Physical Reality – Construction or Discovery?

Michael Grodzicki

Physical Reality – Construction or Discovery?

An Introduction to the Methodology and Aims of Physics

Springer

Michael Grodzicki
Department of Chemistry and Physics
of Materials
Salzburg University
Salzburg, Austria

ISBN 978-3-030-74581-3 ISBN 978-3-030-74579-0 (eBook)
https://doi.org/10.1007/978-3-030-74579-0

Translation from the German language edition: *Physikalische Wirklichkeit – Konstruktion oder Entdeckung?*, © Living Edition/STARNA 2015. All Rights Reserved.

This Springer imprint is published by the registered company Springer Nature Switzerland AG
The registered company address is: Gewerbestrasse 11, 6330 Cham, Switzerland

Preface

Physics owes its numerous practical successes largely to the fact that it segregated from its philosophical roots during the seventeenth century and subsequently developed a self-contained, generally accepted methodology. The preoccupation with questions *about* this methodology and the legitimacy of its foundations ("meta"-physics), as well as considerations about the structure of physical knowledge, are not a prerequisite for gaining and producing physical knowledge, and are considered as unnecessary ballast by many working physicists. However, for anybody who has to teach physics (whether at school, university, or any other educational establishment), these issues cannot be ignored if an appropriate view of physics and its methodology is to be imparted which represents a more current understanding. This concerns primarily the following issues:

- Factual knowledge: what is a physical law (constitutive properties, methodological function) and what is a physical theory (structure, function)?
- Procedural knowledge: what are the characteristic features of physical methodology, in particular, whose methods can be used to obtain structured, secured, and reality-relevant knowledge?
- How can the relations between theory and experience be described in an appropriate fashion?
- What are the aims of physics?

There is no general agreement among physicists concerning these issues, most likely because these problems are rarely discussed in physics. Rather the competence to study and to answer these questions is generally claimed by philosophers of science. Nevertheless, they have not yet succeeded in developing an authoritative conception of theory which provides an appropriate description of the structure of physical theories, and nor have they provided an adequate representation of the relationships between theory and experience.

It is precisely for this reason that this book has come about. In the early 1980s, I started to hold seminars on the methodology of physics that supplemented my courses on theoretical physics, primarily for future high-school teachers. A search of the available literature about the structure of physical theories and the relationships between theory and experience did not yield material relevant to physics, so I was forced to develop my own ideas about theoretical concept formation, the structure of physical knowledge, and the objectives of physics. Initially planned merely as material for the seminar talks with the students, it eventually became a detailed treatise on the methodological issues of physics. Due to this historical origin, the book is a combination of textbook and original contributions. The third and fourth chapters in particular contain material that cannot be found in the available literature to the best of my knowledge. These chapters are based almost exclusively on concrete examples from physics taken directly from my lectures on theoretical, molecular, and solid state physics. These examples either provide the starting point for the derivation and discussion of more abstract issues, or elucidate the general conclusions. Moreover, this extensive use of concrete examples from physics shows that a thorough understanding of physical laws and theories and of the relationships between theory and experience cannot be acquired via formal considerations, as is usually attempted in philosophy of science.

Although there are several connections with philosophical considerations, this book is definitely about the issues of physics itself. It is the translation of the German edition (*Physikalische Wirklichkeit – Konstruktion oder Enteckung?*), which was edited in 2015 by the publishing house Living Edition, Starna GmbH. Note that translations of quotes from other sources are also my own.

Due to the positive resonance among students and colleagues, I was encouraged to prepare an English translation in order to make the text available to a broader audience. In this context, my special thanks go to my editor Angela Lahee at Springer and also to Stephen Lyle for his assistance in improving the style and occasionally clarifying ambiguous formulations.

University of Salzburg, *Michael Grodzicki*
February 2021

Contents

Chapter 1
The World View of Physics I: Presuppositions

The basic requirements for an appropriate representation of the properties of physical methods and knowledge are briefly described, and shown to be in distinct contrast to empiricist views. A comprehensive understanding of physics requires one to include metaphysical and sociological elements in addition to factual and procedural knowledge. The philosophical basis of physics is defined as a "constructive realism" that is subsequently specified and extended by a number of implicit or "tacit" assumptions. A preliminary explication of the conception of a theory is exemplified by Bohr's theory. The method of consistent adjustment, which is of crucial importance to understanding the relations between theory and experience, is elucidated using a simple mathematical example.

1.1 Introduction

Since the foundations of physics as a mathematical natural science were laid in the seventeenth century, the achievements of science have exerted an ever increasing influence on our view of the world. In establishing the scientific world view, physics has played the leading role, and it still does, irrespective of the developments in other branches of science. Firstly, the most striking indications are the directly perceptible technical advances based on physical knowledge. Secondly, physics provides the basis for most of the other natural sciences and influences them accordingly. Finally, the methods of working and reasoning in physics are regarded as paradigmatic for the scientific attitude and rationality. As a result, these methods do not just establish the standard view of scientific methodology, but also constitute the basic pattern for numerous modes of thinking and working in extra-scientific areas. Consequently, these methods are applied to issues for which they were not developed and are not well suited. This has attracted criticism which is justified, at least partially, because such applications are frequently based on concep-

M. Grodzicki, *Physical Reality – Construction or Discovery?*,
https://doi.org/10.1007/978-3-030-74579-0_1

tions of physical methodology dating back to the seventeenth and eighteenth centuries and do not even properly accord with the methodology of classical physics. Certainly, the most widespread prejudice is that (experimental) experience in relation to theory has the distinct priority in the process of gaining and corroborating physical knowledge, as shown by representative textbook citations:

> Observing nature one soon arrives at the insight that it is governed by strictly valid laws. [...] In physics one attempts to read out of experiences relations of as general as possible importance, and then single laws being valid within a subarea to assemble to a whole system without leaving a gap or contradiction. [...] The experimentally obtained facts are untouchable for the physicist as far as the applied observational procedure withstands all criticism. A theory must be changed as soon as a single secured experiment is in contradiction with this theory. [...] Fundamental physical laws can be extracted from nature only by observation (Fleischmann 1980).

> [Physical axioms] are empirical fundamental laws whose correctness[...]emerges solely from immediately given facts (Stroppe 1981).

> Newton's equations [...] were initially empirical formulations of immediately observable quantities (Dransfeld et al. 1974).

> [Physical axioms] emerge as generalization of a finite number of observations. [...] A natural law once established can be refuted (falsified) by a single observation or a single experiment (Fließbach 1996).

> Experiment is the sole judge of scientific truth (Feynman et al. 1964).

> Physics is an experimental science. Everything we know about the physical world and about the principles that govern its behaviour has been learned through experiment, that is, through observations of the phenomena of nature. The ultimate test of any physical theory is its agreement with experimental observations (Sears et al. 1987).

The view of physics emerging from these quotations is as follows. Physics is an empirical science because observations and experiments are the sole, trustworthy source of knowledge and the sole criterion for corroborating theoretical results. Nature is governed by strictly valid laws ("natural laws") discovered through careful observation of nature or by generalization of directly accessible facts. The physicist "asks" nature and obtains unequivocal answers that may, in principle, be scrutinized by any other person. As the aggregation of these laws, physical theories are void of speculative and subjective elements, so that physical knowledge is objective and value-free. Neither the historical process of acquiring knowledge nor individual, social, or economic factors play any role.

A modern philosophical counterpart of this conception is neopositivism or logical empiricism (Haller 1993), developed during the 1920s. Understanding itself as an empiricism supplemented with the results of modern logic, it is characterized by the following assumptions:

- The only trustworthy source of knowledge is pure experience through direct (theory-free) sensory perceptions: all knowledge comes from experience. This assumption, which may be identified as the central dogma of classical empiricism, rests on both the subjectivistic idea of experience due to singular facts and the tabula-rasa fiction of the perceiving subject.
- Scientific concepts and statements may be regarded and accepted as genuine knowledge only if their origin from experience is unambiguously provable: all meaning of scientific terms and statements is derived from experience (empiricist criterion of meaning). The laws of physics are mere descriptions of facts, and theories are the aggregation of laws. Since theories neither possess an autonomous function nor provide any additional knowledge, all "theoretical terms" must be proven to be eliminable. This reflects the anti-theoretical attitude of classical empiricism, considering anything theoretical as a suspect and metaphysical ingredient to be eliminated.
- The relationship between experience and theory rests on the direct correspondence between perceptions or observations and the appropriate conceptual description. As a consequence, this description does not add anything new. The truth of scientific knowledge is guaranteed by the accurate reproduction of facts, followed by their inductive generalization to laws.
- Both the success of physical methodology and the common acceptance of physical knowledge result from avoiding any metaphysical speculations, and restricting assertions to those that are either mathematically provable or empirically testable. The criterion of scientificity is verifiability of those statements advancing a claim of validity.
- Modern logic, as the unifying scientific language, paves the way for the final goal of establishing a unified science. This attitude, in which rationality is identified with logicity, rests on the belief that all problems of content can be reduced to formal problems.

These assumptions lead to the basic features and objectives of the research programme of logical empiricism. Firstly, all theoretical terms must be traced back to direct perceptions ("sense data"). Secondly, the truth of physical knowledge is ensured by inductive generalization of the undoubted sense data to laws. To this end, an inductive logic must be developed. Finally, it must be shown that theories may be replaced by systems without any theoretical terms, but possessing identical empirical content. This would ensure that, at least in principle, all theoretical terms, i.e., all "metaphysical elements" can be eliminated from physics.

Although logical empiricism may be regarded as the first attempt to explain and justify both the success of physical methodology and the intersubjective acceptance of physical theories through a systematic logical analysis, this programme has failed totally owing to its inability to solve any of its problems. The unfeasibility of two steps was recognized rather early on. The intersubjectivity of physical knowledge cannot be justified on the basis of subjective sense data, and a method for truth-conserving generalization of singular facts does not exist (problem of induction). The first problem was believed to be solvable by introducing protocol sentences, understood as verbalized sense data, and after the failure of this attempt, by introducing an observational language. Contrary to the original intention, this resulted in the division of scientific language into an observational language and a theoretical language. This attempt did nothing to solve the challenge of the relationship between theory and experience. Indeed, it just made it harder to understand. The introduction of an observational language meant extending the central empiricist dogma to the conceptual domain, so that not only all knowledge, but also its meaning comes from experience (empiricist theory of meaning).

Attempts were made to get round the insoluble problem of induction by reconstructing theories, not inductively from empirical data, but instead "from the top". Beginning with an uninterpreted formal calculus that is assumed to be independent of any empirical experience, the less general statements are added stepwise until one eventually arrives at the observable facts at the lowest level (Carnap 1966). This conception may be called hypothetico-deductive empiricism because theories are hypothetically postulated systems that are interpreted and justified by the deduction of observable facts. While the empiricist concept of theory is retained with a hierarchical structure of scientific knowledge that has an increasing degree of generality from bottom to top, such an approach implies a slightly different understanding of theoretical knowledge. Theories and laws are no longer a conceptual description of empirical facts, but instead aim to provide predictions and explanations. Accordingly, the need for theories is justified instrumentalistically through their factual indispensability, and theoretical terms are interpreted operationally, in the sense that assuming the existence of a theoretically postulated object, e.g., an atom, means no more than accepting the theory containing this term.

Such an approach is, structurally, not dissimilar to logical empiricism in that its basis is strictly empiricist. Firstly, the empiricist, subjectivistic concept of experience, accepting as real only the perception, but not the perceived, is extended to the theoretical domain through the operational interpretation of theoretical terms. Secondly, the idea that physical knowledge starts with an uninterpreted calculus detached from empirical experience merely replaces the empiricist tabula-rasa fiction with the fiction of an "empirically immaculate conception" of theoretical knowledge. Finally, the inductive logic for the justification of theoretical knowledge is supplanted by the assumption that axiomatization constitutes the decisive tool of the meta-theoretical analysis of physical theories, because this should enable the identification and elimination of metaphysical elements. Therefore, the basic empiricist attitude towards the epistemic status of theoretical knowledge is not substantially dissimilar: the conviction, inherent in logical empiricism, that any kind of theoretical knowledge is suspect, is transformed into the assertion, that all theoretical knowledge is inevitably hypothetical. Consequently, the programme of eliminating all theoretical terms is preserved.

The failure of both logical and hypothetico-deductive empiricism to develop a realistic view of empirical science suggests that this goal can only be attained through a fundamental revision of these empiricist assumptions. Indeed, a programme providing a realistic description of the structure of physical knowledge, physical methodology, the relation between theory and experience, and the objectives of physics must account for a number of facts:

- The assumption of a pure empirical or theoretical stage is a fiction yielding an incorrect account of physical methodology and knowledge. Instead, the development of physical knowledge occurs by theoretical and experimental methods, as two autonomous and equally important parts of physical methodology. Autonomy means that neither can be reduced to the other, but it does not exclude mutual influences. Autonomy means neither theory-free experiments nor invention of uninterpreted calculi isolated from empirical knowledge.
- Similarly, the structure of physical knowledge is characterized by empirical and theoretical autonomy, in that neither type of knowledge is reducible to the other. Theories cannot be inferred inductively from empirical facts and empirical knowledge is not entirely deducible from theories. In fact, the structure of physical knowledge is crucially determined by the kind of relations that exist between theory and experience. Their representation is impossible if understood as a linear, hierarchical relation either inductively from empirical experience to theory, or deductively the other way round. These relations are not unequivocal and can be described only as a process of consistent adjustment between theoretical and empirical knowledge.
- A conception of theoretical knowledge as suspect or inevitably hypothetical totally disregards the crucial importance of theories as the central, epistemically significant, and autonomous parts of physical knowledge about reality. On the contrary, the disqualification of theoretical structures as undesirable metaphysics that must be eliminated precludes the type of physics which is commonly considered as paradigmatic regarding both methods and results. Finally, the identification of rationality with logicity, suggesting that all problems concerning content are reducible to formal ones, implies that the reconstruction of physical theories degenerates to an irrelevant linguistic exercise. The reconstruction of physical theories must rather be preceded by a clarification of the modalities of formation of physical concepts and theories which cannot be carried out on a formal-logical level, but must be oriented toward actual examples.
- The status of physical knowledge as intersubjectively accepted is grounded on the way it has been acquired. Since physical theories are intersubjectively accepted in advance of any reconstruction, the reasons for this must be sought in the way they are formed. Indeed, partition into the context of formation and the context of justification prevents a thorough comprehension of the structure and status of physical knowledge. The context of formation can only be reconstructed properly if sociological and "metaphysical" aspects are taken into account. In particular, the requisite metaphysical elements, in the sense of empirically non-justifiable assump-

tions, must be made explicit in order to better understand their role and function, instead of attempting to eliminate them.

Taking into account these facts leads to a view of physical methodology and knowledge that comprises three structural components that are indispensable for fully comprehending physics as an empirical science:

- Subject-specific or cognitive elements, comprising the explicit part of physics, viz., the factual knowledge in the form of the empirical facts and the language (concepts, laws, theories), and the procedural knowledge, viz., the experimental, empirical, theoretical and consistency generating methods.
- Metaphysical elements, that is, empirically non-justifiable assumptions, implicitly determining the functions and aims of physics. They comprise everything that is not proven, but is accepted as self-evidently valid, and may have heuristic or cognition-guiding functions without themselves belonging to the results of physics.
- Sociological elements, taking into account the fact that modern physics is a highly collective, communicative, and consensus-oriented enterprise. They comprise the structure of the scientific communities, the type of institutionalization, the structures of communication, how the accepted knowledge is imparted, historical background, and finally the position and influence of physics within society as a whole.

This characterization of the basic structures of physics implies that a comprehensive understanding of the methods of gaining, securing, and structuring knowledge, as well as the distinction between presuppositions and results of physical research, necessarily involves considering the influence of the metaphysical and sociological components on the cognitive ones. To this end, the general prerequisites of physical methodology and knowledge must be clarified. These are referred to as "tacit assumptions" because they are not an explicit part of physics, but are learned implicitly. In this respect, these assumptions represent the general, largely unreflected philosophical and methodological basis of physics. The detailed analysis in Sect. 1.6 will show, that they cannot be eliminated, and will thus refute the prejudice that physics is philosophically neutral in the sense of being independent of any philosophical assumptions and, consequently, being compatible with any philosophical doctrine.

1.2 Constructive Realism

Physics in its contemporary form evolved in the seventeenth century by segregation from philosophy. For this reason, numerous connections exist between physics and philosophy, although there exist fundamentally dissimilar opinions about their importance. On the one hand, there is the view that physics, as a special type of human knowledge about nature, definitely possesses a philosophical basis. On the other hand, the prevalent view asserts that physics is philosophically neutral. This, however, is easily refuted because certain philosophical doctrines are definitely incompatible with physical methodology, while others have demonstrably influenced the physical world view. To show this and at the same time to impose some order on the variety of philosophical systems, a suitable classification exploits the fact that physics gains knowledge about nature on the basis of certain methodological rules. Accordingly, the following three classes of philosophies may be distinguished:

- ontological philosophies referring to the nature of reality itself,
- epistemological philosophies concerning the relations between human cognition and reality,
- methodologically oriented philosophies describing how scientific knowledge is obtained or ought to be obtained.

This classification is not disjunct, e.g., ontological philosophies frequently contain epistemological aspects, and methodologies rest on certain epistemological assumptions imparting to them some epistemic character. Since most students of physics are not familiar with philosophy, a brief description will be given of the aspects, relevant to physics, of those philosophical doctrines that are in some way associated with the physical conception of nature (Ritter et al. 1971 ff., Mittelstraß 1995, Batterman 2013).

The most important ontological doctrine for physics is realism. In its general meaning, it comprises all doctrines which assume or postulate the existence of a reality that is independent of human awareness and perception and is, in this sense, subject-independent (an "external world"). Since the universe of discourse of physics is necessarily a reality as object of perception and cognition, but not an unknowable reality-in-itself, all varieties of realism that are relevant to physics contain an epistemic component, in that the objects of perception are assumed to exist independently of the perceiving subject. The existence of a subject-independent reality appears to the layman beyond any reasonable doubt. The problems for realism arise if one asks for criteria or properties that would prove the existence of some part of the external world. The central challenge is the question of whether a phenomenon or property is caused by pure perception or whether it is independent thereof and objective, in the sense that it can be assigned to an object itself. For instance, sounds of music do not exist as sounds in the air, but will be perceived as such only when corresponding regions in the human ear have been stimulated, and the brain has converted these stimuli. The same applies to colours, especially for such colours as purple or brown, which do not correspond to specific wavelengths of the electromagnetic spectrum. Finally, it is by no means obvious in which respect optical phenomena such as rainbows and fata morgana can be assigned an objective existence. Although the experimental methodology of physics that relies on the usage of instruments provides substantial contributions to solving the problem of how subject-independent parts of reality may be identified, such a methodology obviously presupposes a realistic ontology and so does not prove the existence of an external world. For this reason, a realism assuming, against all objections, the existence of a set of properties enabling the identification of constituents of a subject-independent reality constitutes the necessary ontological basis of any concept of nature compatible with physical methodology. The results of modern physics have not changed this.

Suggestions for solving the problem of the relation between subject-independent and perceived reality have led to various epistemic ramifications, as exemplified here:

- Naive or common-sense realism: things are as they appear, i.e., perceptions and the perceived are identified and understood as one and the same. It is also frequently taken to mean that an object and its conceptual description are no longer differentiated. Laws of nature are objective structures of reality and are discovered by careful observation of nature. Acquired knowledge is solely determined by the investigated object, and is independent of the observer. Accordingly, naive realism rests on a belief in the reliability of human perception. In turn, all conceptions assuming direct perceptions to constitute a safe basis rest implicitly on such a naive realism. Such a position is prevalent in physics, too, although physics itself provides many examples that raise severe doubts about the reliability of human cognitive abilities. Further arguments against naive realism as the basic ontology of physics will be given in subsequent chapters.
- Critical realism: the existence of a subject-independent reality is assumed, but it is not taken to be identical to the perceived one. Since any knowledge is affected by the organs of perception and the awareness of the recognizing subject, the existence of an unequivocal correlation between perception and the perceived object cannot be presupposed. For instance, the perception of a tree first requires sunlight, i.e., electromagnetic radiation, to be reflected from it. This radiation hits the human eye and produces an image on the retina, and this image is subsequently processed by the brain. Nevertheless, the perceived is not a purely mental phenomenon, but is caused by a real material object. Realism is then considered to be the simplest hypothesis making the perceptual process comprehensible.
- Scientific realism: in addition to the assumption of the existence of a directly accessible subject-independent reality, it is accepted that objects of scientific research which are not directly perceivable entities, such as atoms, elementary particles, or electromagnetic radiation, can also be ascribed objective existence independent of human cognitive abilities. Associated with this is the assumption that the essential characteristics of these entities are accessible to human cognition and that science enables this access. Depending on additional assumptions, a number of varieties exist (Leplin 1984, Bartels 2009).

- Beside these types, further varieties of realism exist, and the reader is referred to the literature (Ritter et al. 1971 ff., Harré 1986, Mittelstraß 1995, Willaschek 2000, Batterman 2013).

There is also a close relationship between the physical world view and materialism, which claims in its general form that matter is the basis and substance of reality ("primacy of matter"). Most criticism is directed against reductionist versions, attempting also to reduce all mental phenomena and processes to matter. In its moderate version, postulating the primacy of matter without such reducibility, materialism is relevant to physics. A related doctrine that has played a substantial role in the historical development of physics is mechanism. It regards the cosmos as a machine and advances a strictly causal-deterministic world view. At the culmination of classical physics toward the end of the nineteenth century, such a conception was widespread, and not only among physicists, while it is now of minor importance. However, its methodological version still constitutes the basis for the theoretical methodology of physics in the following respect. Without the assumption that nature can be described in a law-like way, knowledge about nature would not exist in the sense of modern natural science. On the other hand, this assumption should not be considered as an intrinsic property of reality, but merely as a heuristic.

The most important opponent of realism and materialism is idealism with its basic assumption that mental categories, such as ideas and consciousness, are primary, while matter is just a form of appearance in the mind. Common to most variants is the notion that reality is constituted by the mind of the recognizing subject. Although one would not expect such an ontology to play any significant role in physics, it enjoys some popularity among particle physicists. For under the extreme conditions of high-energy processes, where particles can be transformed into each other almost arbitrarily, matter appears as a variable element, while only certain symmetries and conservation laws remain constant, and it is these that are then assumed to constitute the proper or underlying reality. In this respect, idealism is of some importance in physics. Moreover, this applies to the epistemological variant which considers as objects of cognition, not the parts of a subject-independent reality, but the ideas or conceptions thereof. According to Kant, these objects of cognition do not exist independently of the ideas, interpreted as the structuring performances of the recognizing subject, i.e., not as things-in-themselves, but they are created only through these structurings. In Sect. 5.5, it will be shown that the physical laws contribute substantially to constituting physical reality. Interpreting the laws, according to the definition in Sect. 4.2, as the ideas of physics or as the structuring performances of physicists, such an epistemological idealism does indeed encompass several aspects of physical reality and is, in addition, compatible with a realistic ontology.

The importance of epistemic philosophies for physics rests on their influence on views concerning the relation between physical knowledge and reality. Historically, empiricism and positivism were initially important. Empiricism is grounded on the general assumption that all knowledge about nature arises from empirical experience and must be justified through it. In other words, concepts and statements are meaningful only if based on empirical experience. The assumption of inherent ideas is rejected, as is the pretense of authority and bare thinking as sources of reliable knowledge. Various ramifications also exist, depending on the meaning of empirical experience. In sensualism, experience is identified with direct sensory perceptions of singular facts, so that human senses are regarded as the only source of experience. According to phenomenalism, the objects of experience are not objects that exist in an external world, but merely forms of appearance ("phenomena"). Empiriocriticism (Avenarius, Mach) may be deemed an attempt to create a systematic basis for this conception. It finds its modern continuation in logical empiricism, already described in Sect. 1.1. Both, however, are of historical interest only.

Closely related to empiricism is positivism, which temporarily exerted great influence on views of the methodology and aims of physics. Its basic assumptions are as follows:

- All knowledge rests on the "positive" facts of empirical experience, which are, therefore, the basis of science. The task of science consists in providing a comprehensive description of these facts and observable phenomena.
- Beyond facts, there is only logic and mathematics.
- All questions decidable neither empirically nor logically are pseudo questions, and therefore meaningless (demarcation criterion for scientificity).

Since positivism disclaims as a part of natural science everything that is not directly perceptible, physical knowledge can only provide descriptions, but not explanations. An explanation of phenomena is solely the subsumption of special cases under general laws. Any considerations beyond this, such as explanation by causes, are epistemically meaningless and irrelevant to practice. This attitude initially made positivism the primary antagonist of the chemical structure theory, and towards the end of the nineteenth century, of any kind of atomic theory. Objections against positivism concern, firstly, the norm of restricting to facts in combination with the question of what is actually a fact, and secondly the demarcation criterion for scientificity. The strongest opposition aims at the implicitly held attitude of classifying all questions beyond the current state of knowledge as meaningless or unanswerable, instead of conceding that this might only be the case temporarily. Among its adherents, there is a persistent tendency to estimate the current state of knowlege as largely definitive. For these reasons, strict positivism would impede research progress.

According to conventionalism, laws and theories rest on agreements among scientists. Founded by Poincaré, such views have been held for example also by Mach and Duhem. As a consequence of the development of non-Euclidean geometries around the middle of the nineteenth century, Poincaré concluded that the geometrical axioms are conventions (Poincaré 1905):

> The geometrical axioms are assessments resting on agreements [...] One geometry cannot be more correct than another one, it can only be more convenient (economical).

The view that economic thinking should guide the choice of a certain representation was widespread among physicists in the second half of the nineteenth century, e.g., Mach, Kirchhoff, Hertz. Concerning mechanics, Poincaré (1905) states:

> Though experiment underlies the principles of mechanics, it can never overthrow these principles.

Although this assessment is entirely correct, as will be argued in detail in Chaps. 3 and 4, and physics contains numerous non-eliminable conventional elements, e.g., the choice of systems of dimensions and units, the claim that theoretical knowledge rests primarily on conventions and arrangements is untenable. Conventionalism had the merit of creating an awareness that even a science as precise as physics must be based on certain well chosen arrangements in order to be an efficient enterprise. Instrumentalism displays a certain affinity with conventionalism, treating laws and theories merely as an auxiliary means to attain practical goals, such as the deduction and prediction of factual knowledge and a mastery of nature. Knowledge is not end in itself, but rather a necessary presupposition and an instrument for successful actions. Laws and theories must work without making any claim to be related to reality. Accordingly, the aim of physical theory is the prediction of factual knowledge, e.g., experimentally comparable measurement data. Instrumentalist views of the status of theoretical knowledge prevail in particular during periods of substantial turnaround. For instance, the early interpretations of quantum mechanics were predominantly of instrumentalist kind.

Most relevant for physics is an approach developed in the 1970s and referred to as radical constructivism. It advances the basic thesis (Watzlawick 1985):

> [...] that any reality in the immediate sense is the construction of those who believe that they are discovering and exploring this reality.

Based on experiments and results in cognitive science and neurobiology, its adherents view the radical dissimilarity of their conception through the relation between knowledge and reality, because (Glasersfeld 1985)

> [...] it develops an epistemology in which knowledge no longer concerns an "objective", ontological reality, but exclusively the order and organization of occurrences in our experiential world.

This may initially be understood in such a way that not only each kind of knowledge represents a construction, but also the scientific object. In this case, radical constructivism contains a remarkable inconsistency. On the one hand, it rests on results of neurobiology, molecular biology and physics which presuppose the existence of a subject-independent reality for methodological reasons, as will be shown later. On the other hand, the existence of such a reality is denied. In order to avoid this inconsistency, radical constructivism may be interpreted as an epistemology that is constrained to the cognitive processes of the recognizing subject, irrespective of the existence of a subject-independent reality. This interpretation is compatible with a realistic ontology and means that reality is to be understood as the object of perception and cognition, as is appropriate for any empirical science. This interpretation is supported by the following metaphor about the relation between lock and key (Glasersfeld 1985):

> A key fits when it opens the lock. The fitting describes the ability of the key, but not the lock.

Moreover, other keys exist that fulfill the same task. Without taking the metaphor too literally, the following remarks are indicated. Firstly, although the various keys do not provide information about the nature and constitution of the lock, the lock exists and is assumed to be part of a given reality. Secondly, the existence of various working keys corresponds to the fact that the relation between reality and its conceptual description is not unequivocal. This insight constitutes the basis of the argumentation in Chap. 4. Thirdly, physics is considerably more radical. As will be shown in Chaps. 2 and 3, a reality is created in a targeted way via idealization and abstraction, not only conceptually, but also materially by experimental methods. This reality is designed in such a way that it becomes accessible to physical and mathematical methods (the "keys") and can be investigated by them. To use the above metaphor, this means that the lock is designed in such a way that it fits the keys. Put like this, radical constructivism is not as radical as it may appear at first sight. In the subsequent chapters, it will be shown that such a moderate constructivism encompasses a number of properties of physical knowledge, and in particular of the relation between physical knowledge and reality.

The primarily methodologically oriented philosophies may be classified with regard to perception, action, thinking, or intuition, these being considered as the most suitable auxiliary means for acquiring scientific knowledge. A common property is a strong normative orientation, reinforced by the tendency to overemphasize one of these four means and to declare it as the general norm for gaining knowledge. In contrast, the methods of physics are so varied and multifaceted that they must be discussed and assessed with respect to physical practices rather than philosophical convictions. Although some aspects of physical methodology may be recorded correctly, their influence on physical practices is therefore modest. The first doctrine is inductivism, which is closely related to empiricism. It understands itself as a method for discovering general laws and pretends to achieve the truth of knowledge through the inductive transition from direct experience to theory. By inferring from a finite number of direct perceptions and observations, one should arrive at the physical laws. The second aim of demonstrating the truth or secureness of knowledge via inductive logic, as attempted in logical empiricism, cannot be realized for reasons described in Sect. 1.1. Although inductive methods constitute an indispensable part of the physical methodology for gaining knowledge, they provide merely tentative and thus hypothetical knowledge. Actually, not only is knowledge secured by other methods, but physical laws are obtained by means other than induction, as will be shown in Chap. 4.

Operations, rather than perceptions, constitute the basis for the physical methodology of operationalism (Bridgman 1928) because measurements, experiments, and the use of instruments are representative of the development of physics. In particular, operational definitions of physical quantities meet important and indeed necessary requirements that will be described in detail in Sect. 2.3. However, the move towards an operationalist theory of meaning, equating the meaning of a term with the defining operations, is untenable. A similar programme is pursued by "Erlangen constructivism", resulting in a hierarchical reconstruction of physics on basic operation-theoretical principles. The result of this approach is termed protophysics, and is considered to be the basis of physics (Böhme 1976). As shown in Chap. 4, such an approach is questionable as a matter of principle.

If pure thinking or reason is ascribed the decisive role in the process of gaining knowledge, this may be considered as a methodological variant of rationalism. The epistemological basis is the conviction that, through sensory perceptions alone, reality cannot be recognized in its full extent, but that thought processes constitute the most suitable, since most adaptive, means to arrive at reliable and thorough knowledge about reality. Indeed, the emergence of physics in the seventeenth century rests essentially on the detachment from direct sensory perceptions, so that even in classical mechanics thought processes, in the form of idealizations and theoretical abstractions, play an important role in gaining knowledge. This is even more relevant in modern physics since numerous results following from theoretical considerations are in radical contradiction with common experience. Further characteristics are the belief in an unlimited human cognitive faculty, i.e., there are only temporarily, but not in-principle, unsolvable problems. Consequently, rationalism is representative of the belief in progress characterizing modern science. The extreme position which assumes that the external world is structured in accordance with the human mind, corresponding to conformity of thinking and reality, is untenable. With positivism, rationalism shares the postulate of value-free science and the programme of the elimination of metaphysical elements. A modern variant that emerged as an alternative to logical empiricism is critical rationalism (Popper 1962), discussed in more detail in Chap. 4. It exhibits some affinity with skepticism, advancing doubt as a principle of thinking. It denies the possibility of achieving definitely proven knowledge. A positive feature is the challenge to engage in continuous critical reflection. Although this may promote research progress in science, strict adherence to this norm would impede efficient science, in the sense of modern natural research, because one would never get beyond discussions on the fundamentals. Finally, intuitionism assumes intuition to be the most efficient source of knowledge. Although the importance of intuition in obtaining knowledge in physics should not be underestimated, this extreme view is untenable when characterizing physical methodology and plays a minor role.

In summary, we have given a first determination of the ontological–epistemic basis of physics that will be deepened and justified in subsequent chapters. The ontological basis is a mixture of scientific realism and materialism with some idealistic admixture. An epistemological consequence is the permissibility of the subject–object separation, with the corresponding methodological implications, but not the naive-realistic belief that the successful working of physical knowledge may be taken as proof of its truth or for the existence of a reality-in-itself. Such conclusions are circular and thus unacceptable. The epistemic basis which determines the basic structure of the relations between physics and reality is formed by a moderate constructivism with conventionalist elements and some admixtures from empiricism and positivism. The influence and weight of these views strongly depend on the historical situation, and reflect the changes in opinions about these relations during the development of physics. The conception corresponding to this philosophical basis is referred to as "constructive realism" and defined by the following properties:

- the assumption of the existence of an external world that is independent of perceptions and conceptual descriptions, so that it is not changed by them,
- the assumption of an at least partial conformity between epistemic and ontic categories,
- the norm of a strict distinction between the subject-independent reality and its conceptual description.

Constructive realism thus assumes the existence of a subject-independent reality, but does not claim to prove its existence. This reality is recognizable, but identity with reality, as constituted by perception, observation, and scientific investigation is not claimed, and nor is the possibility of knowledge about a reality-in-itself. In other words, even the subject-independent reality is that of the recognized objects, but not a fictitious underlying reality. Central importance is assigned to the distinction between the external world and its conceptual description: objects, phenomena, and regularities as constituents of this world are independent of their conceptual description, which may change if the corresponding part of this reality is understood in another or improved mode. Conceptual descriptions, such as physical laws and theories, provide statements *about* this reality, but are not a part of it. This is crucial for a correct assessment of the epistemic status of physical knowledge. Regarding methodology, constructive realism takes a broad view and accepts observation, operation (= experiment), thinking, and intuition as legitimate means to generate knowledge.

1.3 Factual Knowledge: The Structure of Bohr's Theory

The cognitive elements of physics comprise the factual knowledge, viz., facts, terms, laws, theories, and the procedural knowledge, viz., experimental, theoretical, and consistency-generating methods. Among the factual knowledge, theories are of fundamental importance as the most comprehensive constituents. Such a high appreciation of their status is not shared by empiricist philosophies, which consider theories as dispensable, in principle, without any autonomous function in gaining and structuring knowledge, and merely an additional "topping" on the "positive" knowledge of facts. Accordingly, the conception of theory is only vaguely, or not at all, demarcated from other concepts, such as law, model, or hypothesis, and is thus unsuitable as a basis for detailed investigation. Similarly, this applies to views influenced by philosophy of science which largely reduce to formal structure. These theory models assume, on the one hand, an uninterpreted calculus (Hempel 1970) or a mathematical theory (Ludwig 1978) and, on the other hand, empirical data or a part of reality (Ludwig 1978). These are connected through mapping principles (Ludwig 1978), correspondence rules (Nagel 1982), or bridging principles (Hempel 1970). A theory is then represented as an ordered pair (Hempel 1970) or as a mapping (Ludwig 1978) of a formal system onto reality. Actually, such approaches are rarely helpful due to the great differences regarding structure, aim, and function of the various physical theories, which preclude subsumption under a single conception of theory. Instead this diversity requires, in the first instance, the distinction of at least three classes:

- Phenomenological or descriptive theories: examples are physical geometry, kinematics, and geometric optics. Their main objective is the law-like description of relations with relatively direct reference to observation and experiment.
- Deductive or explanatory theories: examples are Newton's theory of gravitation, enabling the derivation of Kepler's laws, or the kinetic gas theory from which Boyle's law may be obtained. They establish deductive structures in that the laws of descriptive theories are obtained from more general theoretical assumptions with the status of patterns of explanation.
- Unifying or grand theories: the most prominent example is certainly Newtonian mechanics with its unifying representation of dynamics. Further examples are electrodynamics, establishing a unified framework for electric, magnetic, and optical phenomena, as well as quantum mechanics providing a unified view of matter.

This classification is not a strict partitioning since descriptive theories also contain deductive structures, and on a higher conceptual level, explanatory theories may be considered as descriptive. A suitable example for a first preliminary explication of theory conception is Bohr's theory of atomic constitution, because it is relatively simple with a well defined domain of application, and because it contains a number of deductive structural elements, despite its predominantly descriptive nature. Furthermore, it is a representative example of a theory that is false, in a sense to be specified later, but nevertheless enables the derivation of many results in agreement with experimental data, which shows that the experiment is not at all the sole judge of scientific truth (Feynman et al. 1964). Finally, concerning didactic aspects, the theory still provides a meaningful heuristic introduction to our views on atomic structure, in particular, since some of its patterns of explanation remained valid. Therefore, its inclusion in physics classes is still justified, provided that its achievements and limitations, and the reasons for its success are presented critically.

Some introductory notes regarding the historical context of its origin are given in advance for the sake of clarity. A psychological moment that was certainly important for Bohr was the failure of electrodynamics when he attempted to calculate the magnetization of atoms in thermal equilibrium (Bohr-van Leeuwen theorem). This experience may have contributed to his skeptical attitude regarding the applicability of electrodynamics to atoms. Another basis was Rutherford's atomic model, which aimed to describe the results of his scattering experiments with α-particles (Rutherford 1911). This aim was achieved with the derivation of Rutherford's scattering formula, while the application to the spectral lines of atoms was not intended. It cannot thus be asserted that his investigations remained at a mere heuristic stage of model formation (Röseberg 1984). Such an interpretation ignores the fact that theory formation is goal-oriented, and that these goals must be accounted for in the assessment of a theory. The sole assumption of Rutherford's derivation which Bohr refers to explicitly is the force law with the distance dependence $1/r^2$. Finally, the empirical basis was provided by the large body of knowledge on spectral lines collected during the second half of the nineteenth century, in the form of both directly measured frequencies and also regularities and empirical correlations, i.e., the constructively processed experimental data. These were mainly the series laws, series limits, the Rydberg–Ritz combination principle, and the Balmer formula for the hydrogen atom. This followed from the "trilogy" (Bohr 1913) in which Bohr emphasizes that he succeeded in explaining the laws of Balmer and Rydberg.

A systematic reconstruction of Bohr's theory contains three dissimilar constituents that are referred to as model assumptions, mechanisms, and correspondence postulates:

(M) Model assumptions defining the system under consideration:

- (M1) The atom consists of a positively charged nucleus, assumed to be point-like, around which the negatively charged electrons revolve, similarly to the planets around the Sun (Rutherford's atomic model). The electrons are ascribed mass m, velocity v, and distance r from the nucleus.

(M2) Stability: on certain, discrete ("permitted") orbits, the "stationary states", the electrons can stay without emitting electromagnetic radiation.

(D) Mechanisms as the assumptions defining the dynamics:

(D1) The motion of the electrons takes place in an electrostatic Coulomb field generated by the nucleus. The laws of classical mechanics and electrostatics are applicable to the dynamics in the stationary states.

(D2) Radiation is emitted only during the transitions of the electron between stationary states. Regarding these transitions, the laws of classical physics do not hold.

(C) Correspondence postulates establishing the connection between the calculable and the measurable quantities:

(C1) The stationary states are defined by the criterion (Bohr's quantization postulate):

$$mvr = nh/2\pi \ ,$$

where h is Planck's constant, and $n = 1, 2, \ldots$. It should be emphasized that this postulate is compatible with (D1) because this restriction only concerns boundary conditions, but not the laws.

(C2) The relation between the energies E_k of the stationary states as the directly calculable quantities of the theory and the measurable frequencies of the spectral lines are given by (Bohr's frequency postulate):

$$E_n - E_k = h\nu_{nk} \ .$$

Hence, the energies themselves do not correspond to the measured spectral lines.

(C3) For very large quantum numbers n, the frequency postulate approaches the classical limit of a harmonically bound electron (Bohr's correspondence principle).

Based on this reconstruction, the key features of the theory conception can be elucidated, which will be developed in more detail in Chap. 3. The analogy with the planetary system in (M1) seems to provide a physical interpretation of the atomic model which, however, has a heuristic function at best, because interpretations based on analogue models are frequently misleading in that they rest on concepts that are not adapted to the theory. In Bohr's theory, this becomes apparent through inconsistencies that follow from relying too closely on the conceptual system of classical physics. Actually, the physical meaning of the basic concepts remains vague. Regarding the concept "electron orbit", this is clear because it does not exist in reality. Concerning the concept of the electron, the inconsistency arises because Bohr's "atomic electron in a stationary state" represents something fictitious. It describes a charged object that does not radiate, despite undergoing accelerated motion. In this respect, it is different from an electron, e.g., in a cathode ray, which will indeed radiate when moving along a curved track. Due to this inconsistency, the concept of the "energy of an electron in stationary state" does not have a clear physical meaning either. Instead, it is only the energy difference between stationary states that inherits physical meaning through the assignment to measurable spectral lines via the correspondence postulates. Consequently, the inconsistencies in relation to classical physics can be removed by considering the Bohr atom as a conceptually constructed model system. Actually, the physical and empirical content of the theory will not be decreased by eliminating the analogy established through (M1) because the analogue model derived from the planetary system neither provides insight into the real physical nature of the atom nor enters the quantitative part of the theory, so it does not provide any knowledge. It is just because such analogue models impart a more concrete access to the abstract part of the theory that they are considered erroneously as essential. Eventually, they even lose their heuristic function, as soon as the theory enables the deduction of experimentally comparable results. Overall, such analogue models can be eliminated without difficulty once the theory has been set up. This proves their epistemic irrelevance.

In distinct contrast, the assumptions (D) concerning the dynamics, i.e., forces and interactions, referred to as mechanisms, constitute the epistemic kernel of the theory. They are the key prerequisites to deduce quantitative statements about empirical facts, directly enter the formal part of the theory, and cannot be eliminated without substantial changes to it. Their central importance rests on the fact that the reduction of physical knowledge to assumptions about interactions forms a basic feature of theory formation in physics. Actually, the success of a theory rests crucially on the construction of the correct mechanism. In addition, it should be mentioned that the assumption of a Coulomb interaction is far from being as self-evident as it may commonly seem. Alternatively, a harmonic coupling between electron and nucleus is conceivable, and was proposed at that time. This interaction yields discrete stationary states as well, and works for certain problems, such as estimating atomic polarizabilities. The fundamental question of the correct mechanism, however, cannot be answered within a theory, and frequently not at all. Essentially, this is the reason for the instrumentalist position regarding the status of physical theories, which denies, although unjustifiably, that any epistemic and ontological significance be assigned to the mechanisms: they neither constitute autonomous knowledge nor correspond to structures of reality.

The correspondence postulates are the assumptions usually referred to as Bohr's postulates, emphasizing their great significance within the theory.

They do not correlate with the correspondence rules or bridging principles of empiricist models because they are theory-internal assumptions that provide the preconditions for the comparison between theory and experiment by specifying the measurable quantities. Establishing such relationships is non-trivial, and forms the second constructive step in theory formation after specification of the dynamics. Indeed, in order to understand the constitution of atoms, the central question to be answered was how to connect the measured frequencies of the spectral lines with the calculable energies of the atomic electrons. This problem was solved by Bohr through postulating that the spectral lines arise due to the transition of an electron between two orbits, rather than by the motion around the nucleus. However, this is a theoretically autonomous assumption that is not unequivocally correlated with experimental data. In conclusion, the correspondence postulates provide a consistent connection between a theoretically autonomous assumption and experimental results.

Mechanisms and correspondence postulates, as the fundamental constituents, show that a physical theory constructively creates autonomous theoretical structures embodying substantially new knowledge. Due to this theoretical autonomy, neither of them are directly correlated with empirical experience. In order to classify a theory as empirical, it must enable the deduction of empirically comparable statements about data, effects, and empirical correlations. Examples of such deductions have already been mentioned by Bohr, viz., the derivation of the Rydberg–Ritz combination principle, the Balmer formula, and the Rydberg constant. Further deductions are the Zeeman effect, the correct interpretation of the Pickering–Fowler series as the spectrum of the positively-charged helium ion, and then, after formulation of the theory, Moseley's law and the interpretation of the Franck–Hertz experiments. Regarding the reconstruction of a theory, it is irrelevant whether or not a fact was already known. This distinction becomes relevant only for historical studies. Overall, physics is not restricted to the description of mere observations, but rather aims at deducing them from theoretical assumptions. Actually, the persuasive power of Bohr's theory rests primarily on such deductions from autonomously constituted theoretical principles.

Hence, there are qualitatively different types of hypotheses within a theory, and these must be carefully distinguished. One type comprises the mechanisms and correspondence postulates entering the theory as assumptions. They cannot be proven within the conceptual framework of the theory: no theory can prove its own model assumptions. For this reason, Bohr's theory does not explain the stability of atoms, in contrast to a widely held belief. Another type comprises experimentally comparable statements that are deducible from mechanisms and correspondence postulates and, in this respect, can be proven within the theoretical framework. Through these deducible hypotheses, a concept of explanation may be defined that corresponds to common use in physics. An observable effect is deemed to be explained by a theory if a corresponding statement is deducible from mechanisms and correspondence postulates. The actual success of a theory, however, becomes manifest less through the deduction of empirical facts than through its ability to provide an efficient structural framework for a broader field of experience. Accordingly, the actual merit of Bohr's theory was not reflected in the mere prediction and explanation of empirical facts, but rather in structuring a large number of previously unconnected observations and experiments through a few unifying assumptions. Eventually, this led Sommerfeld to the conviction that a theory with such a structuring capability must contain a true kernel, despite its apparent deficiencies. A few successful predictions would rarely be enough to conclude in this way. In addition, the use of highly simplifying theories in physics is hardly comprehensible if the central aim of theory formation consists in quantitative predictions, instead of the structuring of knowledge. A representative example is the theory of the free electron gas for the description of the electronic properties of metals. Accordingly, providing predictions is not an objective of theory formation, but the application of theoretical knowledge.

The last characteristic trait of this theory conception is the distinction between the subject-independent reality and its conceptual description by the theory. This reality is theory-independent because it is not changed by conceptual descriptions. Accordingly, effects, data, and empirical regularities representing this reality do not belong to the theory, but only claims about them in the form of the hypotheses deduced from mechanisms and correspondence postulates. In turn, a conceptual system as a theory containing claims about the reality is not a part of it. For these reasons, theories cannot be discovered, but are built up by substantially constructive methods with mechanisms and correspondence postulates as the basic constituents. Consequently, a theory is not a mere summary of empirical facts, but creates new structures, both vertical in the form of deductions and horizontal in the form of crosslinks by connecting previously unconnected experiences. The successes of a theory regarding structuring performance, explanations, and predictions must not obscure the fact that, within the theoretical framework, it is impossible to reach any conclusion concerning the correctness of the underlying model assumptions, beyond stating internal consistency. Therefore, theories cannot be compared with experience in the same way as single statements. Notably, this demonstrates the fact that results can be deduced from a false theory which are in agreement with experimental data.

Finally, it remains to examine in which respects Bohr's theory is false. In the first instance, one must distinguish between the limitations and the inconsistencies of a theory. It is normal that a physical theory has limitations regarding its applicability, i.e., there is no "theory of everything" in physics. As a consequence, the theory cannot be classified as false because it does not properly describe, e.g., many-electron atoms. This just shows its limited applicability. According to such an argument, all physical theories would be false, an attitude perhaps defensible on purely logical grounds, but which makes little sense. According to common opinion, the inconsistency of the theory rests on the inference that the electron as a charged particle should emit electromagnetic radiation due to its accelerated motion. Actually, this contradiction emerges only in relation to the classical conceptual system which identifies the term "electron" with a classical point charge as it seems to be, e.g., in cathode rays. However, there is nothing to say that electrodynamics should be valid for electrons in an atom, and for Bohr the aforementioned reasons were sufficient to doubt this. Thus, postulating the "atomic electron in a stationary state" as a system implicitly defined by the theory, which is not identical with the electron in a cathode ray, was sufficient to remove this contradiction, and makes the theory internally consistent. Such a procedure appears to be unsatisfactory only because either two kinds of electrons should exist or the electric field created by the proton must be assumed to be qualitatively different from the macroscopic electric field in which the accelerated electron emits radiation. None of these considerations are valid arguments to rate the theory as false. Indeed, the actual inconsistencies cannot be assessed relative to electrodynamics, but only through comparison with quantum mechanics, as the theory which also describes the hydrogen atom. Accepting this as correct, Bohr's theory turns out be false for at least the following three reasons:

- Taking seriously the central field as postulated in (D1), the resulting motion should be planar. Schrödinger's equation in two dimensions yields the energy values $E_n = -Z^2/(n + 1/2)^2$, and the eigenfunctions exhibit another shape to the one in three dimensions. Consequently, the theory is either internally inconsistent in relation to quantum mechanics or results in conclusions that contradict experimental results. While a planar motion follows from the model assumptions, agreement with experiment is obtained only for a three-dimensional motion.
- The quantization postulate for the angular momentum contradicts the result of quantum mechanics according to which the angular momentum of the electron in the hydrogen atom vanishes in the ground state $n = 1$. The equation itself, however, remains valid in quantum mechanics if interpreted in a different way. The left-hand side contains the product of position and momentum, whose uncertainties are related to each other by the uncertainty relation. In the framework of quantum mechanics, Bohr's quantization postulate may thus be reinterpreted as a special case of this relation. This ambiguity arises since angular momentum and action have identical dimension.
- Most importantly, the theory implies an incorrect understanding of the dynamics within the atom. The stability of the hydrogen atom is explained, if at all, by an incorrect mechanism, viz., due to the equilibrium between the centrifugal force and the Coulomb force of the proton acting on the electron. This argument is grounded on the implicit assumption of the existence of an electron orbit within the atom, in contradiction to the principles of quantum mechanics. Accordingly, the quantum mechanical explanation is entirely different. While the Coulomb force is preserved as

the attractive interaction, the mechanism of repulsion, requisite to attain a stable equilibrium, turns out to be the increase in the kinetic energy of the electron with decreasing distance between proton and electron. Following from the uncertainty relation, this is a purely quantum mechanical mechanism that can never be obtained from classical concepts.

Due to these contradictions, Bohr's theory is false relative to quantum mechanics. For this reason, it can neither be incorporated into quantum mechanics as a specialization for one-electron atoms, and nor is there a continuous transition between the two theories, as can be constructed in a certain way between classical mechanics and the special theory of relativity. Nevertheless, the theory works in that, within its scope, it reproduces numerous experimental results. The fact that a false theory can yield empirically correct results does not, however, support the instrumentalist conclusion that theories do not represent epistemically significant systems of knowledge, and that the question of the truth of a physical theory is nonsensical. Such a conclusion rests on the implicit assumption that the function of theories is constrained to supplying predictions, and agreement with experiment is the sole criterion of truth. Concerning the question of truth or falsehood, this is not the decisive criterion. It is absurd, e.g., to rate Ptolemy's theory of epicycles as more true than the theory of Copernicus because, at that time, it permitted a more precise calculation of the planetary positions. On the contrary, decisive in rating a theory as false are the mechanisms and correspondence postulates, these being the epistemically relevant constituents. As a theory of spectral lines, Bohr's theory is correct, although limited in its scope, because it postulates the correct mechanism for their origin. In this respect, it is an explanatory theory because it accomplishes the central objective by deducing spectral lines and series. In contrast, as a theory of atomic constitution which aims to explain the stability of atoms, it is false because it assumes an incorrect mechanism. Such a conclusion, however, requires the comparison with an alternative correct theory for the same field of experience. This provides another indication regarding the autonomy of theoretical knowledge. The stability of atoms is deducible from quantum mechanics and can, in this respect, be explained, but not within the framework of Bohr's theory. Here, it is either an assumption, thus not explainable, or it is deduced from an incorrect mechanism, and is likewise not explained. As a theory of atomic structure it must then be classified as descriptive. This demonstrates that the classification of a theory does not depend solely on its formal structure and physical content, but also on the goals being pursued.

1.4 Procedural Knowledge: The Method of Consistent Adjustment

The theoretical and practical successes of classical physics led towards the end of the nineteenth century to the conviction that the methods of physics should be exemplary to gain scientific knowledge. In contrast to the humanities, the conclusion of theoretical development in physics is marked by a consensus about theoretical description that is appreciated as exemplary and has contributed considerably to the reputation of physics to represent objective knowledge. From the very beginning, it was clear that the reason for these successes had to be found in the specific relationship between theory and experience. Therefore, the investigation of this relationship must serve as the starting point for all attempts to understand the successes of physics.

The first systematic attempt to explain how physical theories achieve intersubjective validity can be traced back to logical empiricism. This was restricted, however, to the systematisation and justification of existing knowledge, while its origin was declared as epistemologically irrelevant, totally ignoring the fact that the intersubjective acceptance of physical theories was already established before and independently of any reconstruction. Consequently, the reasons must be identified in the process of formation. As previously mentioned, these attempts failed primarily on the problem of induction. The insight, returning to Hume, that a scheme for the truth-conserving generalization of singular facts does not exist, could not be overcome by means of modern logic. On this insight rests the hypothetico-deductive empiricism which understands scientific theories as propositional systems. In particular, physical laws are propositions that are only valid until revoked, and must withstand the permanent test of experience. The function of experience with regard to theory consists in the possibility of falsifying theoretical assertions, while the function of theory is the explanation and prediction of empirical facts. Similarly, the relationship between theory and experience is not described adequately. The inability of empiricist philosophies of science to achieve this goal caused them eventually to restrict themselves to the reconstruction of theories without taking into account the relation with experience.

In spite of superficial differences, the two conceptions have fundamental structural properties in common. They not only rest on the same empiricist assumptions, but also exhibit an essentially linear structure, either from experience to theory (logical empiricism) or from theory to experience (hypothetico-deductive empiricism). For this reason, neither approach can provide a realistic account of this relationship which cannot be described by a purely additive combination of inductive and deductive methods. Indeed, neither the relationship between theory and experience nor the development of physical terms, laws, and theories can be understood without considering another constituent of physical methodology, viz., the consistent adjustment or adaptation of theoretical and empirical knowledge. Typical properties of this procedure may be illustrated with a simple mathematical example, viz., the solution of the equation $x^2 - 2 = 0$ within the domain of rational numbers. Although it cannot be solved exactly since $\sqrt{2}$ is not a rational number, the solution can be approximated to any desired degree of accuracy. Starting with an approximation x_0, the exact solution is given as $x = x_0 + f_0$ with the deviation f_0 to be calculated. Substitution into the original equation yields

$$(x_0 + f_0)^2 - 2 = 0 \implies x_0^2 + 2x_0 f_0 + f_0^2 - 2 = 0 .$$

The term f_0^2 prevents the computation of f_0 as a rational number. If, however, x_0 is sufficiently close to the exact solution x, f_0^2 is negligible compared with the other terms and can be omitted. The equation for f_0 then reads

$$x_0^2 - 2 = -2x_0 f_0 \implies f_0 = \frac{1}{x_0} - \frac{x_0}{2} .$$

Although the computed $x_1 = x_0 + f_0$ is not yet the exact result, x_0 in the last equation can be replaced by x_1 as a new starting value, and the whole procedure can be repeated with x_1. Generalized, x_{n+1} can be iteratively computed from x_n via

$$x_{n+1} = x_n + f_n , \text{ with } f_n = \frac{1}{x_n} - \frac{x_n}{2} .$$

Depending on the starting value, the following sequences of numbers are obtained:

x_0	1.4	1 or 2	4
x_1	1.4142857	1.5	2.25
x_2	1.4142135	1.4166667	1.5694444
x_3	...	1.4142157	1.4218903
x_4	...	1.4142135	1.4142342
x_5	...	...	1.4142135

It can be seen that, in this example, the starting value influences the number of required steps, i.e., the speed of convergency, but not the final result. This starting value may even be rather far from the exact result, so the initial deviation need not be small compared with expectations. Without overdoing the analogy with a mathematical example, it can be used to illustrate typical structural features of the method of consistent adjustment.

1. In theory formation, the starting value x_0 may correspond to a first theoretical approach. The choice of x_0 is grounded on the personal experience of the acting subject, in this case the previous knowledge that $1 < x < 2$. Since generally valid rules do not exist, a formal-logical reconstruction of this first step is impossible. Even in mathematics, the choice of a suitable initial value is by no means unproblematic, because various circumstances can preclude the success of this procedure:

- the procedure can diverge due to an inappropriate choice of the starting value,
- the solution of the problem need not to be unequivocal, i.e., another choice of the starting value might lead to another final result,
- the procedure ends by going round in circles in the sense that it does not converge to the correct result known, e.g., from other sources.

Analogous cases of failure are encountered in the course of theory development. The first case corresponds to a theory, such as Bohr's theory, leading eventually to inconsistencies or discrepancies during further development, and constitutes according to Lakatos (1974) a degenerative research programme. In most cases, the reason is an incorrect assumption about the dynamics. The second case corresponds to the occurrence of competing theories, and the last to a theoretical attempt that is not extended to a theory due to direct discrepancies with experimental facts. Consequently, the procedure of consistent adjustment provides, at least in a qualitative respect, a suitable approach for understanding certain aspects of theory development.

2. In order to solve the problem within the domain of rational numbers, f_n^2 is neglected without formally exact justification. Such a practice runs like a golden thread through theory formation in physics. Specifically, approximations are made that appear plausible but are not formally justifiable. In the first instance, the justification consists only in some successful applications. As in this example, approximations are frequently made in order to enable mathematical solvability. The advantage of such a procedure is obvious. After establishing such an ansatz, the application of mathematics works by itself, and one is in the deductive phase of theory formation. In this example, this corresponds to the computation of x_{n+1} from x_n.
3. Although the equation cannot be solved exactly within the rational numbers, the result can be obtained to any desired degree of accuracy. Notwithstanding the fact that equality in a mathematical sense is not achieved, this is irrelevant in physics and is analogous to the comparison between experimental and theoretical data. Neither will ever agree exactly, but only within the limits determined by theory and experiment. The limitation of this analogy is encountered when the process of theory formation in its entirety is interpreted as an asymptotic approach towards an "exact theory" or even a "true theory". There is no limiting value for entire theories, but only for single results, such as the numerical values of material or fundamental physical constants.
4. In more complex problems, the iterative procedure must usually be directed in a targeted way in order to achieve convergence. The same is the case when concepts, laws, or theories are purposefully modified or extended in order to reduce deviations from experiment. Steering the iterative procedure requires, in both cases, problem-specific knowledge to be introduced from outside. Hence, the procedure of consistent adjustment is

not an automatic one which requires, for a given algorithm, only a suitable initial value to generate the correct final result.

5. The example demonstrates that approach to a correct final result is possible, even if the initial value is only estimated and also the equation is only approximately valid, so both are false in the sense of (two-valued) logic. This reveals itself to be the fundamental property of self-consistent procedures and establishes their strength. In order to obtain correct results, one requires neither the ultimately secured empirical basis of complete and precise data, as in the case of an inductive reconstruction, nor a theoretical basis of exactly defined terms, as in the case of a deductive reconstruction.

Overall, this method turns out to be the most appropriate for characterizing the relationship between theory and experience with its consistent adaptation of theoretical-deductively and empirical-inductively obtained results, as is typical for the development of physical knowledge by mutual support, which results in extended knowledge. For instance, on the basis of theoretical advances, new instruments and measuring methods are developed, paving the way to obtain new and more accurate experimental data for an improved comparison between experiment and theory. Such an adjustment process is not necessarily circular, but may function as knowledge-extending and problem-generating. Similarly, such feedback, typical of this method, is encountered in concept formation, because the meaning of physical concepts follows only in the context of the entire theory, and as a result of consistent adaptation with empirical definitions. This is essential for the reconstruction of physical knowledge. If concepts acquire their meaning both from theoretical context and from empirical experience, the relationship between theory and experience cannot be described as a linear-hierarchical scheme.

The analysis of theoretical concept formation by means of representative examples in Sect. 2.5 will demonstrate that a linear reconstruction of concepts by explicit and formally exact definitions is not only unattainable, but not even meaningful, since it encompasses neither a transformation of meaning resulting from theoretical developments, nor the model-theoretic aspect inherent in any physical concept formation. Furthermore, important structural features of physical knowledge are ignored, such as the relations and crosslinks between the various physical terms, laws, and theories, as well as the way they mutually support and substantiate each other without ending in a vicious circle. Metaphorically, this may be compared with sticks leaning against each other: none of them is itself self-supporting, but due to the mutual support, the overall construct becomes stable. It is crucial to have this feedback that can be described solely as consistent adaptation, rather than by inductive and deductive methods. Due to this feedback, there are some analogies with the dynamics of self-regulating systems, and science in its development is occasionally interpreted as an evolutionary, self-regulating system. However, scientific development in general, and theory formation in particular, are constructive and substantially goal-oriented processes that are not adequately described by models of this kind.

1.5 The Evolution of Thought Patterns

Since the considerations of Locke, at least, it is apparent that the question of the structure of reality is inseparably connected with the perceptive faculty and thought patterns determining the capability and constraints of human cognition. Since Galileo, physics is also based predominantly on thought patterns for the following reason. Associated with the introduction of mathematical methods by Galileo is the detachment from direct observations, and the construction of an "ideal reality" via idealization and abstraction. As a consequence, physics rests rather on theoretical considerations and reasoning ("thought experience") than on sensory experience (Brown 1986):

> The Scientific Revolution happened because people stopped looking and started thinking.

Accordingly, not only has thought experience been dominant since the development of modern physics in the twentieth century, but it was already so in mechanics with its relatively direct connection to observation. That this is ignored throughout and the importance of sensory experience in relation to thought experience is generally overestimated has roots in both the sociology of science and epistemology. Firstly, it is a matter of fact that, in empirical science, the acceptance of new knowledge is considerably facilitated if legitimated by observations and facts. Secondly, the epistemic root is the missing distinction between perception and cognition, which comes about through assuming that they have essentially the same structure. This is, however, easily refuted (Bayertz 1980, Chalmers 1982). While perception constitutes a physiological process and refers to singular events, the formation of cognition necessitates repeatable or similar events in combination with a memory for storage. Without recognizing repetitions and similarities, cognition cannot be attained. Each perception would then be new, unique, and without connection to previous ones. Therefore, the process of perception is a necessary prerequisite, but not the decisive step toward gaining physical knowledge.

If thought processes play the more important role in acquiring physical knowledge, the question arises as to the relationship between human thought patterns and "structures of reality", in the sense of observable regularities in nature. The spectrum of possible answers is illustrated by the following extremal positions:

- Thought patterns are independent of events occurring in nature: what appears as structure of reality, is actually notional construction. Accordingly, physical laws, for instance, are free inventions of physicists.
- Thought patterns are mere reflections of structures of reality: what is conceivable, exists in reality. Accordingly, physical laws exist as parts of nature and only need to be discovered.

The first view is found in idealism, conventionalism, and among some adherents of radical constructivism, while the belief in the identity of epistemic and ontic structures is characteristic of naive realism. According to this view, Aristotle did not discover Newton's laws merely due to insufficient observations. In spite of their apparent contrariness, both positions have in common a linear structure because, among thought patterns and structures of reality, one is taken to be the presupposition for the other one. Many aspects of this relation, however, correspond to a development through consistent adaptation. In order to describe this evolution of thought patterns, one need not discuss either pure perception as a physiological process or the evolution of the organs of perception. One need only assume that the organs of perception react to external stimuli. Otherwise, only the processing and structuring of perceptions is of interest. Without pretending to encompass all the details, this process may be broken down, in thought, into six steps.

1. There exists a perceivable external world of objects which is, in its essential traits, independent of any individual subject and its perceptions (subject–object separation). In everyday behaviour this is reflected, e.g., in the expectation that something does not vanish as soon as one does not look at it. In this elementary form, subject–object separation appears entirely reasonable and problem-free. The challenges begin, however, when one asks for criteria that would prove the independence of the external world from perceptions.
2. Beside directly perceivable objects, there exist a number of naturally occurring regularities that are independent of human awareness and are recognized as such. Representative examples are the change between day and night or between the seasons, organic life from birth to death, or the periodicity of planetary motion. The recognition of regularities in combination with the ability to collect nonsingular and reproducible experiences necessitates memory. Indeed, recognizing structural similarities or analogies, beyond the mere statement of simple identities, is impossible without stored information as a standard of comparison. This constitutes the central problem for gaining knowledge. On the one hand, a memory is necessary to recognize structures. On the other hand, the human memory works selectively and not always reliably, whence the recognition of structures is flawed with uncertainties.
3. In the next step, based on perceived repetitions, periodicities, or permanences of events, certain patterns for structure recognition are formed despite the existing variability leading to differences and changes. These patterns become manifest in certain attitudes regarding expectations that determine subsequent actions:

   - the expectation that a certain event occurs increases with the number of confirmed expectations,
   - the perception that two events frequently succeed each other increases the expectation that they are associated, irrespective of how this connection is interpreted in detail, e.g., as causal, final, or divine,

- the expectation that two things fulfill the same function increases with the number of common properties,
- each success of an action strengthens the expectation of being successful by the same method in a similar situation,
- differences in perception correspond to differences in the external world.

These attitudes toward expectations have a distinctly normative character and remain unchanged, even in the face of opposite cognitive experiences.

4. Based on this confidence in expectations produced by direct contact with reality, in combination with the patterns of action developed in parallel, elementary thought patterns are formed which serve to interpret already known and new events:

   - inductive or generalizing thinking: universal validity is inferred from a finite number of occurrences of an event,
   - causal and final thinking: connections between events are constructed in thought and considered as truly existing, even when not directly perceptible,
   - classifying thinking: similar things and events are categorized according to certain common properties by abandoning differences in detail,
   - analogous thinking: a few common properties are used to conclude to the commonality of others,
   - equivalence thinking: the equality of events is used to infer to the equality of their causes,
   - forbiddance thinking: a conceivable event that never occurs is unattainable ("forbidden") for certain reasons.

   These thought patterns seem to be universal in the sense that they are independent of any specific culture, although without claiming that they are genetically conditioned. Rather, their universality illustrates how learned actions and practical experiences influence and determine modes of thinking, which is also important in physics. Other more specific thought patterns result from experienced social and cultural structures, e.g., economic and hierarchic thinking or law-like thinking. However, their identification as culturally conditioned would require a comparison between dissimilar cultures, e.g., Western Christian with East Asian or Indian.
5. By means of these thought patterns, a treatment and processing of subsequent perceptions and experiences occurs which goes far beyond any pure process of perception, and is typical of the cognition process. The consolidation of thought patterns due to repeated successes and increasing familiarization eventually results in the transformation of at least some of them into structures of reality (Falk und Ruppel 1972):

   > A sufficiently often and in particular unconsciously repeated experience exhibits the tendency to reinforce itself in such a manner that it is easily considered to be more than pure experience, especially if this experience is

sure to be taken for granted by all other humans, so that reference to it can be accepted without any objection.

This tight coupling between structures of reality and structures of thinking and acting is one of the roots of the difficulties involved in distinguishing between reality and its conceptual description. Law-like thinking and determinism are instances of thought patterns that have played a key role in the development of the world view of physics.

6. Finally, perception, pattern recognition, and interpretation of structures act together in the construction of subsequent order schemes, and these in turn serve to gain new information, insights, and knowledge, in addition to the reinterpretation of old ones. This newly generated knowledge will be used both for constructing new patterns and for collecting new knowledge, so that the entire cycle can start again, although on a higher level.

In summary, the starting point of human thinking is certain attitudes regarding expectations and patterns of action that can be traced back to a number of structures of reality in the form of perceptible and experienced regularities. From these, thought patterns evolve that are already abstractions from and interpretations of these structures of reality, not their mere image. These thought patterns are utilized retroactively to structure and explain phenomena, as well as to create new information and knowledge. The "natural order" constructed in this way then corresponds, in a certain sense, to the order of thought. Consequently, thought patterns are neither a priori nor a posteriori. On the one hand, they go back to perceptions and experiences. On the other hand, they are used to collect and structure new perceptions and experiences, and they affect them accordingly: the handling and processing of perceptions use, in principle, all available knowledge, and what is preferably perceived matches the existing thought patterns. That such a process does not necessarily end up in a logical circle, but acts as feedback, is illustrated by the development of physical theories like the relativity and quantum theories, which lead to results radically at variance with a number of direct experiences. Nevertheless, the problem remains that no ultimate certainty can be attained regarding how far perceptions and thought patterns restrict the abilities of human cognition, and eventually the entire process ends up in a circle that cannot be identified, and thus cannot be escaped. This is a genuine property of processes of consistent adjustment: the possibility of collective errors that also occurred in the historical development of physics can never be excluded.

Analogously, the cognition process in physics contains such feedback, and it is of crucial importance, especially with regard to the relationship between theory and experience, in that it serves to make empirical and theoretical knowledge consistent with each other. For several reasons, this process cannot be reconstructed either as gaining theoretical knowledge inductively or as mere empirical testing of theoretical results. Firstly, empirical data do not constitute the predetermined, invariable basis of physics, but may be improved, revised, reinterpreted, or refuted due to advances in knowledge and development of new instruments. Secondly, since empirical knowledge and theoretical knowledge do not determine each other unequivocally, ambiguities exist in the relation between theory and experience that can be reduced in the course of the adjustment process, but cannot be entirely eliminated. Accordingly, observed phenomena permit dissimilar interpretations, with preference, in general, for those that are compatible with available thought patterns. At the conclusion of this process, structures of reality and thought patterns are no longer differentiated, so they are inadmissibly identified. This tendency is present in all empirical sciences, and constitutes the main source of erroneous notions regarding the structure and status of physical knowledge.

1.6 Tacit Assumptions

The origin of physical methodology may be interpreted as the subsequent development of patterns of action and thought, originally created to master practical problems and to comprehend natural phenomena. Transferring these unreflected experiences of daily life to views on the subject, methods, and aims of physical research yields a number of implicit assumptions which constitute the philosophical and methodological basis of physics as an extension and specification of constructive realism. These assumptions possess the following general properties and functions:

- Apriority: although not a priori in the absolute sense of anteceding any type of experience, they constitute the apriority of gaining knowledge in physics. This does not mean that knowledge about nature is impossible without them, but they are the necessary basis to obtain knowledge in physics, in the sense that a science of nature that does not rest on these assumptions would be structurally different.
- Epistemological status: they comprise largely unreflected views of reality, so that they are to be classified as metaphysical in the sense of empirically unprovable assumptions. Accordingly, they are presuppositions, but not results of physical research. Although they may act to guide cognition in a heuristic sense, they do not possess the status of knowledge about nature.
- Methodological function: they comprise heuristic and pragmatic principles of experimental methodology, as well as ideas and desires about what theoretical methods ought to accomplish and thus what ought to be the objective of physics.
- Normative property: they are assertions without any provable claim of legitimacy since they are not universally valid. Consequently, their justification is not proven by any physical theory, and nor is such a proof the subject or aim of physics. Contradicting experiences require explanation, but do not lead to the correction of these assertions. Accordingly, they have the status of norms that can be made plausible, at best, by

referring to their roots. Nevertheless, in some cases, limited modification may be possible due to new insights. However, such modifications occur rarely compared with changes in the domain of factual knowledge, and then exhibit properties typical of a process of consistent adjustment.

These presuppositions are virtually never mentioned in physics because, due to their claimed self-evidence, most physicists accept and comply with them. For this reason, they are referred to as "tacit assumptions" (Grodzicki 1986). They fall into three groups, viz., assumptions about reality, and also about experimental and theoretical methodology.

The assumptions about the view of reality may be classified as ontological because they are concerned with the properties of nature as the subject of physical research. They comprise three parts:

(1A) There is a part of the world referred to as the external world or subject-independent reality that exists independently of human awareness, perceptions, or experimental and theoretical investigations.
(1B) The external world is structured: there exist structures, independent of awareness and perceptions, that are as real as objects, even when not directly perceptible, but identifiable only through constructive thought processes.
(1C) The entirety of objects and structures in the external world is recognizable: there exists an at least partial conformity between ontic and epistemic categories.

The first assumption is identical to the fundamental assumption of realism and goes back to the pre-scientific experience that many things exist, regardless of whether they are perceived or actually known. Its legitimacy is neither discussed nor scrutinized in physics, i.e., physics *does not* prove the existence of a subject-independent reality, but *takes it for granted*. This assumption is indispensable for reasons of internal methodological consistency. Firstly, as a methodological criterion, the requirement that theoretical results should be empirically verifiable does not make sense if things and phenomena are not part of this reality, but instead results of theoretical reasoning or subject-dependent constructs, or exist only after being discovered. Secondly, the explanation of the mode of operation of instruments, especially in modern physics, rests crucially on this assumption. It does not, however, specify the extent of the subject-independent reality as known at a certain point of time. Concepts, such as phlogiston or caloric as the material carrier of heat vs. atom or electron, demonstrate how the accepted extent has changed over time and depends on the state of research.

The epistemologically more interesting assumption is (1B), returning to the pre-scientific experience that such structures exist, e.g., a living organism is not just an orderless crowd of molecules, cells, or organs. As soon as the status of at least certain structures as a part of reality cannot be questioned, the investigation of these structures belongs to the universe of discourse of physics, irrespective of whether or not they are directly observable. Actually, structures of reality are the main topic of fundamental physical research. Central constituents of physical theories are assumptions and statements about structures, e.g., symmetries or hierarchies. Moreover, the references theories make to reality rest partly on the experience that such structures are a part of reality rather than pure fiction. As in (1A), criticism of this assumption does not concern the existence of structures *per se*, but solely their extent because they are constructed and believed to be objectively existent, even when this is not the case. The constellations of stars provide a representative example. However, there are no methodological criteria or procedures of general validity that would enable one to decide without any doubt whether a postulated structure is fiction or part of reality. Indeed, really existing and fictitious structures are not clearly distinguished in physics, as illustrated by the multifaceted use of the force concept. While the subject-independent existence of electromagnetic interactions is rarely questioned, forces without any truly existing counterpart are sometimes defined ("pseudo forces") because they are useful to treat certain problems. Overall, this assumption is indispensable because it is constitutive of contemporary physics in which the structuring of knowledge is one of the basic objectives, not just the mere collection, classification, and economical description of data and observations. Consequently, abandoning this assumption yields an entirely dissimilar comprehension of the tasks and objectives of physics.

The last assumption regarding the recognizability of the external world is more an epistemic one, but is occasionally reinterpreted ontologically (Einstein and Infeld 1947):

> Without the belief in the internal harmony of our world, science could not occur.

Again this assumption is indispensable for methodological reasons. Without an at least partial conformity between ontic and epistemic categories, observations and experiments could not yield reliable information about reality. That the conformity can be only partial concerns primarily those parts of reality that are not directly accessible to human sense organs, but also misperceptions. In addition, this assumption explicitly expresses the fact that the subject of physics is reality as object of perception and cognition, but not a fictitious underlying reality-in-itself. Otherwise, by analogy with both previous assumptions, (1C) specifies neither the kind of conformity nor the reasons for its origin.

In summary, the assumptions of the first category indicate that the existence of a subject-independent, structured reality is not challenged, even though it cannot be proven. The human organs of perception and reasoning are adapted to reality to such an extent that sufficiently reliable knowledge is obtainable which is consistent with this reality. In contrast, regarding the extent, quality, and constitution of reality and its structures, as well as the type and extent of correspondence and conformity, nothing is declared. In special cases, physical methods and results may contribute both to identifying something as a part of reality and to controlling the reliability of perceptions and observations. However, this proves neither the existence of a structured reality, in general, nor the truth of knowledge.

The tacit assumptions of the second group describe the basis and methodological ideals of experimental practice. Their origin goes back to the seventeenth and eighteenth centuries when experimentation was developed as a systematic activity, disciplined and codified through education and professional rules. Because the development of modern physics in the twentieth century had virtually no influence on the principles of experimental practice, they are still determined by the classical conceptions:

(2A) The entirety of events in the physical world, unique and inseparable in space and time as it may be, can be decomposed into repeatable, isolable elements.
(2B) An experiment repeated under identical boundary conditions must produce identical results (reproducibility), and the experimenter must not be a part of these conditions (intersubjectivity).
(2C) An experiment in the laboratory as a "constructed phenomenon" has the same ontological status and epistemic significance as a natural phenomenon.
(2D) Among the physical properties measured on any system, some of them may be assigned to the system itself, independently of any experimental studies ("strong objectifiability").

The first assumption provides us with decomposition and isolation (the "analytic method"), the methodological foundation of physics. It constitutes the indispensable condition to obtain reproducible, controlled, intersubjective knowledge and to establish the connections between theory and experience. Both require repeatable, dynamically isolable processes because regularities and law-like relations can neither be recognized nor scrutinized by examining singular events or processes that are too complex. As a consequence, theory formation in physics cannot be oriented toward processes that are either not reproducible or last too long to be factually reproducible, like the evolution of the cosmos or the evolution of life from protozoans to *Homo sapiens*. Of course, this does not exclude the possibility that singular events may give an impulse to theory formation or that physical knowledge may in turn be applied to the study of nonreproducible processes. However, the application of physical theories is quite different from their development and is of marginal interest in this context. The assumption of repeatability goes back to the pre-scientific experience that regularities exist in nature in the form of periodically recurring events. This leads to the expectation that two events should occur identically if they differ solely by occurring at different times or places. In other words, space and time are assumed implicitly as homogeneous. Isolability means that certain events or parts of the world are sufficiently well separable from their environment that they may be assumed to be isolated, irrespective of the possibility that everything may interact with everything else. More precisely, this part may be termed dynamical isolability. Its success rests on the fact that in many physical processes only a small number of influences play a role. It is important to emphasize that the assumptions about homogeneity of space and time and dynamical isolability must not be misunderstood as claims about the constitution of nature. They merely express the expectation that such a decomposition is feasible and admissible. Although contradicting experiences necessitate explanation and will not simply be ignored, they do not lead to the revision of this attitude regarding expectations ("normative"). Most difficult to estimate is the dynamical isolation. Actually, it may happen that the assumption of closure or the neglect of interactions turns out to be improper and must be revised. For example, many molecular properties cannot be described by decomposing the molecule in thought into the constitutive atoms and attempting afterwards to investigate the associated atomic properties, because most molecular properties only emerge due to the mutual interactions, and yet other properties may emerge from the much weaker intermolecular interactions. In particular, dynamical isolation necessitates extensive expert knowledge, and depends on the systems and properties under study.

Assumption (2B) is the extension of (2A) towards the indispensable methodological norm of experimental science: experiments must be performed and described in such a way that they may be reproduced by every scientist possessing comparable devices and knowledge at a later time or at another place, but under otherwise identical conditions. In contrast, the uncontrolled production of singular facts is not science. The norm of reproducibility cannot be fulfilled without theoretical knowledge, because performing reproducible and controlled experiments is impossible without information on both the experimental conditions and the physical state of the system under study. This becomes apparent when one reaches the limits for realizing this norm, e.g., the location dependence of pressure or boiling point measurements, the repeatability of experiments with radioactive sources, or experimental investigations on real crystals where the influence of dislocations, vacancies, etc., becomes difficult to control. The production of reproducible data can then require considerable effort. Similarly, the norm of intersubjectivity must not be understood in the sense that the scientist is a negligible variable in the process of gaining knowledge, but rather in the sense that experiments must be performed in such a way that this condition is fulfilled. This norm holds without any restriction both in classical and modern physics. Experiments on quantum systems are performable in a manner in which the results are subject-independent. However, the possibility of intersubjective knowledge does not permit the conclusion that an intersubjectively recognized phenomenon corresponds to an element of reality. If all scientists make an identical observation, this could be caused by a general deficit of human perceptive capability or by some feature of the experimental method. Even reproducible and intersubjectively accepted knowledge does not exclude the possibility of collective errors.

Although the assumption (2C) appears so self-evident to contemporary physics that it hardly seems necessary to mention it, it is not at all dispensable. It establishes the assertion that physical knowledge is valid not only for the reality of the laboratory, but also for the external world. This corresponds to the assumption that, under laboratory conditions, physical systems do not behave differently than in nature. Consequently, the legitimation of laboratory experiments becomes dubious when these conditions are no longer fulfilled, as for laboratory experiments in psychology. It may also be mentioned that Goethe's basic criticism of Newton's optics was exactly this "unnatural manipulation of nature". Similar objections were raised against the transfer of results of Darwin's breed selection experiments to natural selection. Finally, laboratory physics has substantially contributed to the abolition of the differences between nature and technics. On the one hand, technics became "natural". On the other hand, the result was a technical-instrumentalist attitude against nature. Both directly influence the physical view of reality.

The last assumption (2D) is crucial for the identification and recognizability of physical systems. Substantial doubts about its justification, raised in quantum theory, show that it is far from being as obvious as it might initially appear on the basis of experience of classical physics. Actually, every measured property is, in the first instance, defined as the combination of the measured system, measuring instrument, and their mutual interaction. The subsequent conclusion from the relation between two systems to a property of one of them is a substantial, additional step appearing unproblematic only relative to the background of classical physics. It was only with the development of quantum theory, where dynamical properties such as position, momentum, or energy cannot be assumed, in contrast to classical mechanics, to be strongly objectifiable, i.e., pre-existent system attributes independent of the measuring process, that it became clear that this step is not at all unproblematic. Nevertheless, even in quantum mechanics, some strongly objectifiable properties exist, such as mass, charge, and spin, thereby enabling the identification of quantum systems. Consequently, analogously to the ontological assumptions, the existence, in general, of strongly objectifiable properties as a prerequisite for identification cannot be questioned. It is just that the number of such properties remains undetermined and depends on the mode of theoretical description.

The last group of tacit assumptions constitutes the implicit basis of theoretical methodology, and reflects expectations about the tasks and objectives of physics. These expectations are dominated by views with roots going back as far as Ionic natural philosophy, e.g., the desire for omniscience, the search for truth, or the belief that nothing happens without cause. Despite numerous contrary experiences which should prompt modification or revision of these views, this is rarely the case. In contrast to the assumptions of experimental methodology, they act as a more or less naive heuristic guideline for theory formation:

(3A) Nature is "governed" by laws ("natural laws"): the structures of reality, assumed to exist objectively according to (1B), can be described by strictly valid ("true"), mathematically expressed laws.

(3B) Everything can be explained: for each observed phenomenon some mechanism ("cause") can be found from which it can be deduced and explained in this respect.

(3C) If two or more events are correlated, i.e., occur largely at the same time or in succession, a mechanism exists, e.g., in the form of a physical interaction that is the cause of this correlation.

(3D) The type of admissible mechanism is constrained by the condition of being mathematically representable, and divine, magical, animist (paranormal), and anthropomorphic explanations of phenomena and correlations are excluded.

(3E) Among the various theoretical descriptions of the structures of reality, the simplest is to be preferred (principle of economy).

(3F) A unified conceptual theoretical description ("theory of everything") of the subject-independent reality is possible.

The common attribute of these assumptions is the mathematical presentability of knowledge: physics is the mathematized form of natural research, but to conclude subsequently that nature itself is mathematical would be naive. The assumption (3A) regarding the law-like descriptiveness of reality defines the fundamental, non-eliminable heuristic of theoretical methodology. Only under this assumption can observable phenomena, regularities, and correlations be considered to be in need of explanation, and the patterns of explanation are then the mathematically expressed physical laws. Accordingly, the strategy of physics for describing observations and experiments consists in the thought decomposition into a general constituent assumed to be strictly valid, viz., the physical law, and into particular components comprising system-specific and contingent conditions. Such a decomposition rests on theoretical considerations and is metaphysical in the sense that it is not enforced by mere empirical experience, let alone by the constitution of nature. In contrast, a science constrained to describe phenomena and observations rather than considering them to be in need of explanation would be structurally dissimilar. The possibility of such an option, however, brings out the fact that (3A) is actually an assumption. Neither the existence nor the unequivocal form of a law follows solely from the observation of a phenomenon, regularity, or correlation. Even the most precise experiments are affected by error margins, and are subject to influences from the environment that can never be suppressed entirely. Consequently, the belief in the existence of strictly valid physical laws must be rated as the choice of a particular method to describe nature which is based on the anthropomorphic thought pattern "lawful thinking". For these reasons, even the convincing successes of the strategy of a law-like description of reality do not prove that physical laws are a part of reality and could be discovered or inductively inferred from observations and experiments alone. Overall, physical laws as mathematical relations are conceptual descriptions of structures of reality, i.e., statements about reality but not a part of it. Consequently, natural phenomena and processes are not 'governed" by laws because conceptual descriptions cannot influence natural processes. Nevertheless, in many textbooks, formulations are still found that assert that nature obeys strictly valid "natural laws" or is governed by strictly valid laws. Formulations of this type must be rated merely as "textbook lyricism".

With the belief that nothing occurs without cause, assumptions (3B) and (3C) emphasize even more forcefully the autonomy of a theoretical component of physical methodology that is linked not unambiguously to empirical experience. As demonstrated by Bohr's theory, physics is by no means constrained to the description and classification of factual knowledge. The central subject is rather the investigation and construction of law-like describable structures from which the empirical facts can be deduced. Therefore, the persuasive power of Bohr's theory rests predominantly on the circumstance that numerous empirical facts could be deduced from autonomously constructed, theoretical assumptions. It is precisely the construction of suitable mechanisms which provides the decisive, innovative, genuinely theoretical aspect of fundamental research in physics. Although not free from errors, (3B) and (3C) establish an extremely successful heuristic that can be taken as the basic characteristic of physics since Newton. In addition, it is obvious that these assumptions rest decisively on a concept of reality determined by the assumptions (1A)–(1C).

Besides the occasional occurrence of errors, the fundamental problem with such a heuristic is the lack of any unequivocal correlation between model assumptions and mechanisms on the theoretical side, and observable phenomena and regularities on the empirical side. Therefore, the methodological function of (3D) consists in preventing uncontrolled theoretical speculation. Firstly, the requirement of mathematical representability is the presupposition that one can deduce quantitative, experimentally comparable results. While mechanisms and laws constitute the patterns of explanation in physics, the mechanisms must not be identified with causes. Accordingly, physics neither provides explanations by causes nor pursues the goal of studying causal connections. As will be shown in the subsequent chapters, physical laws are not causal laws. Actually, the law-like description reveals itself as considerably more universal because it is applicable irrespective whether a causal connection exists. Overall, with its laws, principles, and theories, physics provides a set of patterns of explanation that is entirely independent of how these are interpreted. Any explanations going beyond this are not supplied, i.e., physics does not provide explanations for its patterns of explanation. It is exactly this that Newton was referring to in his dictum "Hypotheses non fingo" when asked for the cause of the gravitational force.

Another option to restrict the ambiguity of model assumptions and mechanisms is provided by (3E) as an additional selection criterion between competing theoretical approaches. The concept of simplicity has two dissimilar meanings, viz., mathematical and conceptual simplicity. Mathematical simplicity means that the quantitative description of correlations or data sets ought to be as simple as possible. This does not constitute a criterion of ontological or epistemic relevance in selecting a representation, but arises for merely pragmatic and psychological reasons. Firstly, empirically comparable results are more easily deduced from simple mathematical equations. Secondly, a simple description is sensed as being more elegant than a complicated one. Therefore, among the various options to describe the non-equivocal relation between theory and experience, the simplest one is preferred, but this choice is not enforced by nature. Conceptual simplicity represents the principle of economy in its genuine meaning, viz., a theory should be based on a minimum number of unprovable assumptions. A procedure introducing a new theoretical assumption for each new fact counts as unscientific. In turn, this results in the heuristic principle of postulating no diversities in the theoretical domain that are not amenable to experiment. In contrast, (3E) must not be understood as a statement about a property of nature. A postulate of simplicity does not exist for nature itself, in the sense that nature does not do anything surplus or useless, although such a belief can be traced back to the Ionic philosophy of nature, and is thus deeply rooted in human thinking. To conclude from the success of simple descriptions to the simplicity of nature again corresponds to the identification of ontic and epistemic categories. Actually, there is no guarantee that a strategy oriented toward a principle of economy must be successful a priori. The frequent success of such descriptions rests, in analogy to dynamical isolability, on the circumstance that in many physical processes only a limited number of influences play a role.

The last assumption defines, as one of the central objectives of physics, the unification of knowledge by reducing the diversity of phenomena to a minimum number of fundamental principles. Although prominent physicists in the nineteenth century advanced the opinion that the aim of physics ought to be a complete economical description of observable phenomena, this was never the case. This is already demonstrated by an example as simple as Kepler's laws, because the observed planetary orbits are not ellipses. In contrast, the directly observed paths are described by the epicycle theory of Ptolemy. If physics were actually constrained to the description and prediction of observable facts on the basis of arbitrarily changeable assumptions, the efforts of physicists to construct unifying theories are not comprehensible, and nor is the fact that physics has changed and established world views. How unification should be realized remains undetermined. A prevalent attitude advances the view that, due to the "oneness of reality", a conceptually uniform, theoretical description of reality must exist. Such a conclusion is again structurally nothing else than identifying an object with its conceptual description, and just represents a faith in naive realism. In particular, the analysis of the various types of physical theories in Chap. 3 will show that a conceptually uniform, theoretical description of reality cannot be expected. The modified ultimate aim of modern physics is instead the construction of a consistent, unitary world view (not theory). If unification is understood accordingly as subsumption under a few general structuring principles, it remains undetermined whether these principles, constituting the fundamental patterns of explanation in phyics, are conventions or fiction, or indeed to what extent they correspond to structures of reality.

In summary, the tacit assumptions demonstrate the extent to which physical methods and results are based on metaphysical, i.e., non-empirical elements in the form of normative beliefs, pre-scientific experiences, and more or less naive attitudes toward expectations. Although their significance for physics is ignored throughout because things regarded as self-evident are rarely scrutinized, these assumptions are indeed indispensable in that a science of nature not relying on these assumptions would not be identical with contemporary physics. In this respect, they form the basis of the type of research about nature which has developed under the socio-cultural influences of the Christian Occident. On the other hand, this does not justify the conclusion that any type of knowledge about nature is impossible without these assumptions, so they must not be misunderstood as a priori or transcendental requirements for gaining knowledge about nature. Accordingly, these assumptions are modifiable within certain limits. Due to their normative character, however, such changes occur comparably rarely and considerably more slowly than in the domain of factual knowledge. In addition, such changes are accompanied by profound changes in the world view of physics which are frequently considered to be revolutionary. This became apparent when certain assumptions which characterize classical physics were revealed to be untenable, e.g., views about space and time, determinism, the relevance of causality, and ultimately the belief in an unrestricted, law-like description of nature. In the subsequent chapters, it will be shown to what extent these tacit assumptions influence the world view of physics and the external image of physics.

References

Bartels A (2009) Wissenschaftlicher Realismus. In: Bartels A, Stöckler M (eds) Wissenschaftstheorie. Mentis, Paderborn

Batterman R (ed)(2013) The Oxford Handbook of Philosophy of Physics. Oxford UP

Bayertz K (1980) Wissenschaft als historischer Prozeß. W Fink, München

Böhme G (ed)(1976) Protophysik. Suhrkamp, Frankfurt

Bohr N (1913) On the constitution of atoms and molecules. – Part II. Systems containing only a single nucleus – Part III. Systems containing several nuclei. Phil Mag **26**: 1, 476, 875

Bridgman PW (1928) The Logic of Modern Physics. McMillan, New York
Brown JR (1986) Thought Experiments since the Scientific Revolution. Internat Stud in the Phil of Science **1**:1
Carnap R (1966) Philosophical Foundations of Physics. Basic Books, New York
Chalmers AF (1982) What Is this Thing Called Science? The Open UP, Milton Keynes
Dransfeld K, Kienle P, Vonach H (1974) Physik I. Oldenbourg, München, Wien
Einstein A and Infeld L (1947) The Evolution of Physics. Cambridge UP, Cambridge, Mass.
Falk G and Ruppel H (1972) Mechanik, Relativität, Gravitation. Springer, Heidelberg
Feynman R, Leighton R, Sands M (1964) The Feynman Lectures on Physics. Vol I, London
Fleischmann R (1980) Einführung in die Physik. Physik Verlag, Weinheim
Fließbach T (1996) Mechanik. Springer, Heidelberg
Glasersfeld E von (1985) Einführung in den radikalen Konstruktivismus. In: Watzlawick P (1985)
Grodzicki M (1986) Das Prinzip Erfahrung - ein Mythos physikalischer Methodologie? In: Bammé A, Berger W, Kotzmann E: Anything Goes– Science Everywhere? Profil, München
Haller R (1993) Neopositivismus. Wiss Buchges, Darmstadt
Harré R (1986) Varieties of Realism. Blackwell, Oxford, New York
Hempel CG (1970) On the Standard Conception of Scientific Theories. Minnesota Studies in the Phil. of Science. Vol IV:142. U of Minnesota Press, Minneapolis
Lakatos I (1974) Falsification and the Methodology of Scientific Research Programmes. In: Lakatos I and Musgrave A (eds) Criticism and the Growth of Knowledge. Cambridge UP, Cambridge, Mass.
Leplin J (ed)(1984) Scientific Realism. U of California Press, Berkeley
Ludwig G (1978) Die Grundstrukturen einer physikalischen Theorie. Springer, Heidelberg
Mittelstraß J (ed)(1995) Enzyklopädie der Philosophie und Wissenschaftstheorie. Stuttgart-Weimar
Nagel E (1982) The Structure of Science. Routledge & Kegan Paul, London
Poincaré H (1905) Science and Hypothesis. The Science Press, New York. Reprint 1952 Dover, New York
Popper K (1962) The Logic of Scientific Discovery. Hutchinson, London
Ritter J, Gründer K, Gabriel G (eds)(1971–2007) Historisches Wörterbuch der Philosophie. 13 vols. Schwabe AG, Basel, Stuttgart
Röseberg U (1984) Szenarium einer Revolution. Akademie Vlg, Berlin
Rutherford E (1911) The Scattering of α- and β-Particles by Matter and the Structure of the Atom. Phil Mag **21**:669
Sears FW, Zemansky MW, Young HD (1987) University Physics, 7th edn. Addison-Wesley, London
Stroppe H (1981) Physik. Hanser, München, Wien
Watzlawick P (1985) Die erfundene Wirklichkeit. Piper, München, Zürich
Willascheck M (ed)(2000) Realismus. F Schöningh, Paderborn

Chapter 2
Methodology of Physics

The structure of physical knowledge and methods comprises three domains, viz., an empirical one with predominantly inductive methods, a theoretical one with predominantly deductive methods, and the relations between theory and experience characterized by the methods of consistent adjustment. The various experimental methods establishing the empirical basis will be discussed in detail. This empirical basis consists among other things of reproducible events and representative objects instead of singular experiences and direct observations ("sense data"). The dissimilarities between empirical and theoretical methods of concept formation are elaborated and contrasted with each other. Theoretical orders are defined as the conceptual reproduction of structures of reality and as the epistemological kernel of reality cognition in physics. In this context it is shown that Newtonian mechanics is not a causal theory. The diverse meanings and uses of models in physics are classified and elaborated in detail.

2.1 Scientific Methodology

The subject-specific or cognitive constituents of empirical science are the factual knowledge and the methods constituting this knowledge. Although there is broad agreement that something like a scientific method exists, distinct from the problem-solving strategies of everyday life, rather the opposite follows from descriptions of scientific methods in many textbooks which illustrate only the pragmatic attitude of everyday problem-solving. This impression is enforced by metaphors comparing science with detective stories or the exploration of unknown continents. One conceivable reason might be the prevalent view that the objective of science supposedly consists in providing explanations and predictions. Accordingly, science is understood merely as a particularly sophisticated type of "puzzle solving" (Kuhn 1970), so that the differences between scientific and extra-scientific methods are, at best,

M. Grodzicki, *Physical Reality – Construction or Discovery?*,
https://doi.org/10.1007/978-3-030-74579-0_2

of a gradual, but not of a qualitative nature. This raises the question: what are actually the aims of science, and according to which criteria are certain modes of knowledge rated as scientific?

The emergence of a scientific methodology could ideally be described as a process of consistent adaptation of the desire for knowledge to the actual capability of scientific methods. Starting from the primordial aims of the human quest for knowledge, viz., the desire for omniscience and the search for truth, normative ideas can be enunciated about ideal forms of knowledge, properties of scientific knowledge, and objectives of science. Subsequently, methods should be developed to realize these ideas. Experience gained by practical handling of these methods, in combination with general epistemological considerations, should uncover the capabilities and limits of these methods. Finally, this should lead retroactively to correction of these primordial ideas. In fact, this is not the case because normative convictions change very slowly or not at all, in spite of contrary cognitive experiences. As a consequence, concerning both the status of physical knowledge and the aims of physics, views still prevail that correspond to these primordial ideas, instead of the capabilities of physical methodology.

This is also demonstrated by the historical development of scientific methodology (Losee 1972, Detel 1986). The onset of science is basically associated with the insight that the desire for omniscience requires the structuring of knowledge and, in addition, does not only comprise the recognition of perceivable phenomena, but consists of knowledge of their causes. Accordingly, in ancient Greek philosophy, it was considered that the essential feature of science was not merely ascertaining facts, but searching for the underlying causes and general principles. Scientific knowledge is then true, necessary, and largely provable knowledge of these structures. The first systematic representation of a scientific methodology goes back to Aristotle and consisted of two parts, viz. induction and deduction. The characteristic feature of science was induction as the first step. Based on observations and factual knowledge, this was supposed to lead to knowledge of their causes. It consisted of two components, viz., enumerative and intuitive induction. Enumerative induction comprised the enumeration of those attributes of objects that served as representative properties to define a classification term. This resulted in empirical generalizations of the type "all swans are white". Since every entity has, in principle, an arbitrary number of properties, the number of class members will decrease as the number of properties included in the description increases, until finally each class contains just a single individuum. In a second step, enumerative induction had therefore to be supplemented with intuitive induction, which consisted of a selection and evaluation, and separated the significant properties from unessential or accidental ones. As the central function of the inductive method, this was supposed to lead eventually to the discovery of the general principles and the primary causes imagined as the invariable structures inherent in nature. The search for these was taken to be the genuine topic of science. In the last step, further factual knowledge would be gained and explained by deduction from premises containing among other things these general principles. The rules according to which the deduction was carried out were summarized in the two-valued logic developed also by Aristotle. In this context, he recognized that the logical truth of a statement is determined exclusively by the relation between premises and conclusions so that, in terms of purely logical criteria, deductions exist that do not explain anything according to common sense. In particular, this is the case when the premises are either false or unrelated to the explanandum. Therefore, with regard to their content the premises of an explanation had to fulfill four requirements, viz. to be true, unprovable, better known than the explanandum, and to be its cause(s). The postulate of unprovability goes back to the insight that science is necessarily grounded in some unprovable principles, in order to avoid an infinite regress of explanations. Consequently, any science contains parts that are unprovable.

From the deductive aspect of this methodology evolves soon afterwards one of the fundamental principles used in the structuring of scientific knowledge, viz., the ideal of deductive systematization which aims at the deduction of knowledge from a few assumptions, the axioms, and comprises three postulates:

- axioms and theorems must be deductively related to each other,
- axioms must represent evident truths,
- theorems must be in accordance with the corresponding observations.

While opinions concerning both the status of the axioms and the empirical relevance of the theorems diverged considerably, the first postulate was never questioned and may be considered as the defining property of deductive systematization. The first scientific theory formulated along these lines was the geometry of Euclid, with the consequence that mathematical methods became the ideal of scientific methodology due to their stringency and exactness. Overall, scientific methodology comprises two components, viz., the inductive–deductive method and the ideal of deductive systematization. Knowledge of the true and necessary first principles is obtained through intuitive induction, factual knowledge is gained either empirically through observation or by deduction from the principles, and structuring is realized by deductive systematization. According to Aristotle, scientific knowledge is thus true and structured knowledge. In comparison to physics, physical knowledge is also structured, but physical methods cannot furnish a proof of the truth of knowledge, as will be shown in the following chapters. What physical methods can accomplish is to provide the *secureness* of knowledge in the sense that this secured knowledge enables one to obtain reliable and corroborated predictions. Accordingly, physical methods comprise the procedures for gaining, structuring, and securing knowledge.

Subsequent developments until about the end of the sixteenth century were dominated by attempts to extend induction to a logic of discovery of the first principles, because the belief prevailed that proof of their truth must be furnished by the inductive process of gaining knowledge. Since such an objective is unrealizable, these attempts did not exert any significant influence on the development of methodology, as relevant for physics. In particular, mathematical methods played a marginal role. Their application to nature was considered pointless because natural phenomena are influenced by numerous disturbing factors that should inhibit any representation through strictly valid laws.

This attitude changed only with Galileo, who established physics as mathematized natural research by the introduction and efficient utilization of mathematical methods to describe physical phenomena and processes. This created not only the basis for the quantification of physics, but also extended the tools available to back up knowledge by mathematical methods of proof. The necessary requirement for this mathematization is the second essential innovation, viz., idealization, conceived by Galileo as supplementing intuitive induction because assumptions about idealization, similarly to selection and assessment of properties, cannot be obtained through enumerative induction, but necessitate creative imagination. While selection and assessment are rather taxonomical methods, providing the presupposition for concept formation and classification oriented toward real objects, idealization aims to create an "ideal reality" that can be described and studied using mathematical methods. Accordingly, idealization is carried out with the requirements of the mathematical description in mind, so it is an essentially theoretical type of concept formation dominated by constructive aspects. Finally, the third innovation is the amendment of bare observations by experiments during both the inductive stage of obtaining knowledge and the deductive one in the form of experimental confirmation. In Sect. 2.3, it will be argued that, due to the first two innovations, Galileo was undeniably the founder of the mathematical method in physics, but not the founder of the experimental method, at variance with common opinion. The first attempt to systematically implement experimental methods in a theory of physical methodology goes back to Bacon. Only later did the view of experiments as the sole source of secure knowledge, and the final judge over theory, develop into the normative ideal of physical methodology.

According to Bacon, science must start with the collection of facts about observations and experiments, as complete and precise as possible, which should serve as the secure basis of knowledge. In this process, experiment plays a more important role than bare observation. In the next step, all possible correlations between these facts must be found in order to ascend subsequently by induction from correlations with a minor degree of generality to more general ones. The most universal correlations will eventually turn out to be the universal principles of our knowledge of nature. The challenge of discriminating between accidental and significant correlations should be solved with a number of methods summarized in 27 "prerogative cases", among which the most important ones are the methods of exclusion and decision (*experimentum crucis*). The novel aspect here was not only the attempt to establish a systematic connection between generalizing induction and experimental methods, but the reduction of scientific methods essentially to just these two aspects. Therefore, both the intuitive–constructive aspect of induction and the deductive part are largely eliminated from Bacon's conception of scientific methodology. Although he does not deny that science contains a deductive state, it has a substantially different significance compared with Aristotle, in that the ideal of deductive systematization is absent and is replaced by a hierarchy of scientific generalizations. Consequently, the deductive state is not an essential constituent either for gaining knowledge or for structuring and securing it. This conception is also reflected in Bacon's diffuse metaphysics, understanding the world as a collection of substances that interact with each other and are provided with certain properties. Since his notion of scientific law is formed by analogy with judicial law, which means that natural laws govern nature and regulate its behaviour, neither his metaphysics nor his methodology reflect the standards achieved previously in ancient Greek science. Although the enduring merit of Bacon is to have established the philosophical basis of experimental methodology in physics, his views overall appear more like a cartoon of scientific methodology and might have been largely ignored if he had not so enduringly influenced the views of empirical scientific methodology in the Anglo-Saxon world that these views still prevail almost unaltered.

The first temporary endpoint in the development of physical methodology was accomplished by Newton. In its basic features, his view resembles those of Aristotle and Galileo, but with somewhat different emphasis and amendments due to the increasing importance of experimental methods. He refers to the inductive–deductive method of Aristotle as the method of analysis and synthesis, and the deductive systematization is realized by axiomatization along the lines of Euclidean geometry. His method of analysis does not correspond to Bacon's reduced version of induction as inductive generalization, but rather resembles the intuitive induction of Aristotle, supplemented by idealization in the sense of Galileo. This is apparent from those places in the *Principia* and *Opticks* where Newton himself uses the term induction. For instance, the law of inertia is not an inductive generalization of observable motions of real bodies, but a thought abstraction, in the sense of detachment from direct perceptions, because there are no uniform rectilinear motions and no force-free bodies. His method of synthesis transcends Aristotle's in that Newton emphasizes the significance of deduced statements that have not been used in the inductive derivation. This entails an important modification. Since novel knowledge can also be gained by theoretical deduction, experiment serves an entirely dissimilar purpose in the process of securing this knowledge. As far as the mere subsumption of given knowledge under general principles is concerned, an experimental demonstration just serves as an illustration, rather than possessing a real function in securing knowledge. Overall, the procedures of idealization and abstraction, gaining novel knowledge through theoretical derivations in combination with their experimental verification, constitute the basis for a theoretically autonomous methodology which did not exist previously in this specific sense. This demonstrates both Newton's *regulae philosophandi*, which may be considered as heuristic principles of theoretical methodology, and his axiomatic method, made up of three steps. The first step is the setting up of the axiomatic system, the second is the specification of correspondence rules that aim to map between theorems of the axiomatic system and observation, and the third is the experimental verification of the deductions from the physically interpreted axiomatic system. The introduction of correspondence rules as an independent constituent of the axiomatic method underscores the fact that Newton differentiated precisely between the abstract meaning of the axiomatic system and its application to experience. As a consequence, besides methodology, scientific knowledge also contains a theoretically autonomous constituent.

Once the theoretical aspects of physical methodology had been assigned their own weight by Newton, the way was paved for the insight that theory formation in physics need not necessarily start with perceptions and observations. This was probably pointed out explicitly for the first time by Herschel (1830), when he recognized from his historical studies that numerous scientific discoveries and developments cannot be subsumed under an inductive scheme. He concluded that two essentially different ways exist to obtain scientific knowledge, viz., through induction or by setting up hypotheses. Consequently, both inductive and hypothetico-deductive methods are legitimate procedures for gaining knowledge, and the way it actually takes place is irrelevant for its acceptance. This created the basis for the subsequent insight that induction constitutes a method for gaining, but not for securing scientific knowledge. This had an important implication. The central goal of formulating and extending induction since the Middle Ages had consisted in the attempt to settle the normative claim about the truth of scientific knowledge, and thus its acceptance, directly through the method used to obtain it. Herschel, on the contrary, declared this to be irrelevant, because the experimental verification of hypothetico-deductively obtained knowledge can also lead to acceptance. Firstly, the securing of knowledge was shifted to the deductive part, and secondly, the acceptance of the premises was no longer linked to their confirmability in a logical–mathematical sense. Only the renunciation of this pretension to confirmability makes the distinction between the context of formation and the context of justification comprehensible and meaningful. This distinction does not mean, however, that they are independent of each other because the strategies of justification depend on whether a certain result is obtained by inductive or by hypothetico-deductive methods. It only matters that the securing of knowledge does not necessarily take place during the process of gaining it, as implicitly assumed in all inductive methods, but also ex post.

This historical overview has recorded the roots and basic aspects of physical methodology. To conclude, let us give a brief introduction to the methods of gaining, structuring, and securing physical knowledge in combination with a preliminary structural model resulting from these methods. Firstly, it should be recalled that physical knowledge is not only determined by the scope of physical methods, but also through general conditions of the human cognitive faculty. Among these are a physiological presupposition concerning the possibility of perception by our sense organs, combined with the ability to store perceptions in memory, at least for some time, and to retrieve them. A good memory is therefore the essential requirement for structuring perceptions and eventually acquiring cognition. Secondly, a central feature of modern science consists in communicating knowledge (Ziman 1978), so that it must be possible to transform observations and results into a communicative form accomplished through language. Altogether, these are physiological, epistemological, sociological, and linguistic issues that are not directly related to physical methodology. For this reason, they do not need to be discussed in detail, except for the reducing aspects, inherent in every step towards more detailed knowledge which any scientific method has in common with perception, structuring, and linguistic specification. For instance, every act of perception selects a single possibility among those available, and this reduces the spectrum of future perceptive faculties. This applies even more to structures that have been stored in memory on the basis of previous experiences and constitute the background influencing further perceptions. More generally, it is a common phenomenon, not restricted to science, to perceive preferably what fits with existing structures. Finally, the linguistic specification of a perception accompanied by the exclusion of non-verbal modes of communication implies, on the one hand, a reduction of the complexity of direct perceptions, but creates, on the other hand, the precondition for generating precise knowledge that is essential for intersubjective communication. The fact that a certain interpretation of the world is already unconsciously assimilated during language acquisition is thus accepted in view of the advantage of an unambiguous mode of communication through language. Given these preliminary remarks, the constituents of physical methodology and knowledge can be itemized, although the presentation here should not be misunderstood as in any way chronological:

1. The first constituent is gaining factual knowledge primarily by experimental methods, but also by observations and theoretical deductions. The result is defined in Sect. 2.2 as the empirical basis comprising objects, data, and effects that must satisfy the requirement of reproducibility as the most important criterion because only reproducible facts are testable and suitable for checking general relations such as physical laws. A result can thus be considered as scientific fact only if reproducibility is confirmed: physical knowledge rests on reproducible experience, but not on singular perceptions. The criterion of reproducibility not only implies a reduction of that part of reality forming the universe of discourse of physics, but cannot be fulfilled without experimental methods. Any type of experimental activity, however, contains a noneliminable, theoretical component because it is goal-oriented and rests crucially on the utilization of instruments.

For these reasons, the elements of the empirical basis exhibit a degree of theory dependence that resides far beyond any theory-ladeness of pure perceptions.

2. Since the empirical basis does not consist of singular perceptions, the conceptual description of its elements is not an unproblematic verbalization of direct observations and experiences. While the colloquial description of a phenomenon already contains an interpretation restricting the potential connotation, this applies even more to reproducible facts and processes. Their description must contain information regarding the experimental conditions of their production, because they are neither reproducible nor testable otherwise. Next, objects are not described colloquially by directly perceivable attributes, but by quantifiable and operationally definable properties, i.e., physical quantities. Many of these properties, like solubility or conductivity, occur only under certain external conditions that must be included in an intersubjectively traceable description. Consequently, the conceptual description of the empirical basis consists neither of protocol sentences nor of the vocabulary of an observational language, because it is based on a physical terminology.
3. Although every description is already associated with an interpretation, the proper interpretation of data, effects, and regularities involves the constructive search for organizing principles enabling the structuring of factual knowledge. The construction of such empirical orders does not merely pursue the pragmatic goal of more clearly arranging factual knowledge so that it becomes more easily accessible. These organizing principles also aim to constitute "natural" orders in the sense that they correspond to structures of reality. For this purpose, one must recognize the essential properties of those objects and phenomena that are to be classified. However, there are no universally valid criteria for distinguishing between essential and nonessential properties, because they depend on the goals pursued within a certain classification. In addition, any assumption concerning the status of the organizing principles as structures of reality implies an unproven claim of validity, and is thus hypothetical. Accordingly, the construction of empirical orders may be described formally as the setting up of hypotheses serving as patterns of structuring and explanation. This reduces the information that is potentially contained in the bare facts, but also means that new experiences are perceived and assessed through the spectacles provided by these patterns. During the production of knowledge, there thus occur selection processes that may contain errors, as in any human activity.
4. According to empiricist views, physical laws and theories are obtained from empirical facts and orders through increasing experimental confirmation, stepwise specification, and generalization. As shown in Sect. 2.5, this is unattainable, largely because empirical concepts do not constitute an appropriate basis for theory formation. For a conceptual specification oriented toward real objects means a more detailed description by more

properties. In contrast, the concepts establishing the basis of physical theories are precise in an entirely dissimilar sense, because model systems are *defined* through properties that are selected with regard to theoretical issues rather than real systems. Only in this way can concepts be obtained that are precise enough for the development of an abstract theory like a physical theory. The loss of flexibility associated with the definition of these theory-oriented terms implies a further reduction of the many more potential representations than empirical terms. Therefore, theoretical concept construction is not the conceptual description of pre-existing objects, but theory-oriented modeling that constitutes the central part of theory formation in physics, because it is oriented toward the objectives of the theory, so that the respective terms acquire their meaning only in the context of this theory. This type of abstraction through the theory-oriented, axiomatic definition of properties is structurally dissimilar from the object-oriented idealization of empirical concept formation by the selection of certain properties assumed as pre-existing attributes. It represents a construction of reality in the literal sense, because it is not certain a priori whether such a constructed model system corresponds to an element of reality.

5. The second step in theory formation, following the theory-oriented construction of terms, is mathematization. It is not at all identical to the bare mathematical representation of results that occurs in the empirical description, e.g., of data sets. Mathematization in theoretical physics means, firstly, the structuring of physical knowledge according to the formal structure of mathematical theories, and secondly, the securing of knowledge by adopting the associated proof-theoretic methods. The mathematical theories are assumed to be axiomatized throughout, and to be pre-existing, self-contained structures, entirely independent of any possible physical interpretation. Apart from the fact that the language of mathematics constitutes the most precise form of communication, mathematization not only helps to avoid logical errors, but is the indispensable prerequisite for quantitative comparison between theory and experiment. Although contributing nothing to the acquisition of factual knowledge, it may exhibit explicitly structural connections and provide new insights. Mathematization, however, is not an end in itself. Since the diversity of a description is reduced with increasing precision, in any particular case, one must ask whether the gain, regarding insights by specification and structuring, compensates for the loss of information compared with a richer though less precise description.
6. While mathematization largely determines the formal framework of the theory, the physical content is not yet specified; in particular, the question of whether a physical theory is empirically relevant in the sense that it possesses any relation to reality. Accordingly, the third essential constituent of theoretical methodology is the physical interpretation of the theoretical framework which takes place in two steps. The first, described

in Sect. 2.6, is the construction of theoretical orders through structural principles. With regard to universality, these exceed by far the ordering principles of empirical classifications, which represent systems of knowledge of rather limited scope due to their orientation toward real objects and direct experiences. In contrast, the world view of physics is primarily constituted by theoretical orders representing autonomous and significant knowledge, since they are based on structural principles which are entirely independent of particular systems and conditions. Accordingly, the theoretical framework does not contain any statements or data that are directly comparable with experimental results. Therefore, in the second step of physical interpretation, the theoretical framework is supplemented by system-specific assumptions as system parameters and initial or boundary conditions that contain the necessary information for comparison with experiment. This transition from the abstract theoretical framework to the empirically relevant realizations can be described formally by the addition of hypotheses. These empirically relevant realizations of a theory are referred to as scientific applications. This decomposition is exactly in the spirit of Newton's axiomatic method, which distinguishes between the abstract part of the theory and its application to reality.

7. The last constituent of physical methodology establishes the relationships between theory and experience that are not at all limited to the bare comparison between measured and calculated data, but concern rather the general relationships between theoretical and empirical knowledge. In Chap. 4, it will be shown by means of representative examples that this comprises among other things the adjustment between operationally defined properties at the empirical level and the associated theoretical definitions. In fact, these are not necessarily compatible a priori due to the dissimilar procedures of empirical and theoretical concept formation. The essential feature of these relationships is their ambiguity which cannot be removed, as a matter of principle. For this reason, these relationships can be described only as the result of a consistency-generating process of mutual adjustment or adaptation.

In summary, the methods for gaining, securing, and structuring physical knowledge comprise the following parts:

1. Constitution of the empirical basis by experimental methods.
2. Conceptual description of the empirical basis.
3. Interpretation and classification of empirical results by hypothetical patterns of explanation.
4. Theoretical formation of concepts through axiomatically postulated properties.
5. Construction of a formal theoretical framework ("theory frame") by mathematization.
6. Physical interpretation by theoretical structuring principles and formulation of scientific applications.
7. Establishment of the relationships between theory and experience.

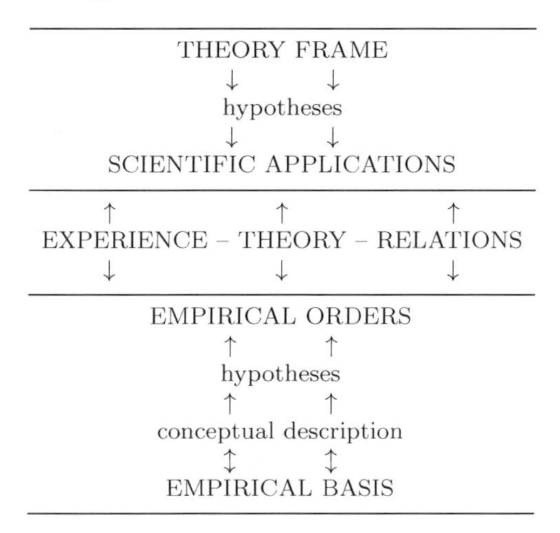

Fig. 2.1 Simplified schema of the structure of physical knowledge

It should be emphasized again that this enumeration does not constitute a chronology because, in most cases, the individual steps do not occur successively. In general, empirical and theoretical knowledge develop in parallel and with mutual adjustment. The first three items comprise the empirical methodology, with largely inductive methods. The acquisition of knowledge is mainly a cumulative process, and structuring methods are distinctly hypothetical and tentative in character. Experimental methods constitute a substantial part of empirical methods, but the two are not identical. Items (4)–(6) comprise the theoretical methodology, with largely deductive methods. The predominant goal is the structuring of knowledge, which itself may lead to new insights and represents an essentially noncumulative process. Mathematical methods constitute part of the theoretical methods, but theoretical physics is not the mere application of mathematics. Finally, item (7) comprises the methods of consistent adjustment between theory and experience with the predominant function of securing knowledge. The resulting schema of the structure of physical knowledge, depicted in Fig. 2.1, provides a guideline for subsequent investigations.

Just as the different steps (1)–(7) do not occur successively, this representation must not be misunderstood as a hierarchical order from concrete to abstract or from particular to general. Theoretical and empirical knowledge are two self-contained constituents of physical knowledge that are neither hierarchically related nor reducible to each other. Grossly simplified, physical methodology can be described as follows. Knowledge is acquired primarily by empirical methods, structured predominantly by theoretical methods, and secured by the methods of consistent adjustment. The methodology of physics comprises the methods themselves, their classification and function with regard to gaining, securing, and structuring knowledge, and also their mutual relationships. Further constituent parts are the methodological norms, as expressed in the tacit assumptions, the explanation of its successes, and the way physical reality is constituted.

2.2 Empirical Basis

According to common opinion, physical knowledge receives its status as reliable, sound, or even true largely, if not exclusively, because it rests on directly given facts rather than on metaphysical assumptions, and because theories are scrutinized through permanent control by observation and experiment. Facts are considered as individual things and direct perceptions (sense data) which together should establish the theory-independent basis of physics. Different views exist as to whether what is directly given are the perceived entities and phenomena as a part of an external world, or solely the single acts of perception. In the first case, one must clarify how to characterize them and to what extent they are modified through appearing as perceived objects, instead of unobservable things-in-themselves. This results in the question of criteria for the existence of this external world, its constitution, and its relation to perceptions. Assuming perceptions as primary, it must be asked to what extent they are a matter of spontaneous acts, independent of previous experiences and acquired thought patterns. This problem is treated in philosophy of science under the heading "theory-ladeness of perception" (Chalmers 1982, Hanson 1972), a designation that is somewhat misleading in that it concerns the generally accepted fact that perception does not occur in a nonbiased manner as a spontaneous act according to the empiricist tabula-rasa fiction, but is intrinsically tied to structuring against the background of previously acquired experiences. This obvious fact of structured perception, understood as the perception of differences and similarities, and associated with the relevant division of the environment, already occurs among the lowest developmental stages of life and is unrelated to the theory-dependence of scientific facts.

The question of the existence and constitution of reality and the challenge raised by the theory-ladeness of perceptions lead to dissimilar views regarding the basis of empirical science. The broad spectrum of opinions is expressed by the following extreme positions (Jung 1979):

- Things and phenomena, as the facts that are accessible to everybody, are parts of an objective, real world which are not influenced by perceptions and theoretical views. These public domain facts constitute the firmly given foundation which all empirical scientific theories rest upon: theories come and go, the facts persist. Without phenomena, theories are empty thought constructions.
- Only descriptions of facts are relevant in science, but not the facts themselves. Descriptions utilize terms that are on hand, not prefabricated, but

constructed: things and phenomena depend on conceptual constructions, and the objective reality of the world of things and phenomena is fiction. This world is changed permanently by science and cannot constitute the firmly given foundation of theories.

Due to the basic significance of these questions to epistemology, it is frequently presumed that they should also be relevant to the analysis of the methodology and knowledge in physics. That this is an error has its reason in the specific structure of that part of reality which is the universe of discourse of physics. This neither consists of "a chair in a room, a cup of coffee on a table, or a stone having shattered a window glass" (Ludwig 1974), nor of singular observations of individual entities, as presumed in logical empiricism (Carnap 1966):

> What has been observed is not thermal expansion in general, but the thermal expansion of this iron bar when it was heated at ten o'clock this morning.

Actually, unspecified entities and singular statements do not provide a suitable basis for obtaining intersubjective and testable knowledge, because a singular event can neither be confirmed nor refuted, but merely stated. Instead, the recognition of regularities and errors is possible only for reproducible events, so that both theory formation in physics and the comparison between theory and experience can be carried out solely on the basis of reproducible data and events. As a consequence, a phenomenon can be an element of the empirical basis of a theory only after its reproducibility has been confirmed. In order to be reproducible, an event must occur under known and controlled conditions, because only this enables its repeatability. This emphasizes the significance of the tacit assumption (2B), postulating the reproducibility of experimental data as a norm.

The methodological norm of reproducibility requires that phenomena and data not be included indiscriminately in the empirical basis; they should instead be subject to assessment and selection which eliminate questionable and nonreproducible results. A representative example are Millikan's experiments to prove the existence of an elementary electric charge (Holton 1973). His published data did not comprise all of the measured results, but rather a "net" data set omitting widely deviating results. According to empiricist principles, such a selection is an inadmissible manipulation, and thus nonscientific. Rather the complete data set must be considered as equally relevant, irrespective of any theoretical prejudices, and must be submitted to the scientific community as the basis for subsequent discussion. Accordingly, Millikan violated scientific norms because he published only those data that were consistent with his assumption of an elementary electric charge. In fact, due to the requirement of reproducibility, the assessment and selection of data are compatible with the norms of physical methodology. Millikan attempted to argue that some of his data were non-reproducible results, adulterated by perturbations or systematic errors. In principle, such a selection is admissible, and frequently proves justified after improving the experimental conditions, but universally valid criteria justifying such a procedure do not exist, so the decision is left to the discretion of the experimenter.

The norm of reproducibility also concerns the constitution of the object domain because, in contrast to the examples above, physics investigates neither arbitrarily selected entities, nor properties of a particular iron bar. Regarding its use as an item for practical purposes, its properties may be important. With regard to physics, it is relevant that the experimentally investigated piece of iron should be representative of the physical system of iron, in that its crystal structure and, if necessary, further physical properties are specified, in order to obtain reproducible measured data. For instance, the experimental investigation of physico-chemical processes on solid surfaces would not yield reproducible data without detailed characterization, e.g., normal direction, reconstruction, vacancies, or contaminants.

Finally, direct perceptibility is not a criterion for elements of the empirical basis. On the one hand, this would exclude any quantitative, measured value as empirical datum: measured data never correspond to direct perceptions because quantitative data are obtained only through the use of instruments, and are never free of conventions. On the other hand, entities like atoms, molecules, or electric currents exist without any doubt, irrespective of their theoretical description, and may be an element of the empirical basis of a theory, in principle. Consequently, objects and facts count in physics as a part of reality, even when they are detectable only by instruments or amenable only via theoretical knowledge. This corresponds to the position of constructive realism that the extent of reality does not depend on the current state of knowledge. Otherwise, something could have come into being as really existent only after being recognized. This means that, e.g., the far side of the moon or the cosmic background radiation came into existence only in that moment when it was first photographed or detected, respectively. Such a view definitely contradicts both methodological principles and the research practice of physics, and is thus untenable. For instance, any question about the origin of the cosmic background radiation would be meaningless, and it could not be used to assess cosmological theories if it actually had come into being only after its detection. Analogous criteria, regarding objectivity, apply to objects and phenomena produced in laboratories. Artificially produced elementary particles, atoms, or crystals are considered as real as naturally occurring ones. This corresponds to the everyday notion that a house or a car, designed by humans, is as real as a tree or mountain, and is expressed in the tacit assumption (2C) on the identical ontological status of "natural" reality and the reality created in laboratories.

In a first summary, the elements of the empirical basis of a physical theory must fulfill the following requirements:

1. Phenomena and data are possible elements of the empirical basis only when their reproducibility is guaranteed; a reproducible phenomenon will be called an effect.

2. An object, as a possible candidate of the empirical basis, is not an arbitrary, unspecified element of reality, but representative in the sense that experiments performed on it must be reproducible.
3. Direct perceptibility is not a requisite criterion for an object, datum, or effect to belong to the empirical basis.

Accordingly, the reality of objects, data, and effects studied in physics is largely a reproducible and representative reality because physical knowledge rests on observations, measurements, and data resulting from goal-oriented experiments. Consequently, the concept of experience in physics is based on experimental, i.e., reproducible, representative, intersubjective, and in this respect law-like experience, in distinct contrast to the subjectivistic concept of experience in empiricism which rests on singular sense perceptions and unspecified entities. Therefore, the establishment of the empirical basis is accompanied by a "theoretization" which is admissible and not put into question according to the tacit assumptions of experimental methodology. For these reasons, both challenges mentioned above, viz., the existence of a subject-independent reality and the theory-ladeness of perception, are largely irrelevant for the discussion of physical methodology:

- The general question of the existence of an external world is irrelevant since what is of primary concern are the challenges regarding the physical characterization of objects already presumed to exist. This is another instance of the tacit assumption (1A) that physics does not prove the existence of an external world, but takes it for granted. This is not in contradiction with the fact that the proof of existence of a specific, theoretically postulated object may well be a matter of physical research because this merely concerns the extent of the external world that is not specified by (1A).
- The theory-ladeness of perception, understood as the problem of structured perception, is irrelevant because physical knowledge does not rest on singular sense data, but on reproducible experience. For this reason, the elements of the empirical basis are much more "theory-laden" than due to the mere fact of structured perception. The problem in physics is rather centered on how to ascertain reproducibility and representability.

Accordingly, the actual challenges associated with establishing the empirical basis are of an essentially different type than generally presumed in the philosophy of science, because physics is not concerned with a given reality of singular phenomena and unspecified entities that is studied from a non-biased point of view, but rather with a reality that is designed for its own purposes. This has profound consequences for the relation between the empirical basis and the associated theory. According to empiricist views, based on the dogma of the priority of empirical knowledge, the empirical basis as the representative and foundation of firmly established knowledge should be independent of the associated theory. For the secureness and reliability of the factual knowledge, contained in the empirical basis, should rest precisely on their independence from any theoretical assumptions. According to such an a priori conception, formative retroactions of the theory on the empirical basis are forbidden for methodological reasons, because the secureness of the empirical foundation would be destroyed. Actually, an aprioristic, theory-independent definition of the empirical basis is impossible because its extent cannot be determined without regard to the associated theory. Since without theoretical background all empirical facts have the same status, an assessment as to whether something is significant or irrelevant for the theory is impossible without knowing the theory. In order to circumvent this problem, the empirical basis must contain the whole body of known factual knowledge. Even under the restriction to reproducible facts, this is neither possible nor meaningful because the empirical basis must not be too extensive. On the one hand, it contradicts the conception of a basis containing a multitude of data that are totally irrelevant for the associated theory. On the other hand, the theory will be at variance with numerous facts, because every physical theory has a limited scope. Overall, the empirical basis does not represent a cumulative collection of facts. Instead, its extent must be adapted consistently with the theory since it is determined by the scope of the theory.

In the first instance, the selection of objects to be included in the empirical basis is theory-dependent. This is exemplified by the question of why an electric current, understood as an entity of moving electrons, does not belong to the empirical basis of classical electrodynamics. In 1820 Oersted discovered that a magnetic needle near a current-carrying wire deflects when the voltage source is turned on or off. From this discovered effect, the property of electric currents was constructed conceptually to generate a magnetic field in its environment. This property is not an element of the empirical basis because it is already an interpretation of the observed effect, instead of its bare statement. Subsequent experiments revealed that electric currents also exhibit chemical and thermal effects. Correspondingly, the conceptual description was augmented with chemical and thermal properties where again only the effects, but not the associated properties, were elements of the empirical basis. Therefore, in electrodynamics as a macroscopic theory, an electric current is solely the *conceptual* summary of its magnetic, thermal, and chemical properties, irrespective of its physical nature. In electrodynamics, only these macroscopic effects and properties are studied, so the electric current is not an element of the empirical basis, but only a concept leaving undetermined whether it corresponds to an element of reality.

In this respect, classical electrodynamics is a descriptive theory because it is constrained to the description of the effects and the associated properties, and does not contain any model assumption on the material carrier, the "electric current", as an element of the external reality to which these properties can be assigned. A microscopic theory $\mathrm{T_{mic}}$ goes beyond this level of description. The macroscopic properties are independent of $\mathrm{T_{mic}}$ in the sense that they are invariant against changes in the microscopic description. Of course, they nevertheless remain conceptual descriptions and are not elements of the empirical basis of $\mathrm{T_{mic}}$ either: in the microscopic theory as well, the macroscopic properties assigned to the electric current do not belong to the empirical basis; only the associated effects do so. Additionally, T_{mic} comprises those data and effects that are experimentally provable deductions from the model assumption that the electric current corresponds to an entity consisting of moving electrons. In contrast to electrodynamics, T_{mic} contains an existence hypothesis concerning the material carrier of the electric current. In the case of the successful experimental verification of this existence hypothesis, the empirical basis EB of T_{mic} can be augmented with this material carrier. This modality for constructing reality may be represented schematically by the sequence:

$$\text{effect } (\in \text{EB}) \longrightarrow \text{property } (\in \text{T}) \overset{?}{\longrightarrow} \text{carrier } (\in \text{EB})$$

Due to the discovery of effects, certain properties are constructed conceptually and assigned to a hypothetical carrier, initially on the conceptual level. If later on a material carrier is discovered, the empirical basis may be augmented correspondingly because one has switched from the conceptual to the object level. In the case of the electric current, however, this extension concerns only the empirical basis of T_{mic}, but not that of electrodynamics. This does not mean, of course, that electrodynamics as a physical theory decides on the existence of electric currents, but solely that their constitution as an entity of moving electrons is irrelevant in relation to this theory. Otherwise, Maxwell could not have developed electrodynamics more than 30 years prior to the discovery of the electron. Admittedly, the two levels of description are rarely distinguished from each other, and electric currents are also described as an entity of moving electrons in electrodynamics, even though this is irrelevant in that context and is not part of that theory. The electric current has different conceptual descriptions in terms of either a macroscopic or a microscopic theory. In this respect, the mode of theoretical description essentially determines the extent of the empirical basis, in the sense that the theory defines what is observable within its scope. It is, however, obvious that this must be understood in an epistemic sense, i.e., the theory defines its subject of research by delimiting that part of reality and selecting those properties which constitute its universe of discourse. This neither permits the conclusion that a change in the theoretical description will imply a change in the physical nature of the electric current, nor that the theory decides on its (material) existence.

Accordingly, the theory-dependence of the object domain results from defining model systems through certain properties on the theoretical level, irrespective of whether such a model system corresponds to an element of reality. This theoretical concept construction will be described in detail in Sect. 2.5. As a consequence, all empirically comparable deductions within the framework of the given theory are strictly valid solely for these model systems. Hence, the theory remains empirically irrelevant as long as no real systems exist that realize these model systems sufficiently well. For this purpose, objects must be fabricated that possess the required properties as precisely as possible. Accordingly, one of the functions of experiments, described in Sect. 2.3, consists in the material realization of these theoretically introduced model systems, e.g., the fabrication of crystals, as ideal as possible, by crystal growing procedures, the generation of homogeneous fields or of monochromatic light by appropriate experimental designs, or the production of elementary particles in accelerators. After the experimental creation of an object according to the theoretical requirements, a transition has taken place from the conceptual to the object level, i.e., the model system possesses an empirical counterpart that may be included, as a part of the external reality, into the empirical basis, if appropriate, notwithstanding the fact that this creation has presuppposed knowledge of the associated theory. Subsequent experimental investigations are then performed on these material counterparts generated in a goal-oriented way because, due to their representativeness, they enable one to obtain data that allow a considerably more precise comparison between experiment and theory. The intention behind this practice is to have available a representative collection of data that is as precise as possible, instead of an empirical basis of indiscriminately amassed facts. In so doing, it is inconsistent to postulate that the empirical basis may be augmented by new effects and data, but not by objects, although they can be defined as conceptually precise only within the theory. Indeed, most of the new data that enable an improved comparison with theory are obtained just on these objects that are created according to the theoretical requirements, e.g., artificially grown crystals or surfaces prepared in a goal-oriented way. Even when initially the experimental investigations in a certain research field are carried out on naturally occurring entities, they will successively be replaced with the material counterparts of the model systems which then constitute the basis of subsequent experiments. This substitution due to advances both on the instrumental and theoretical side represents one aspect of the progress of knowledge in physics. For instance, early investigations in high energy physics were carried out on the naturally occurring cosmic radiation, while today elementary particles are generated in a selective and goal-oriented way in accelerators. Only this can enable extensive and systematic experimentation. In short, in parallel with the development of the theory, the extension and reshaping of the empirical basis occurs in accordance with the theory.

This stepwise exchange of natural entities with laboratory-generated empirical counterparts, designed according to theoretical knowledge and thus adapted to the requirements of the theory, has crucial consequences for the reconstruction of the structure of physical knowledge. Its goal consists in representing the theory as transparently and compactly as possible, along with its relationship with the relevant aspects of the external reality. This is unattainable without including in the empirical basis the theory-oriented designed material carriers of the model systems, because it is only in this way that a quantitatively provable relationship between theory and experience can be established. Without such a constructive improvement of the objects of investigation, which aims to approach the theoretical ideal as closely as possible, discrepancies between theory and experiment could be far too easily ascribed to the imperfectness of the systems, instead of serving as indications of deficiencies in the theory. If such a constructive improvement of systems is impossible, the theoretical description remains one of scarcely compelling "model theories" because the objects of investigation rarely resemble the model systems of the theory. In a caricatural form, it could be said that physical theories resemble reality for the reason that their statements refer to a reality constructed in accordance with their requirements. Irrespective of whether one shares this view, it remains clear that establishment of the object domain of the empirical basis must occur in accordance with the theory, and must remain subject to reshaping and revision due to improved knowledge.

Analogous conclusions apply to the data basis. According to common prejudice, the data basis must be independent of the theory due to its function of establishing a firmly given basis. This assumption is untenable, since quantitative measuring results are generally subject to improvement or revision, either through repetition of measurements attributable to the background of novel theoretical knowledge, or due to novel measurement procedures, or due to more representative systems. Moreover, experimental data are obtained by instruments and, especially in modern physics, by sophisticated measurement and evaluation procedures. Before the instruments used for generating the data can be included in the empirical basis, theories must be available for both the measuring procedures and the mode of operation of these instruments. Otherwise, the corresponding data cannot be considered as secured. For instance, the use of data obtained with microscopes or telescopes necessitates optics, which is the theory underlying these measuring instruments, in order to be secured. This condition cannot usually be met since the measuring procedures and instruments required to establish the empirical basis of a theory T can be substantiated only by T itself, or indeed by theories going beyond T. This applies to all procedures for measuring the fundamental quantities of a theory, such as length in physical geometry or temperature in thermodynamics. For example, the measurement of distances cannot be theoretically justified before precisely defining the terms "rigid body" and "inertial system", and the measurement and definition of temperature is impossible without knowledge of thermodynamics. As a consequence, the theoretical justification of the relevant measuring procedures can be described only as a process of consistent adjustment.

The theory-guided production of material counterparts of model systems, gaining and improving experimental data on the basis of theoretical knowledge, as well as their selection and assessment as relevant for the theory, preclude any definition of an empirical basis that is independent of the associated theory. Consequently, the empiricist a priori conception is a fiction, suggesting an erroneous belief in an ultimate, theory-free foundation to which the whole of physical knowledge could eventually be reduced. Adhering to this fiction would be justified, at best, if it could contribute to an improved comprehension of the structure of physical knowledge. Actually, the opposite is true because this conception rests on a view of experience relying on direct perceptions and observations, instead of on the use of instruments. Without using them, empirical experience may be considered as unchangeable from a certain state on. In contrast, an empirical basis relying on the application of instruments can always be extended and improved due to the advancement of new instruments and measurement techniques. Accordingly, the assumption of an a priori determinable empirical basis leads to a severely distorted view of the structure and development of physical knowledge. Actually, the theory dependence of both the extent of the empirical basis and the possibility of improving experimental data are typical structural properties of physical knowledge. Without accounting for these features, an appropriate comprehension, e.g., of scientific progress in physics is impossible. In conclusion, the conception of the empirical basis is meaningful, despite its theory dependence and the impossibility of its a priori determination, and corresponds to a realistic ontology of physics that assumes the existence of a subject-independent reality. As the representative of this reality, the empirical basis can be characterized as follows:

Definition 2.1 The empirical basis of a theory T represents a certain part of the subject-independent reality. It is established through consistent adjustment with T, and consists of objects of investigation (object domain), effects, and data (data basis), and the instruments required to obtain them. The following conditions must be fulfilled:

- the objects of investigation must be representative,
- effects and data must be reproducible,
- the relevance of the empirical facts for the theory must be confirmed,
- due to possible instrumental and theoretical developments the empirical basis must remain open to:
  - extension and improvement of the object domain due to construction of new objects according to theoretical requirements,
  - changes in the data basis due to newly discovered effects, new or more precise measured values, and the revision of data,
  - development of novel instruments, even when their mode of operation can only be justified within T.

Due to these theory dependencies, the question arises whether the separation between theory and empirical basis is meaningful, or whether it should be assumed instead as a part of the theory, as in other theory models (Carnap 1928, Ludwig 1978). Against these approaches, it must be argued that they lead to both methodological inconsistencies and an incorrect understanding of the relationship between theory and experience. Firstly, the experimental examination of theoretical results reveals itself to be a logical circle, because a component of the theory is being used for its own justification. However, secureness and acceptance of theoretical knowledge in physics rest precisely on the fact that experimental data possess the status of *external* instances to control theoretical results, so experimental examination is not circular, but rather corresponds to the consistent adjustment of theory and experience. Secondly, taking into account the fact that dissimilar theories may exist for the same part of reality, and that they may even be mutually exclusive like atomistic and continuum theories of matter, the inclusion of the empirical basis in the theory might lead to the inference that a theory change is identical with a change of reality. To abandon the strict separation between conceptual and object level could then suggest that this reality is alterable by theoretical descriptions and is no longer subject-independent. This corresponds to a radical form of constructivism, judging theory dependencies as so essential that the assumption of a subject-independent reality is untenable. However, the transition from one theory to another does not correspond to a change of reality, but merely to another description. Otherwise, two adherents of such dissimilar theories would actually be unable to engage in a discussion, not only because their theoretical points of view were different, but also because they would even be talking about different facts, so there would not even be a common ground for their discourse. Actually, a proof by examples from the history of physics for the existence of such incommensurabilities postulated by Kuhn (1970) has never been provided. Conversely, the inclusion of the empirical basis in the theory might support a naive-realistic position that attributes the theory to reality and believes that theories were discovered as objects and effects. Consequently, radical constructivism and naive realism are actually two sides of the same coin, viz., an undue identification of reality with its conceptual description. They differ solely by the result, in that naive realism ascribes everything to the object level, and radical constructivism to the conceptual level.

For these reasons, differentiation between empirical basis and theory is not only justified, but is absolutely necessary for methodological, epistemological, and ontological reasons. Concerning methodology, this differentiation corresponds to the fact that experimental and theoretical research represent two methodologies for gaining physical knowledge that are autonomous in the sense that they are not reducible to one another. The empirical basis is theory-independent in that it is established, not by theoretical methods, but instead through results of practical activities resting on methodological principles that are basically dissimilar to those of theory formation. This is exemplified by the different tacit assumptions of experimental and theoretical methodology. This methodological autonomy is the prerequisite for the comparison between theory and experience not to be a circular procedure, in spite of the theory-determinateness of the empirical basis.

Regarding the ontological aspect, the conception of the empirical basis represents the assumption of the existence of a subject-independent reality, but is a proof neither of the existence of a reality-in-itself, nor of the truth of physical knowledge. The separation of theory and empirical basis conforms to the ontology of constructive realism which strictly distinguishes between the levels of concepts and objects. A theory provides statements *about* reality, but is not a part of it. The necessary distinction between objects, effects, and data, on the one hand, and their conceptual theoretical description, on the other hand, is absolute in that it is independent of the description by a particular theory, and may be exemplified through three criteria:

- effects are discovered and demonstrated in experiments, but not invented,
- objects are discovered or manufactured, while their properties are constructed conceptually,
- data and effects are stated without providing an interpretation.

These criteria reflect the intuitive attitude that an object or effect as a part of reality appears whenever the necessary conditions are realized, so that they can in principle be discovered by accident and without any theoretical knowledge. The knowledge serving as the basis for planning and performing experiments solely determines the purpose of the experimenter to deliberately provide the conditions for the occurrence of the desired effect. This type of theory-determinateness of experimental activity is considered to be unproblematic because the goal and motivation of an acting subject do not have any relevance to the status of the subject-independence of the external reality. The same applies to the theoretical knowledge needed to manufacture the counterparts of the model systems and to construct the instruments. Although the reality may become only partially accessible through goal-oriented or theory-guided experiments, those parts that become known are, according to the tacit assumption (2C), subject to the same criteria of objectivity as those that are known without any theoretical preconceptions. Overall, the ontological status of objects and effects as elements of reality does not rest on whether something gets known through direct perception, goal-oriented experiments, or theoretical considerations.

Regarding the epistemological aspect, the empirical basis is assigned the function of providing the basis for the intersubjectivity of physical knowledge, and thus a consensus-qualified basis that can be resorted to in cases of doubt. Indeed, it is precisely for this reason that the experimental data are included in the empirical basis. Although, in a strict sense, they are not a part of a subject-independent reality, they are highly consensus-qualified. The conception of the empirical basis embodies the idea that a communicative undertaking like modern physics necessitates an intersubjectively accepted foundation, in order to avoid endless, unproductive discussions about methodological and foundational issues. In addition, the empirical basis refers to an actively constituted, representative, and reproducible reality, instead of a passively received one made up of singular, subjective perceptions. As a result, it contains a conceptual component because, in a strict sense, factual statements like instrument readings or pointer positions are judgements and are verbalized as such. Accordingly, facts are evident, unquestioned judgements that are unproblematic regarding perception and verbalization, and are thus particularly consensus-qualified.

In summary, the conception of the empirical basis represents in methodological respects the autonomy of experimental research, in epistemological respects, the particularly consensus-qualified component of physical knowledge, and in ontological respects, the existence of a subject-independent reality. The impossiblity of an a priori definition emphasizes the openness to novel experiences and improved data which both constitute essential aspects of research progress in physics. The non-eliminable instances of theory-determinateness are solely epistemological, but not ontological, and concern the extent of knowledge about reality, the delimitation of the theory-relevant parts, and the mode of conceptual description. Accordingly, they are unproblematic for the status of the empirical basis as representative of a subject-independent reality, and merely correspond to the fact that the constitution and view of this reality is affected by thought patterns and theoretical knowledge.

2.3 Experimental Methodology

Although the utilization of experimental methods is regarded as the characteristic feature of physics *par excellence*, the performance of experiments is not an invention of seventeenth century science, but can be traced back to ancient Greece, although with another meaning. According to a definition due to Aristotle, *experimentum* was used in the sense of "emerged from many reminiscences", thus denoting the entirety of acquired experiences. Hence, *experimentum* and *experientia* were not distinguished, and the two terms were used synonymously approximately until the Renaissance (Ritter et al. 1971ff.). Accordingly, experimentation did not possess any self-contained function within scientific methodology, and in particular, it was not considered as a method for gaining new knowledge about nature. Galileo's attitude was similar: experiments served primarily to illustrate insights derived from previous theoretical considerations. Apart from being mainly thought experiments, the structure of the arguments in his dialogues demonstrates that experiments were just an auxiliary means for those who were not convinced by the conclusiveness of his theoretical argumentation (Grodzicki 1986). In contrast, the change in the function and significance of experiments, especially as an important tool for obtaining new knowledge, emerged in England, and developed in several stages over a period of more than a century:

- In his studies on magnetism at the end of the sixteenth century, Gilbert used experiments for the first time in a systematic fashion to obtain new knowledge (Gilbert 1600). In this way, he demonstrated that it was possible to arrive at new knowledge through experimental methods.
- Twenty years later, in his *Novum Organum*, Bacon set the decisive basic attitudes for subsequent development. By postulating experimentation as a legitimate means for the systematic production of new knowledge,

he established the ideological foundation of natural science in the subsequent centuries. In contrast to Greek and scholastic natural philosophy, the following attitudes were particularly important:

  - It is possible to obtain new knowledge by experiments and individual experiences.
  - It is not experience per se that is decisive, but controlled and goal-oriented, i.e., experimental, experience produced in laboratories.
  - An experiment is not an act designed by humans to outsmart nature. The experimenter is instead in concordance with nature, exploiting its principles and acting accordingly. As a consequence, the technique of experimentation is freed from any artificial and magical character, and becomes a legitimate constituent of scientific methodology.
  - The sole source of safe knowledge is experiment. Consequently, the sole route towards true scientific knowledge is via induction from experiment. It is worth mentioning that Bacon himself was not an experimenting natural scientist. Given the limited experimental capabilities at that time, it would rarely have been possible for experimentation to establish the methodological foundation of natural science.

- In the wake of these approaches towards an experimental methodology, initially restricted primarily to England, empiricism emerged as the dominant English philosophy of science (Hume, Locke, Mill).
- The implementation and acceptance of an experimental-inductive methodology on the European continent took place only during the eighteenth century with the dissemination of Newton's *Principia*, in which, indirectly or directly influenced by Bacon, he propagated this methodology. The successes of Newtonian theory have been ascribed primarily to the application of these methods.
- Due to the increasing application of experiments thereafter, methodological rules were developed whose implicit content is contained in the tacit assumptions, and whose explicit embodiment became the foundation of experimental methodology.

It is remarkable that, until the end of the eighteenth century, the use of instruments was not ascribed great importance. It was only after the application in particular of telescopes and microscopes, which led to a large number of novel results, that the implementation and extension of experimental methods became closely linked to the development of new instruments. The successes thus achieved led in the nineteenth century, in combination with the representation of experiment as a theory-free factual statement, to an increasing and uncritical belief in facts, one that prevails in principle right up until now. Accordingly, in this view, the acquisition and securing of knowledge are both achieved by experimental methods, and experimentation is considered as a largely unproblematic question put to nature that supplies unequivocal and objective information about nature because it should be unaffected by theoretical prejudices. Such an attitude overlooks a number of important properties of experimental methods:

- All experimental methods possess a theoretical component: design, evaluation, and interpretation of experiments is impossible without theoretical background knowledge.
- Instruments are indispensable constituents of experimental methods: their mode of operation and function cannot be understood without a theoretical basis.
- When a science manufactures its own objects of investigation in laboratories, the question arises as to whether new knowledge is actually obtained in this way, or whether one has merely reproduced the theoretical ideas that have entered the experimental design.
- Any knowledge obtained exclusively by experimental methods has only approximate character and is incomplete (Duhem 1908): every measuring result is determined only within certain error margins due to measurement accuracy, and the fact that only a finite number of measurements can be carried out. Therefore, unequivocal conclusions regarding the quantitative form of a relation between two measured quantities cannot be drawn solely from measurements: $n+1$ measuring points can always be precisely reproduced by a polynomial of n th degree. In short, a science based exclusively on experimental methods can provide neither precise statements nor unequivocal conclusions concerning the form and validity of general relations like physical laws.

These theory dependences become especially challenging when the experiment is intended to serve to provide the secureness of theoretical knowledge. Due to these theory dependences, any experimental strategy for scrutinizing theoretical knowledge is inevitably subject to the objection of circularity. Actually, the acquisition of experimental knowledge is also concerned because the idea of experiment as a theory-free factual statement is untenable. It is thus not clear a priori to what degree the experimentally obtained elements of the empirical basis may be considered as secured. To begin with, the answer must clarify both the abilities and limitations of experimental methods by critically analyzing the various functions of experimentation. They can be classified into three qualitatively dissimilar categories: methods to establish the empirical basis, methods to establish the relationships between theory and experience, and application-oriented methods for extending system-specific knowledge. Discussion of the two latter categories necessitates the conception of a physical theory, and is postponed to Chap. 4. In this section, those procedures will be described that are relevant to establishing the empirical basis.

(1) The first function consists in the objectification of perceptions and observations. A number of everyday experiences demonstrate that direct sense perceptions are not reliable. This is exemplified by heat sensations and op-

tical illusions. When attempting to estimate the temperature of water by dipping one's hand into it, the result will depend on the previous treatment of the hand. If it was previously in a colder environment, the water will be estimated as warmer than if the hand came from a warmer environment. Similarly, numerous optical illusions result from the experience that objects of the same size appear unequal with regard to dissimilar backgrounds. Indeed, the reason for many erroneous perceptions is that they do not happen in isolation, but relative to some spatial and temporal environment. Observation methods using exclusively human sense organs are referred to as subjective methods. In order to assure oneself that a perception does not involve an illusion, the following alternative exists. The first option to make oneself sure of a perception is communication with other people, a method also applied beyond science. The second is the fundamental constituent of experimental methodology, and consists in the transition to procedures of objectification by the application of instruments. Due to this detachment from bare sense perceptions, which aims to reduce the influence of erroneous sensations, physics is frequently considered to be an objective science. In this context, one has to bear in mind first of all that even the use of instruments never entirely eliminates human senses: at the end of an experiment, it always remains to read off a number, pointer position, etc., fixing the result by the experimenter. Therefore, a total elimination of subjective methods is unattainable. All one can do is reduce to those that are deemed to be unproblematic, because one is convinced not to be subject to sense illusions. Secondly, any use of instruments contains a non-eliminable theoretical component because the reliability of measured data is not secured without a corresponding theory of the instrument and the measurement procedure. Due to these inherent constraints, the methods of objectification establish neither a strictly theory-free basis of physics, nor access to knowledge about a thing-in-itself in a verifiable manner. Accordingly, objectification does not pursue the ontological goal of proving the existence of a reality-in-itself, and nor does it prove the truth of knowledge, but it has the epistemic function of controlling and correcting direct perceptions. Finally, from the application of instruments for objectification of human perceptions results the classical view of instruments as the refinement and extension of human sense organs. In this respect, measuring device and observer are considered as a unity, so instruments do not possess an autonomous status. In addition, perturbations to the measured system due to the measurement process are assumed to be negligible.

(2) The second function of experiments consists in extending the objectification towards the quantitative definition of properties by their metrological realization (operational definition). This concerns, in the first instance, fundamental quantities like length, time, mass, temperature, and electric charge, but also operational definitions of derived and theoretically introduced properties. This is frequently associated with the development of new devices, such as the thermometer for temperature measurements. The simplest ex-

ample of an operational definition is the one for length carried out with the following rules (Döring 1965):

- The length of a line segment between two (marked) points on a body remains unchanged under translational displacements of the body.
- Two line segments on different bodies have identical length if the two initial and end points can be superposed simultaneously.
- When p line segments of identical length can be attached along a straight line in such a way that the end point of one coincides with the initial point of the subsequent one, the length of the whole line from the initial point of the first segment to the end point of the last one equals p times the length of the single line segment.

Analogous procedures apply to other quantities. First, one states some invariance properties of the quantity to be defined, then one defines the equality of two quantities of the same type, and finally one describes how the ratio of two quantities of identical type is determined via the previous definitions. Explicitly, the operational definition of a physical quantity ("observable") comprises the specification of the system "possessing" the property to be measured, the measuring device together with the instruction for use, and the measurement unit. Beyond the concept of space, all quantitative, experimental methods implicitly presuppose that an algebraic structure and a metric are defined on this space since neither are part of nature. This measuring space, forming the space of all experimental experiences, is described mathematically by a three-dimensional Euclidean vector space. This has not changed due to the results of modern physics.

Even without great demands on operational definitions, their deficiencies are obvious. Disregarding the fact that, in the case of length, notions must exist about terms such as "shorter" and "longer", one requires the background knowledge that solid, fluid, and gaseous bodies exist, and that among the solid ones some do not perceptibly change their form under ambient conditions, and are thus suited for the fabrication of measuring rods. In a strict sense, however, there is not a single body whose shape and length is entirely independent of external conditions like temperature. Consequently, the precise operational definition of length also requires a procedure for measuring temperatures. This, in turn, usually occurs by measuring a length, e.g., by determining the expansion of the column of a liquid in a thermometer, if not through techniques presupposing considerably more elaborate theoretical knowledge. Finally, an important objective when introducing the concept of length is the determination of the distance between two bodies, or between the positions of a moving body at different times. This requires not only the concept of an inertial system, but also a quantitative concept of distance, since the distance between two real bodies cannot be precisely defined due to their finite extent. A definition such as "distance = distance between the centers of mass" again necessitates previous theoretical knowledge from mechanics itself, and a distance defined in such a way would not be directly measurable, in general. Overall, a precise operational definition of length is impossible without borrowing from theories whose basis this definition ought to constitute. Even this simple example demonstrates that operational definitions of physical quantities are not at all unproblematic and theory-independent determinations. In addition, elementary limits exist regarding concept formation via operational definitions. Firstly, it is impossible to specify all the conditions under which an operation is performed. For the sake of reproducibility this would be necessary. Secondly, the reduction of operations to other, more elementary ones, cannot be continued indefinitely. Consequently, in any operational definition, there enter implicit assumptions of the type that only a limited number of factors will influence the results, and that certain operations, e.g., the metering of a scale do not require further analysis. Thirdly, all operational definitions rest on conventions, and as quantitative determinations, they are never precise, but are valid only with finite accuracy. Hence, on the one hand, operationally defined quantities are not theory-free, and on the other hand, they provide definitions that are not sufficiently precise for theoretical argumentation and they do not encompass the full extent of the physical meaning of a quantity.

Due to these deficiencies, operational definitions are occasionally considered as provisional, i.e., something which ought to be eliminable once a theoretically precise definition has been established. However, this overlooks the fact that operational definitions fulfill a number of important and necessary functions when introducing physical quantities. Certainly, the best known is the feature that empirically meaningless conceptions are being identified and excluded. For instance, Mach (1919) pointed out that Newton's wording "without reference to anything external", occurring in the definition of absolute space and time, excludes any possibility of a measurement. In the special theory of relativity this criticism is substantiated by the proof that Newton's conception of space-time cannot be realized through measurements, although it had been commonly accepted as self-evident. Accordingly, Newton defined his ideas of space and time intuitively, and not by exclusive reference to operations with measuring rods, light signals, clocks, etc. Regarding Newtonian mechanics, this was innocuous only because the theory did not make explicit use of it. Operational definitions furthermore warrant (Bridgman 1928):

- Clarity: acts are less ambiguous than words. The meaning of a term is learned through observing what somebody does, but not what they say about it. For example, when the length of an object must be determined, every physicist knows exactly what practical operations must be carried out. Although consensus about the precise length need not exist, subjective descriptions of the type "as long as my finger" or "rather small" are not used.
- Consensus: every measurement constitutes a comparison of the quantity to be determined with a standard accepted by all members of the scientific community, together with certain commonly accepted methodological

rules. This is one of the reasons for the success and the effectiveness of physics, as well as for the rapidity with which disagreements are removed.

- Criterion of meaning: physically meaningless terms, statements, and questions can be identified and removed by checking their operational feasibility. According to current knowledge, this may be the question of what physical laws would look like in a universe without matter. In particular, ideas like a thing-in-itself or a reality-in-itself are physically meaningless abstractions that do not belong to the universe of discourse of physics.

Application of the operational meaning criterion necessitates some caution because such judgements may depend on prejudices or on the current state of knowledge. For instance, during the nineteenth century a number of physicists and chemists considered the questions of the existence of atoms or the geometrical structure of molecules as meaningless. Furthermore, physically meaningless questions may well make sense in another context. Finally, the methodological necessity of operational definitions must not be extended to an operational theory of meaning which identifies the physical meaning of a term with the defining operations. Apart from being an unjustifiable reduction of the meaning of physical quantities, a number of empirically meaningful quantities exist, like instantaneous velocity or vertical ionization potential, which are initially defined only through mathematical operations. The operational determination of such quantities then involves the combination of an experimental measuring instruction and a mathematical calculation rule.

(3) Apart from objectification and operationalization, the necessity of instruments becomes even more obvious for the exploitation and exploration of fields of experience that are not amenable to direct perception by human senses. This concerns both the expansion and the refinement of perceptible reality. Expansion means, on the one hand, that research fields such as electromagnetism are made accessible. On the other hand, it also concerns fields whose investigation presupposes constructive experimental techniques like the generation of extreme pressures and temperatures. Refinement involves the detection of differences that are not perceptible by human senses, e.g., the fact that the yellow light of sodium vapour does not arise from a single line, but from two spectral lines 6 Å apart. This experimental potential inevitably enforces a revision of the concept of observability. The pre-scientific view, shared with empiricist philosophies of science, is largely identical with requirement of direct perceptibility by the sense organs. In this case, ultraviolet radiation, electric currents, atoms, or molecules would be unobservable, in contrast to the common view in physics. Consequently, admitting the use of instruments, the definition of observability must be substantially extended because it depends on both instrumental development and the state of theoretical knowledge.

(4) Another function relating to the establishment of the empirical basis concerns experiments for the discovery of effects, especially if occurring by accident, such as Oersted's discovery of the reaction of a magnetic needle when switching on an electric current, or the discovery of radioactivity and superconductivity. Initially, however, an accidental discovery corresponds to a singular fact, so it is not yet a scientific datum and thus cannot be an element of the empirical basis. In order to confirm reproducibility, the singular event must be explored and analyzed by systematic experimental studies which transform it into a reproducible event. Only with this prerequisite can nonreproducible "incidental discoveries" due to poor experimental performance be distinguished from real physical effects. Therefore, in this context the actual function of the experiment consists in the transformation of an incidental discovery into an effect, i.e., a reproducible and controllable phenomenon. Although this may require some theoretical ideas, they do not play any role with respect to the ontological status of the effect, so the view that effects can be discovered by experiments is justified, regardless of whether it happens by accident or is well-directed due to theoretical considerations.

According to a widespread belief, not only effects, but also physical laws can be discovered by experiment. This belief rests on the implicit assumption that physical laws exist as a part of nature, because in a strict sense something can be discovered only if it exists independently of any experimental and theoretical investigations. Since physical laws as conceptual relations between physical quantities rest on concepts that get their meaning only within the framework of a theory, this presupposition is not justified. Hence, physical laws cannot be discovered by experiment. What may be discovered, in a certain way, are empirical regularities, but these must not be identified with physical laws. This challenge is discussed in Chap. 4 in the context of the definition of the conception of law in physics.

(5) The experimental laboratory fabrication of objects, as ideal as possible, is of huge importance for technical applications and also for fundamental research. Examples like the generation of a vacuum, the design of homogeneous electric fields, or the growth of crystals demonstrate how in physics items are constructed in accordance with the model assumptions of a theory, not only at the conceptual level, but also at the object level. In addition, the optimal realization of the model systems of a theory paves the way for more precise empirical testing of deductions from the underlying theory. This emphasizes the strong theory orientation of this type of experimental activity, because the properties of the entities to be manufactured are given by the theory.

(6) Similarly, the function of the theory-oriented development of instruments is to make properties measurable that are introduced within the framework of a theory, e.g., measurement of kinetic energy in mechanics or the enthalpy of a gas in thermodynamics, or the determination of the wavelength of light. The necessary development and improvement of instrumental techniques is

usually guided by theoretical knowledge of the same theory that provides the definition of the given quantity. This is the case for virtually all spectroscopic techniques in modern physics. Through metrological realizations, a property defined initially only theoretically acquires an empirical connotation, but it does not become theory-free, of course. To anybody who is not convinced of the wave nature of light, any attempt to build a device to measure the wavelength of light is nonsensical. This theory dependence bears important consequences for the status of the instruments, because they can no longer be considered as epistemologically unproblematic refinements of human sense organs, but play a constitutive role in the definition of the given property. Accordingly, the question of criteria of observability comes up in a tightened form because observability of a property, and thus the constitution of reality, depends on the underlying theory (Planck 1983):

> The decision as to whether a physical quantity is observable or whether a certain question bears a physical meaning can never be made a priori, but only from the viewpoint of a certain theory.

(7) Finally, the goal-oriented production of quantitative data is an indisputable function of experiments. Extending and increasing the accuracy of the data basis using novel theoretical and instrumental developments are characteristic features of experimental methodology. Representative examples are the precise specification of quantitative data regarding the material properties of solids, e.g., density and conductivity, as well as the continuous improvement of the numerical values of the fundamental physical constants. Increasing this accuracy is not only important in order to improve the quantitative comparison between experiment and theory, but proves conclusively that quantitative experimental methods provide the basis for progress in research only in combination with mathematical methods. In addition, the production of more precise data regularly necessitates an improvement, if not a new definition, of standardized measuring procedures, especially for the fundamental units. The historical development of the operational definition of the meter is a representative example that will be described in detail in Sect. 4.5.

This selection of functions of experiment emphasizes, on the one hand, its outstanding and indisputable importance in establishing the empirical foundation of physics. On the other hand, the common structural feature of all experimental methods is a more or less pronounced theory-dependence. For this reason, experimental data and operationally defined properties are never strictly theory-free. The first four functions may be rated as relatively weakly theory-determined because they do not presuppose the existence of a theory. In contrast, the function of the latter three consists in the adjustment of the empirical basis to theoretical presettings, and is therefore directed by a strong theory orientation. Accordingly, these functions are representative of the methods of consistent adjustment between theory and experience: based on the assumptions and results of a theory, objects and instruments are designed that subsequently serve to extend and increase the quality of the empirical basis of that theory.

The second common feature of all experimental methods is the extensive use of instruments because the transformation of subjective perceptions and singular events into reproducible, controllable, representative, and intersubjective experience is unattainable otherwise. This close link with instruments is not only the constitutive feature of all experimental methods, but also leads to conceptions of observability, experience, and experimental methodology which differ significantly from empiricist views. Moreover, experimental methodology by no means consists merely in the collection of experimental techniques developed and used in physics, but constitutes a comprehensive network of assumptions, views, convictions, and action instructions:

- The tacit assumptions of experimental methodology: firstly, they set norms showing that the experimental methodology is grounded in a set of methodological principles, generally accepted in physics. Secondly, they are representative of the attitude and comprehension of nature in physics: the directly perceptible reality ("nature") is not more highly ranked regarding its ontological status (it is not "more real") than the reality created in laboratories.
- Experience, observability and the status of the experiment: experience in physics relies on experimental, i.e., instrumental, reproducible, representative, and intersubjective knowledge, rather than on direct sense perceptions. Consequently, observability is not identical to direct perceptibility since it depends crucially on both the development status of experimental techniques and the current state of theoretical knowledge. Experiments, on the one hand, contribute considerably to the acquisition of knowledge about nature and are indispensable tools for establishing the empirical basis of physics. On the other hand, they are neither unbiased experiential sensations, nor an approach to nature without any presuppositions, nor the unproblematic questioning of nature they are frequently considered to be. Knowledge and experience obtained by experiment are never theory-free in an empiricist sense.
- Status of instruments: the instruments are not epistemologically unproblematic refinements and extensions of human sense organs, but play a key role in experimental methodology, with an autonomous function in establishing physical reality. Firstly, without instruments, it would be impossible to create material reality, in the sense of manufacturing representative systems as ideally as possible. Secondly, the extension of knowledge through exploitation and exploration of those parts of reality that are not amenable to direct perception necessitates instruments. Thirdly, the measurable properties (observables) of physical systems can be defined only through the use of instruments, in fact, initially only as the unit comprising system, measuring device, and mutual interaction. Although under certain conditions the observable can be assumed to be an attribute of the

system itself so that one may neglect the instrument and its influence on the system, the definition of quantifiable properties is impossible without instruments.

- Experimental methods: the cognitive significance of the instruments becomes apparent in the specific measuring procedures and experimental techniques of physics. The investigation of the mode of operation of instruments, the acquisition and interpretation of data, and the connection between the properties of a system and the registered data of specific examples provide the basis for judging the epistemological significance of instrumentally obtained knowledge.
- The idea of scientific progress: both the expansion of reality by experiments, with respect to extent and knowledge, and the potential for improving experimentally obtained data represent central aspects of the idea of progress in physics. Consequently, experimental experience is never ultimately closed, and the fiction of an aprioristic empirical basis of fixed, final, and non-revisable facts and data goes back to normative beliefs from the period of pre-instrumental natural philosophy.
- Methodological autonomy: the categorical theory determinateness of experimental methods does not indicate that experiments merely reproduce the theoretical assumptions they rest upon. Although the possibility of such a vicious circle cannot be entirely excluded, experimental methods and results are autonomous in that they are not reducible to theoretical methods and results. In particular, the empirical basis is the result of operations and practical actions, but not of theoretical considerations and preconceptions. Although the methods for establishing the empirical basis are never theory-free in an empiricist sense, theoretical methods are not used.

This definition of experimental methodology rests primarily on the function of experiments to establish an intersubjective foundation of experience. In contrast, the possibility of securing theoretical knowledge by experimental methods has not yet been discussed. Due to the theory dependencies, the extent to which experimental methods are capable of accomplishing this task is by no means obvious. This challenge will be the central topic of Chap. 4.

2.4 Empirical-Inductive Methodology

Once the empirical basis has been established by experimental methods, its conceptual description adds physical meaning to its elements. According to empiricist views, physical meaning should be assigned to the terms for objects by referring to directly perceivable entities (referential aspect). This meaning should be acquired, firstly, by the immediate and unequivocal correlation between term and object established through showing (ostensive definition), in accordance with the naive-realistic identification of an object and its conceptual description. Secondly, the meaning should arise from the description through directly observable attributes, such as fast, heavy, or short. The "empirical terms" formed in this vein are supposed to be theory-free, directly comprehensible, and completely interpreted. In contrast, even the object-oriented description of physical systems and processes does not rest on colloquial attributes, but on quantifiable properties, i.e., physical quantities. Although directly observable properties exist in an empiricist sense, similarly to singular perceptions, they are not a part of physics as long as they are not measurable. However, a quantitatively defined property cannot be designated as perceptible because it does not correspond to a direct perception. For this reason, physical concepts receive their meaning, not via reference to reality, but through the properties assigned to them, so ostensive definitions do not play any role in physics. Furthermore, due to the reproducibility and representativeness of its elements, the empirical basis contains numerous structures that must be included in the conceptual description. Regarding data and effects, for example, the experimental conditions of their production must be specified, not only because their reproducibility must be accounted for, but also because their meaning partially arises from these specifications. Accordingly, the conceptual description of the empirical basis is grounded in a scientific conceptual system, so it is never theory-free in an empiricist sense. Consequently, it does not consist of protocol sentences, and it is not reducible to an observational language whose terms are directly correlated with perceptions.

The conceptual description of physical systems, i.e., the object domain of the empirical basis, already occurs through the definition of physical quantities and may be referred to as empirical concept formation. These methods go far beyond any unproblematic verbalization of direct perceptions and experiences and comprise various types of definitions. A nominal definition is the explicit definition of a new concept in terms of known concepts, so it is bare convention or naming. Examples are the names electron, proton, neutron, and pion for elementary particles defined, i.a., by the numerical values of their mass, electric charge, and spin, or phonon, magnon, and exciton for the quanta of elementary excitations in solids. Another type of feigned nominal definition is the definition of a physical quantity through other, known quantities. A simple example is the mass density $\varrho = M/V$ of a body. Since formally the quotient of two known quantities does not contain any additional knowledge, this concept formation looks like a nominal definition, so the concept of density merely arises as an abbreviation for the expression M/V. Such a reduced view immediately raises the question of the significance of this concept. As a mere abbreviation, this term is eliminable without problems, and thus dispensable. Moreover, it is not comprehensible why just the quotient is introduced as a new term, instead of the product or any other combination. Actually, the significance of this concept does not follow from the formal definition through two already known and measurable quantities, but from the circumstance that the density constitutes a property which is characteristic of the given material, while mass and volume depend on the size and shape of the particular body. In this respect, the density is an "invariant" with respect to the size and shape of a body. This is not self-evident: it is conceivable that a larger body of the same chemical composition might have a higher density because an increasing number of parts might result in an increased internal pressure due to mutual interactions, and this might reduce the volume. Therefore, the concept of the density of a body acquires its meaning and empirical significance because it expresses an invariance, corresponding to an empirically provable regularity. Another example is the definition of the momentum $\mathbf{p} = m\mathbf{v}$. Its empirical significance rests on the circumstance that $\mathbf{p}$ is one of the fundamental conserved quantities. In the formal sciences many concepts arise from such nominal definitions, but in physics most new concepts contain more meaning and knowledge than can be read off the formal equations of definition. Since, by elimination, this additional meaning, and thus empirical content, is lost, concept formation in physics does not rest on nominal definitions. For didactic reasons, it follows that physical concepts cannot be properly understood if they are just introduced as nominal definitions without pointing out the knowledge that has led to this concept.

Consequently, the conceptual description of the empirical basis is grounded less on nominal definitions than on empirical and analytic definitions, together referred to as real definitions. An empirical definition represents the description of an object or phenomenon by empirical facts. For instance, the object "air" may be empirically defined as a quantitatively determinable mixture of various gases. An analytic definition is the explanation of a term through other terms that are better known or more precisely definable. Unlike a nominal definition, the term to be defined already has a meaning, albeit not precisely specified. Examples of such generic terms are tree, animal, model, or theory. The conceptual description of the empirical basis frequently refers to colloquial terms that are so diffuse that an analytic definition does not yield a sufficiently detailed one. This necessitates either a restriction of meaning through a new, clear-cut meaning (explication) or differentiation into different meanings by attributes. Accordingly, an analytic definition aims to determine a meaning that is specific enough for demarcation from similar objects or phenomena, but is at the same time general enough to cover a sufficiently broad range that its use makes sense. For this reason, the conceptual description of objects does not take place as an additive aggregation of "all" properties (enumerative induction), which would eventually have the consequence that a different term must be introduced for each individuum. The aim is rather the selection of a *small* set of attributes, considered as significant with respect to a certain context (intuitive induction). As a consequence, dissimilar descriptions and meanings are obtained, depending on the selected properties. Overall, the conceptual description of the object domain of the empirical basis corresponds to the construction of representative terms for classes of objects, so that an univocal correlation between term and object no longer exists. This also militates against the empiricist theory of meaning, which implicitly assumes the existence of such a correlation and largely ignores the aspects of idealization and classification inherent in any concept construction.

A physical quantity acquires another part of its meaning through its operational definition, described in detail in Sect. 2.3, in combination with the measuring instruction. Firstly, the measurability of a property like the density extends its empirical content. Knowing, for example, the mass of a body, its volume can be determined by measuring its density. Afterwards, it can be checked whether the result is in agreement with values obtained by other methods. Secondly, measurability extends the meaning because it makes the concept largely independent of the model assumptions underlying the formal definition. Thirdly, the measurability is a necessary criterion for a physical quantity, just as representativeness and reproducibility are for objects and effects: solely by means of measurable quantities, a description of objects is accomplished that meets the criterion of representativeness. All in all, formal, operational, and contextual definitions, due to observable effects and regularities, contribute to the meaning of physical quantities. Just as a theoretically postulated object becomes a part of reality through its discovery, and thus becomes independent of theoretical assumptions, a physical quantity acquires a theory-independent aspect of meaning through the latter two definitions. This demonstrates that even quantities introduced and defined on the basis of theoretical considerations are not at all pure conventions.

The conceptual description of the empirical basis by real definitions and measurable physical quantities constitutes the first step towards a structuring of factual knowledge. This results in the classification of the object domain with respect to selected properties, and becomes manifest in representative, classifying terms, and in generic names like "metal", "halogen", or "red giant". The structuring aspect, beyond a pure description, is immediately apparent because the result of the classification depends on the selected properties. Similarly, the conceptual description of effects does not consist of the mere statement of an observation. In order to guarantee reproducibility, the description must comprise the experimental conditions and procedural steps that are necessary to produce the effect. If these conditions are known, this description implicitly contains a causal interpretation as structuring adjunct. Thse structures, which provide interpretations of the factual knowledge contained in the empirical basis, are called empirical orders. They range from the presentation of individual functional relations up to comprehensive ordered systems, and can be subdivided into rules, algorithms, empirical generalizations, causal hypotheses, empirical correlations, and empirical classifications. With the exception of rules and algorithms, the central common feature is their hypothetical character. Hence, their formulation can be described as the formation of hypotheses. A hypothetical statement or structure is one that postulates new knowledge and advances an unproven claim of validity, and may be therefore subject to refutation. With respect to this definition, rules are not hypotheses, but heuristic descriptions that do not advance any refutable claim of validity because exceptions are generally permitted and are thus not falsifying instances. Similarly, algorithms as mere computational instructions do not advance any claim of validity because they are purely instrumentalist procedures for the generation or description of data sets.

A dissimilar structure is exhibited by empirical generalizations such as "all metals can conduct an electric current" or "all things fall to the ground", which can always be transformed into a statement of the type "every object G has the property E". Their claim to validity consists in the idea that they represent a universally valid relation. Formally, they cannot be definitively verified by quoting corroborating instances, but they may at best be falsified by counterexamples, although even this turns out to be impossible in most cases. For instance, the assertion "all metals can conduct an electric current" is not falsifiable as long as there is no precise definition of metal which is independent of the property of electrical conductivity. Such a definition is provided by the electronic theory of solids, whereby this empirical generalization is transformed into a deducible theorem. What is decisive here is that, firstly, assertions with an unconditional claim to validity are not put forward in physics: the claim to validity of physical laws, as the representative scientific statements, is always tied to certain conditions. Secondly, a physical law is not an assignment of a property to an object, but constitutes a functional relation between properties. Due to these structural dissimilarities, empirical generalizations are of minor importance in physics.

The interpretation and explanation of effects occurs, at the empirical level, by the assumption of a material carrier or a physical interaction. In the first case, an existence hypothesis is obtained, but in physics the bare existence of an object is never postulated because this does not offer any possibility for testing. Instead, properties are given which enable the identification of the object in question. Accordingly, existence hypotheses are always substituted by hypotheses about properties, as illustrated by the existence hypothesis for the chemical element "ekasilicon" (germanium) that was replaced by a set of interpolated physical properties permitting the verification of its existence. Analogously, the hypothesis "there is another planet beyond Uranus" was realized by providing the approximate position at a certain point in time. Thus, if an effect results from the existence of an object, such as Neptune as the cause for the anomalies in the orbit of Uranus, an existence hypothesis may be interpreted as a causal hypothesis. However, this commonly involves an explanation of an effect by a physical interaction. For instance, the splitting of certain spectral lines in an external electric field (Stark effect) is explained by the interaction of the atomic electrons with this field. In fact, the distinction between existence and causal hypotheses is blurred, as demonstrated by the effect, known since antiquity, that water in a lift pump never rises higher than about ten meters. Galileo still assumed as cause the idea that "nature abhors a vacuum", and it was only in the middle of the seventeenth century, due to experiments by Torricelli and Pascal, that air pressure was identified as the cause for this effect, i.e., the explanation is once again provided by a material carrier, viz., the air. In general, a feature of many causal hypotheses is that the postulated cause is either initially or not at all directly observable. As a consequence, the construction of causal hypotheses to explain effects passes through a highly speculative stage and is prone to errors. Although the arbitrariness of such speculations may be constrained by methodological norms, stipulating, e.g., that the postulated interaction ought to provide additional empirically testable effects besides the one it was constructed for, the options for verifying hypotheses of this type through purely empirical methods are so limited that they are not transformed into secured knowledge. For this reason, the explanation of effects in physics does not take place through the proof of hypothetical causes, but at the theoretical level by deduction within the framework of a theory. Overall, causal hypotheses and empirical generalizations may be ascribed some heuristic function, but they do not play a role in structuring physical knowledge.

The most comprehensive empirical orders are the empirical classifications (Böhme et al. 1977). Although the idea of empirical orders returns to ancient Greek natural philosophy, they have played a subordinate role because even comprehensive empirical classifications are unsuitable to constitute world views, the main objective of preclassical science. Only with the rise of classical physics in the seventeenth century, and the development of an inductive methodology, in combination with the program for eliminating metaphysical elements from natural science during the eighteenth century, did the construction of empirical classifications become the dominant research program of science. The idea that this is a meaningful scientific objective rests on the experience expressed in the tacit assumption (1B), namely, that nature is structured and that these structures can be discovered. The representative example is Linne's classification of plants, deemed at that time as exemplary among the sciences. Examples of empirical classifications in physics and chemistry are the periodic system of the elements, the Hertzsprung–Russell diagram for the classification of stars, and the classification of crystals according to symmetry. In the latter case, solids included in the class of face-centered cubic crystals are physically as dissimilar as argon, sodium chloride, and copper. Similarly, the periodic system groups the elements according to similar chemical behaviour, irrespective of their rather different physical properties. Overall, empirical classifications are characterized by subsequent properties and functions (Böhme et al. 1977):

- They arrange the many facts more clearly and thereby improve access to existing factual knowledge by facilitating the retrieval of information.
- They create theoretical knowledge, in the sense that it is not necessary to observe everything in order to become aware of it. In the ideal case, this may enable empirical deductions like the prediction of new elements suggested by the periodic system.
- Through the selection of properties, they define those features that are considered to be significant. This selection is essentially determined by

the objectives pursued by classification, so that a diversity of possible classifications results.

- By structuring the object domain through grouping, they create an autonomous conceptual system with generic names like alkali metal and halogen in the periodic system, or supergiant and main sequence star in the Hertzsprung–Russell diagram.
- The flexibility of empirical terms is transferred to empirical classifications, in that they remain subject to supplementation and modification. Accordingly, their function does not consist in advancing claims about the completeness of the object domain under consideration.

The practical objectives pursued by a classification, such as improved access to factual knowledge, do indeed play a certain role, but it is far from being the most important one. In this regard, alphabetical order, the ideal type of purely conventional classification, is the most efficient. Concerning observable regularities, interpreted as structures of reality, the construction of empirical classifications rather has the goal of achieving a "natural classification" (Duhem 1908) that is adapted as well as possible to the object domain and contains a minimum of conventionalist elements. However, the attribute "natural" is neither a unique criterion nor a universally valid quality characteristic. For example, crystalline solids can be classified according to symmetry in crystal systems, crystal classes, Bravais lattices, and space groups. Such a classification is "natural" to describe crystal structures, optical anisotropy, or the directionality of elastic behaviour. For the description of other physical properties, such as cohesion or lattice dynamics, the classification according to types of chemical bonding into ionic, metallic and covalent crystals is "more natural", and for the description of electric or magnetic properties, it is a partition into conductors, semiconductors, and insulators or into dia-, para-, and ferromagnetic substances, respectively. Due to these dissimilar possibilities, even the best empirical classifications contain distinct conventionalist elements and are afflicted with a certain arbitrariness. Finally, they are purely descriptive, even when they represent natural classifications. For instance, without knowledge of the atomic structure, the periodic system can provide the reasons neither for the chemical affinity of the elements of the same group, nor for the fact that no additional element exists, e.g., between carbon and nitrogen or between oxygen and sulphur. Consequently, it was possible to infer the existence of new elements from gaps within the schema, but the existence of rare gases could not be predicted and neither could the exact number of transition metals and rare earths. This shortcoming is grounded in the fact that, solely by means of empirical methods, the basic ordering principles cannot be justified because they are not entirely understood. On the other hand, the attribute "natural" is at least implicitly connected with the hypothetical, empirically unverifiable ontological claim of validity that the basic order principles correspond to structures of reality. Finally, since in most cases the structural principles do not possess a mathematical representation, empirical classifications play a subordinate role in physics, and are more a transitional phase on the way to theoretically well-founded knowledge. In contrast, they may dominate in non-experimental sciences like astronomy or in sciences that are not quantifiable.

Empirical correlations play the most important role in physics, since virtually every goal-oriented conducted series of measurements yields such a correlation, but also because they exhibit an identical formal structure to physical laws. Typical instances are:

- Ohm's law: representation of the connection between the voltage applied on a conductor and the induced electric current.
- Gas laws: representation of the connection between two of the state variables pressure, temperature, and volume for a gas.
- Titius–Bode sequence: ratios of the average distances between the Sun and the planets.

Further examples, mentioned in Sect. 1.3, are Moseley's law and Balmer's formula. The properties and functions of empirical correlations are, as far as they are reasonably transferable, similar to those of empirical classifications:

- The data volume is structured: extended series of measurements are clearly arranged and reduced to their essential parts.
- By interpolation or extrapolation, they produce theoretical knowledge in the same manner as empirical classifications, and they may provide the heuristics for empirical deductions like Moseley's law predicting new elements, or the Titius–Bode sequence suggesting the search for another planet between Mars and Jupiter.
- In the ideal case, experimental studies may serve to identify significant properties, e.g., the electrical resistance of a conductor can be studied with respect to temperature, pressure, or size and shape of the sample.

Since they are obtained by experimental and inductive methods, empirical correlations possess a meaning that is independent of the conceptual system of a specific theory. For this reason, they are invariant against theory changes and are sometimes called experimental laws (Nagel 1982). The Balmer formula, for example, is valid regardless of whether it is derived from Bohr's theory or from quantum mechanics. The hypothetical elements of empirical correlations are, in the first instance, the range of validity and the kind of functional relation. Regarding Ohm's law, it is assumed valid for all metals, and the relation between voltage and electric current is assumed to be precisely linear. Additional hypothetical elements concern the selection of variables, but above all, the additional and for the main part implicitly advanced claim that this constructed relation is not accidental, but significant in the sense that it corresponds to a structure of reality. Without this ontological claim, an empirical correlation would merely have the status of an algorithm, i.e., a purely formal relation that would be useless for gaining knowledge about nature. Due to these hypothetical elements, one is faced with the challenge that, solely by means of empirical methods, it is impossible to provide the secureness of knowledge postulated by empirical correlations. The significant correlations cannot be separated from the accidental ones, nor can the mathematical form of the relation be justified, and nor can any claim to validity. This is illustrated by the procedure for arriving at such correlations.

The first step consists in searching for properties correlated with each other. To this end, one uses a procedure called correlation analysis, based on the following idea. For two variables a correlation coefficient is defined that represents a degree of confirmation for a supposed connection. This coefficient is equal to unity if there is an exact correlation and zero if there is none at all. When the correlation coefficient is equal to or at least approximately equal to unity, this suggests that the correlation represents a natural law, provided that one believes that laws can be "discovered". However, the actual situation turns out to be far from unambiguous. On the one hand, exact correlations do not exist due to the finite accuracy of experimental data, and on the other hand, accidental correlations cannot be excluded. Despite these constraints, the belief, that one can arrive at general laws by starting from single observations and experiments has formed the basis of inductivism since Bacon. More generally, the transition from experience to theory ought to be established by purely inductive methods. This should take place in the following steps (Chalmers 1982, Ritter et al. 1971ff.):

- Since perceptions, observations, and experiments must be considered as the sole sources of true knowledge, because they are independent of any theoretical prejudices and metaphysical speculations, physics must start with a collection of facts, as complete as possible, which provide a theory-free and safe basis.
- Subsequently, one must search for possible correlations between these facts. By means of inductive procedures like correlation analysis, methods of comparison, exclusion, etc., a limited number of single observations are used to arrive at special laws with differing degrees of confirmation that are also true and represent objective knowledge.
- With increasing empirical confirmation and generalizing induction, one ascends from special laws to increasingly general ones until one finally arrives at the most general laws of nature.
- The formulation of physical theories as true propositional systems is achieved by summarizing the (true) laws in the form of theories. Even though actual theory development does not rest solely on induction, the theory must be reconstructed in such a way that all terms without direct connection to observation ("theoretical terms") can be eliminated.
- As the completion and support of inductive methods, true statements about particular events and phenomena are gained by deductive methods, in accordance with the aim of physics to provide predictions and explanations.
- The truth of physical knowledge is guaranteed by the circumstance that the observational statements can be confirmed as true through direct use of the sensory organs. By induction, this truth is transmitted to the laws

and theories. As a result, the predictions and explanations deduced from these laws are also true if the deduction was logically correct.
- The objectivity of physical knowledge rests on the fact that both observations and experiments, as well as the inductive argumentations, are objective: they are verifiable by any other person, and they are independent of any suspect theoretical assumptions or individual convictions.
- Both the development of science and scientific progress are essentially cumulative processes: scientific knowledge grows to the extent to which the amount of factual ("positive") knowledge increases.

Although this is admittedly a somewhat simplified description of inductive methodology, a feature of all variants is that the actual execution of this program is faced with insurmountable challenges. Disregarding the fact that perceptions are not at all reliable, they are intrinsically tied to structuring due to previous experiences. Experiments a fortiori are never theory-free in an empiricist sense. Next, the complete data basis is unsuitable as a starting point for gaining physical knowledge. Since the data must not be separated into significant and irrelevant data, which would be viewed as an inadmissible assessment involving theoretical assumptions, the significant data are obscured among a wealth of irrelevant and useless information, whence it is impossible to recognize general structures. Actually, the recognition of general structures necessitates sufficiently simple systems and phenomena. For instance, neither the exceedingly fruitful concept of the chemical bond with well-defined valencies, nor the concept of a definite molecular structure could have been developed if a complete basis of empirical facts had actually been available, e.g., if beside CH_4, the compounds CH_3, CH_2, and CH had also been known, or if the intermetallic compounds, already known at the time, had been taken into consideration. However, while the inductive construction of structures necessarily relies on sufficiently simple cases, the transfer to complex systems and phenomena may lead to erroneous interpretations and may seem artificial since the description of reality is confined to an ordered schema that is much too limited. This is precisely the reason for numerous scientific errors that have been corrected only through theoretical knowledge, rather than by new empirical facts. In addition, the generalization of empirical experience faces tight limitations. Firstly, generally valid relations cannot be induced from and justified by singular perceptions. This necessitates reproducible experience. Secondly, it is impossible to unambiguously determine the quantitative form of the relation between two quantities solely by measurements, because measurement results are never exact, and the conclusion from a data set to an empirical correlation requires additional assumptions, e.g., the principle of economy, which cannot be justified empirically. Finally, no proof of induction as a truth-maintaining generalization of singular statements has ever been provided (problem of induction): the principle of induction is justifiable neither by experience nor by logic, and it is generally acknowledged that there are no other schemes of justification. Even in a reduced form, the problem of induction cannot be solved: one cannot specify universally valid criteria for the required quantity (number) and quality (diversity, variety of conditions) of observations that would turn an empirical correlation into a physical law.

In summary, the capability of inductive methods for gaining, securing, and structuring physical knowledge are subject to fundamental constraints and are indeed so limited that all attempts to reconstruct physical knowledge in this way have failed. Consequently, these difficulties and the problem of induction, in particular, have led to the following alternative. If one assumes that the securing of scientific knowledge rests essentially on induction, there is no rational justification of science as a whole due to this unsolvable problem. This conclusion was drawn by Hume who declared that the belief in natural laws and scientific theories is merely a consequence of custom and not justifiable by experience. If, in contrast, it can be shown that in physics neither theory formation nor the establishment of the relationship between theory and experience rest on induction, the inductive reconstruction of physical theories is totally useless. Precisely this is the case because inductivism is based on an entirely erroneous view of experimental science and in particular of physics:

1. The status of perceptions as the source of true knowledge: sense perceptions are not structurally different from dreams or hallucinations so they are by no means a reliable source of knowledge. As singular, irreducible elements of experience, they do not constitute the basis of physical knowledge which, in contrast, rests on reproducible experience and is never theory-free. The assumption of a theory-free, empirical basis as the safe foundation of inductive theory formation is a fiction that does not serve as a suitable starting point, either for the reconstruction of theories, or for the reconstruction of the relationship between theory and experience.
2. The status of physical concepts: concept formation in physics is not the mere verbalization of direct perceptions. While even perception works in a selective and structured way, concept formation is always a combination of abstraction and constructive additions so it cannot be described as purely cumulative or selective in the sense that it summarizes all possible perceptions or records the invariant or intersubjective aspects of sense data. Consequently, there is no unequivocal correlation between concept and object. Every conceptually described reality constitutes a partially constructed reality. In addition, the methods of empirical concept formation are unsuitable for establishing a conceptual basis which is precise enough to serve as a starting point for theory formation in physics.
3. The status of physical laws: there is no inductive pathway from empirical propositional statements to physical laws because they are fundamentally dissimilar in terms of both structure and methods of justification. Physical laws are not highly confirmed empirical generalizations, and nor are empirical propositional statements representative of physical laws. The implicit assumption of logical empiricism, that a statement of the type "all swans are white" exhibits the typical structure of a scientific statement, is

fallacious. In particular, proving the truth of empirical propositional statements is not a part of physical methodology, and is therefore irrelevant for physics.

4. The status of theoretical knowledge: laws and theories of physics are neither the mere summarization of empirical facts, nor epistemologically suspect constructs. On the contrary, they represent significant, autonomous knowledge and are indispensable constituents for structuring and securing physical knowledge. The "theoretical terms" in the laws enlarge the empirical content of theories so the empiricist program of eliminating theoretical terms reduces the facility for gaining even factual knowledge.
5. The status of predictions and explanations: it is an error, and not only of inductivism, that the central aim of physics ought to be to provide these. Predictions and explanations of particular events are applications of physical knowledge, rather than an objective that determines theory formation. This has direct consequences for theory conception in physics. Assuming them to be the central aims of physics, the theories, rated anyway as suspect constructs, would be dispensable. In this case, they are not required for this task because specifically adapted algorithms or empirical correlations fulfill this function equally well or even better. Physical knowledge constitutes primarily secured and structured knowledge, so the central objective of physics consists in constructing a consistent and coherent system of findings, as well as securing and structuring it, rather than predicting and explaining single facts and events.

Notwithstanding the fact that inductive methods are an indispensable constituent of physical methodology because they provide novel, although not secured knowledge, any inductivism which pretends to accomplish the securing of knowledge, and to reconstruct both the transition from experience to theory and the relation between them, is obsolete. Actually, it refers to a model of science that exhibits a primarily classificatory character and rests, marginally at best, on the use of instruments. In this case, the central goal would indeed be the generation of a natural classification, optimally adapted to certain object domains. In a science constrained in such a manner to obtain, describe, and classify factual knowledge, direct observations have clear priority in the acquisition of knowledge, the methodology is primarily characterized by inductive procedures, and scientific progress is essentially cumulative. On the other hand, such a science represents a comparatively weak form of knowledge because the knowledge contained in the empirical orders cannot be secured and no proof of the ontological significance of the underlying structural principles can be supplied. The limitations of any type of empirical knowledge can no longer be overcome within the framework of an empirical-inductive methodology. Such limitations can only be overcome by explanation of the structural principles through deduction from a theory. This reduces the dependence of knowledge on a hypothetical classification schema because the theoretical deduction, e.g., of an empirical correlation assigns to it a semblance of logical necessity. The clear starting point to achieve this goal is a kind of theory formation relying on largely autonomous theoretical methods for forming concepts, models, and hypotheses, and also structural principles that are substantially more comprehensive, and thus more formative for the knowledge and the world view of physics than any type of knowledge resulting from the application of inductive methods.

2.5 Theoretical Concept Formation

The historical development of the periodic system of the elements demonstrates that the construction of efficient empirical orders is not at all simple, because structures are not identifiable without constructive cognitive efforts such as recognizing analogies, abandoning differences, and selecting relevant properties. Nevertheless, the construction of empirical orders is not the initial stage of theory formation in physics, since theories are not inductively developed from them. Since empirical orders are constructed with regard to certain object domains, they provide terms that cannot be made more precise in a way that would make them more suited for theoretical reasoning. A more precise description, in an empirical sense, would involve including further properties in the characterization, thus making the term less general and representative due to its covering a less extended domain. Therefore, concept formation in theoretical physics does not consist in attempts to specify empirical terms, but in constructing concepts that are precise because right from the outset they do not refer to real objects. Due to the fundamental importance of the dissimilarities between this kind of theoretical concept formation and an empirical one, three representative examples are discussed here to elucidate these dissimilarities.

The empirical definition of a mass point starts from the pictorial idea of a tiny piece of matter. The applicability of this term to real systems is paraphrased by postulating that sizes must be negligible compared with mutual distances. A conceivable specification of this idea might be to use the δ-function, exploiting a mathematical limiting process. Apart from some formal doubts about such a "precise" definition, it is not meaningful for basic reasons because it does not embrace the physical significance of this conceptual construct, as may be illustrated by the Earth–Sun system. The mutual distance is definitely large compared with both sizes, so applying the mass point model seems to be justified from an empirical point of view. This is the case, indeed, when calculating the orbit of the Earth around the Sun. In contrast, as soon as conclusions about the total energy of the system are concerned, this condition reveals itself to be irrelevant, since the energy contains a nonnegligible contribution from the intrinsic angular momentum of each body. A mass point, however, does not possess intrinsic degrees of freedom, and thus has no intrinsic angular momentum, because that depends on the size, shape, and mass distribution of the body. Accordingly, the applicability of the concept “mass point” is not determined by system properties but by the problem under consideration, and thus through the theoretical context. Hence, it must be interpreted as a theory-oriented concept rather than an object-oriented one. A definition taking this feature into account might declare that a mass point in mechanics is defined through the properties mass, position, and velocity at a certain instant of time. The contextuality is expressed through the dynamical variables position and velocity. This definition still contains a redundancy that can be traced back to the empirical origin of this term. With respect to the definition within the framework of mechanics, the momentum $p = mv$ is the decisive property. Taking this into account, the mass point is defined as follows as a term in the physical theory of mechanics:

Definition 2.2 A mass point as a physical system of classical mechanics is a non-existent model system whose dynamical state is, irrespective of size and shape, exhaustively described by the $2d$ parameters $x_i(t_0)$ (position coordinates) and $p_i(t_0)$ (momentum coordinates) at a certain instant of time t_0, where d is the dimension of the manifold containing the motion.

This definition of systems through the dynamical state and its properties is typical for classical physics, and does not involve any pictorial reference to real objects, in contrast to the empirical term. Instead, the concept “mass point”, as defined in this way, acquires its meaning only within the framework of the physical theory “mechanics of the mass point”, so it is no longer comprehensible in itself. For this reason, this definition might appear circular when it presupposes the theory whose conceptual basis this term is thought to belong to. Consequently, the alternative of an object-oriented definition through a mathematical limit pursues the goal of establishing a formally precise conceptual basis for the theory without presupposing it. This is, however, impossible. Taking an object-oriented definition seriously, it is not just that it is faced with several more serious challenges, but it is not even meaningful for the following reasons:

1. The definition of the the mass point through an idealizing thought process with reference to real bodies of finite extension must specify the point where the mass is thought to be concentrated. Without such a specification of this point (and the corresponding momentum), it is impossible to establish a dynamics which is quantitatively comparable with experimental data, because the precise positions and distances cannot be determined. Although this position can be correctly defined through the assumption that the mass point standing in for the real body is identified with the center of mass, this definition presupposes at least the definition of a rigid body, and this in turn cannot be formulated without referring to concepts of mechanics. All in all, such an attempt of reconstruction through an object-oriented idealization process does not produce a precise, theory-independent definition: theoretical precision and theory-independence cannot be realized simultaneously.
2. Only the definition of the mass point through its dynamical state determines the scope and the empirical content of the theory: the mechanics of

the mass point is applicable to any physical system whose dynamical state is exhaustively described, with respect to a certain context, by positions and momenta at a certain instant of time, and the theory accomplishes nothing more. It thus reveals itself as the theory of the dynamics of systems without intrinsic degrees of freedom, but is in no way related to point-like masses, or to the complexity of real systems. For instance, the spatial motion of a binary star whose center of mass does not necessarily correspond to a material point can be described by the concept of the mass point, as can the motion of an entire galaxy containing a huge number of stars. In the context of an object-oriented definition, it appears absurd to consider a galaxy as an instance of a mass point. On the other hand, if further properties are required to specify the dynamical state, e.g., angular momentum, shape, or rigidity, another theory must be chosen which contains definitions of the relevant properties, irrespective of the geometrical dimensions of the system under consideration. This demonstrates convincingly that the applicability of the concept "mass point" is determined by the particular context, and must therefore be interpreted as a theory-oriented term.

3. Any object-oriented reconstruction ignores the fact that the mass point in mechanics does not simply correspond to a tiny piece of matter. It is defined not by mass and point-likeness, but through the dynamical properties position and momentum. Without this dynamical definition, the difficulties associated with the transition to quantum mechanics are not comprehensible. Since according to the uncertainty relations quantum systems cannot be defined through their dynamical state, it has sometimes been concluded that quantum systems do not exist without reference to a measurement process. Finally, the dynamical definition yields another definition of equality. From an empirical perspective, two bodies represent the same mass point if they have identical masses. In contrast, with regard to the dynamical point of view, two mass points are identical if they are in the same dynamical state. Under identical external conditions, the reproduciblity of the trajectories is then guaranteed. The mass does not play any role in this context: an external force has the same effect on a body of a given mass and velocity, viz., the same change of momentum, as on a body that is half as massive, but twice as fast. Altogether, mass and mass point are not the basic concepts of the theory, and any theory-independent definition is obsolete.

A similar situation is found by analyzing the concept of "ideal gas". A common object-oriented description paraphrases it as an entity made up of non-interacting mass points. Again a precise definition cannot be object-oriented since the ideal gas is not a part of reality. In this case, the dynamical properties position and momentum are replaced by the two intensive state variables P and T, which completely determine the state of the ideal gas in thermal equilibrium, and which are related to each other via the ideal gas equation. Corresponding to the theoretical definition of the mass point one obtains:

Definition 2.3 The ideal gas of classical thermodynamics is a non-existent model system whose state in thermal equilibrium is completely described by the two intensive state variables temperature T and pressure P, and which is defined by the state equation $PV = nRT$, where n is the amount of matter and R the universal gas constant.

Intrinsic or system-specific properties do not occur, and the theory referred to as "thermodynamics of the ideal gas" is entirely independent of any assumptions regarding its physical nature. In this respect, this theory is more universal, although less specific, than a theory based on a microscopic model, such as the kinetic gas theory. Accordingly, within the framework of thermodynamics, only a single ideal gas exists, and an object-oriented "definition" by a microscopic model is not a part of this theory.

The last example, illustrating the dissimilarity of empirical and theoretical concept formation, is provided by the periodic system of the elements (PSE). This can be interpreted, on the one hand, as an empirical classification, according to its historical origin, and on the other, as a classification system described within the framework of quantum theory. Each of the two variants leads to a different concept of the chemical element. In the framework of the empirical PSE, an element is defined through its macroscopic physical and chemical properties under ambient conditions, e.g., aggregation state, electrical properties, or homologous chemical behaviour. This results in statements of the kind: in the top right-hand corner, the elements are gaseous; down to the left, metallic properties are more pronounced; in the upper right half, one finds the insulators; and along a diagonal from upper left to bottom right, one finds the semiconductors. The sequence of the elements is given by their masses, but in some cases the chemical affinity leads to exceptions as the dominant property. In contrast, the concept of "element" in the quantum-theoretical PSE is based on the properties of the electronic states in an electrostatic central field. In combination with the Pauli principle, this leads to the typical shell structure of the radial electronic density distribution. The sequence of the elements is determined by the atomic number, which corresponds to the number of protons and is a pure counting parameter without any physical meaning in the empirical PSE. No macroscopic properties are used as defining attributes: the statement that an iron atom possesses metallic properties is obviously meaningless. Another example illustrating these dissimilarities is given by nitrogen. Within the framework of the quantum-theoretical PSE, it is an atom with seven electrons, among which the five valence electrons determine the chemical behaviour. In the empirical PSE, nitrogen is a largely inert gas consisting of diatomic molecules, at variance with the quantum-theoretical concept of an element, viz., the atom. This is the only reason for the inertness, while atomic nitrogen is quite reactive. Overall, the empirical and theoretical concepts of a chemical element are distinctly different. Although in a certain sense the quantum-theoretical description provides an explanation of the empirical classification, it does not make it obsolete. For instance, the arrangement in groups of chemically homologous elements is not deducible solely from quantum-theoretical principles, e.g., the empirical term "halogen" does not correspond to a well-defined Hamiltonian. Nevertheless, the two conceptual levels are rarely distinguished and it is common practice both in chemistry and physics to consider the PSE as a combination of the empirical and theoretical classification systems.

In summary, these examples demonstrate the crucial structural difference between empirical and theoretical concept formation. The empirical approach is object-oriented in that it starts from real objects and describes them as complete as possible by all properties observed under various conditions. Subsequently, among all of the possible properties assigned to the object itself, those that are considered to be essential are determined and selected. This selection of properties may be referred to as idealization: a real system is transformed, in thought, into an idealized object that exclusively possesses the selected properties. Theoretical concept construction goes exactly the other way round. Oriented toward the objective and the context of a theory, a set of mathematically well-defined properties is introduced that defines the dynamical or physical state. This state is described by an equation, e.g., an equation of motion or a state equation which, in turn, conceptually defines a non-existent model system. Since this is not a part of reality, this term does not in the first instance have any meaning that is independent of these defining properties. The relationship with reality ("referential aspect") is rather established only within the framework of system-specific applications. This type of concept construction may be called abstraction, in the sense that it rests on the detachment from direct observations and experiences. However, the two approaches must not be viewed as isolated from each other. In the object-oriented idealization, the assessment of relevant properties takes place in relation to a certain theoretical background. Conversely, the selection of the properties in the theory-oriented abstraction rests on systematic experimental studies. Together they constitute one of the multi-faceted relationships between theory and experience, discussed in detail in Chap. 4.

Consequently, the conceptual basis of a physical theory is established not by idealized objects like mass point or ideal gas, but by properties. The definition of properties is therefore the first step in theory formation in physics, rather than the definition of idealized objects. Accordingly, one may ask whether a set of properties can be defined that would serve as the theory-independent conceptual basis of the theory. In logical empiricism, these are directly observable properties like short, fast, or liquid. However, conceptual descriptions in physics rest on physical quantities, i.e., measurable and quantifiable properties, so only these come under consideration as the basis of a theory. In order to show that their theory-independent definition is unattainable, first of all, the ontological status of properties must be clarified, i.e., one must ask whether they can be assigned to a system irrespective of any experimental and theoretical context, or whether they come into existence only due to observations and experimental investigations, or whether they are, as theory-dependent constructs, merely conventions, or even fictitious. Actually, examples exist for each of these three possibilities.

A property of the first type is definitely the existence of macroscopic bodies: that a tree, a mountain, or the moon exists when nobody looks at it, is rarely put seriously in question. At least, I do not know any physicist who stays in front of his apparatus during a longer measurement because he is afraid it could vanish if he were not looking at it anymore. Existence can thus be taken as an attribute that is assigned to a body independently of perception and human consciousness, and may be called, in this respect, an objective property. Other properties of this type are the mass and moment of inertia of a body, or the length of a bar. The measurement of such a property merely yields its quantitative value, and common consensus agrees that it is not generated by the measurement. This may serve as the constitutive feature of such properties: the system possesses this property as a "pre-existent" attribute, and a measurement just fixes its quantitative value. Only under this precondition is it meaningful to ask how a system behaves when it is not undergoing a measurement. In this sense, in classical mechanics all dynamical properties such as position, momentum, or energy may be considered as objective.

Properties of the second type are those which occur only under certain external conditions like water solubility: the solubility of a substance becomes manifest or observable only if it is actually put into water. The status of such relational or dispositional terms may be interpreted alternatively as follows. According to the first view, such a predicate corresponds to a potential or latent property of the system which is "actualized" only under certain conditions. This corresponds to the assumption that any behaviour, observed sometime, is correlated with a property that the system itself possesses as a pre-existent, though latent attribute. This would imply, e.g., that not only each atom potentially possesses the properties of all its compounds, but even that elementary particles are already potentially intelligent. This view obviously ignores the fact that, due to the interaction between two systems, new properties may emerge which can be defined in a meaningful fashion only by including this interaction, but which cannot be assigned to the isolated systems themselves. This insight leads to the alternative position that such properties must be interpreted, not as potentially or latently existent, but as newly formed so that they may be referred to as emergent. This interpretation not only avoids the nonsensical consequences of an "actualization of potential properties", but is also of fundamental importance to any interpretation of quantum theory. Since the dynamical properties cannot be assigned to quantum systems as pre-existent system attributes without encountering inconsistencies, these properties can be defined only by including an appropriate measuring device. In contrast, the view of the dynamical properties as "potential" system attributes that are "actualized" in a measuring process results in interpretations of quantum theory that have been revealed as untenable. In addition, the emergent view leads to an appropriate understanding of the construction of conceptual hierarchies in physics. For instance, the hierarchy of the conceptual levels to describe many-particle systems comprises four stages:

- Free particle: a system is assumed to be isolated from the environment, i.e., independent of the properties of the latter. In this approximation, the properties of the associated many-particle system are additively composed of those of the individual particles. The many-particle system does not exhibit additional properties.
- "Effective" or quasiparticle: a part of the interaction of the particle with its environment is included in the particle. This results in the modification of some of its properties. The definition of the effective mass of electrons in the periodic potential of a solid may serve as a representative example.
- Particle in an external field: the environment and its influence on the particle is accounted for explicitly, but not the influence of the particle on the environment.
- Interacting particles: the particle, the environment, and their mutual interactions are explicitly included. Only at this stage of the hierarchy do substantially new properties emerge. An example is the electric dipole moment of a molecule that is defined only for the entire molecule, but not for the separate, non-interacting atoms.

While the second stage of this hierarchy corresponds to the projection of influences of the environment on the isolated system, an identical procedure yields meaningless constructs in the case of the emergent properties defined in the last stage. To speak about the latent or potential dipole moment of an atom with respect to the various dipole moments of its compounds seems nonsensical, but such a "potential dipole moment" is not even well-defined, quantifiable, or measurable, in contrast to the effective mass of an electron, for example. As a consequence, emergent properties are non-existent without the interaction of an object with its environment, so they are defined solely for the whole system "object + environment + interaction", but not for the isolated parts: emergent properties are defined only in a higher stage of complexity.

The antithetic interpretation of dispositional predicates either as potential or as emergent properties demonstrates once again the crucial dissimilarity between empirical and theoretical concept formation. The empirical description of a real system which claims to be as complete as possible also comprises properties arising from interactions with its environment. Through "dynamical isolation", these properties are detached from the experimental context and projected onto the system, whereupon they are transformed into properties that are assigned to the system as potential attributes. Theoretical concept construction goes the other way round. In the first step, properties are introduced that define a model system conceptually. The introduction of a dispositional predicate corresponds to the definition of a new model system that comprises the original system together with both its environment and their mutual interaction.

The third type concerns properties that seem initially to be meaningful only within the framework of a certain theory. From an empiricist perspective, such properties are at best conventions, or even fictitious. That this is not at all the case is shown by the property "orbital period of a planet around the Sun", which is defined only with respect to a heliocentric theory, but is totally meaningless in the framework of a geocentric theory. Another example is the property "mass density at a point" within a body. It is definable in a meaningful way only within a continuum theory of matter: the term "density of an atom" is obviously meaningless. Within the framework of a continuum theory, the density at a point can be defined conceptually by imagining continuously decreasing volume elements ΔV around this point, and looking for a limit in the ratio between mass and volume. Under certain conditions, this limit yields a well-defined value that is defined as the density ρ at point $\mathbf{r}$:

$$\rho(\mathbf{r}) := \lim_{\Delta V \to V_0} \frac{\Delta M}{\Delta V} , \quad \mathbf{r} \in V_0 .$$

In order to apply this conceptual construction to atomic-like structured matter, the volume V_0 must be chosen distinctly larger than the order of magnitude of atomic dimensions, otherwise this ratio will begin to oscillate and will not approach a well-defined value. In that case, the term "density at a certain point" ceases to make sense. The same applies to terms like "charge density" or "current density", which are likewise meaningful only with respect to a continuum theory. Since such properties rest on an incorrect model of real matter, it is not clear, in the first instance, to what degree such theory-dependent properties must be assessed as pure fiction, or whether they correspond to some property of reality.

These examples of the status of physical properties show that each of the three types occurs. There are properties like existence, volume, or mass that may be assigned to the system as pre-existent and objective attributes, in the sense that they seem to be independent of any context. Next, there are properties that are dependent on the experimental context and can be defined in a meaningful fashion only at a higher level of complexity, e.g., as the result of the interaction with an environment. Finally, there are properties depending on the theoretical context which seem to be fictitious because their carrier does not correspond to a part of reality, like the material continuum as the carrier of the density of a body. Actually, even the objective properties are not entirely context-independent, but they are linked to external conditions, e.g., the existence of ice is constrained to a certain range of temperature. As a consequence, the assessment of a property as objective contains implicit assumptions about these external conditions. Concerning ambient conditions which constitute the general frame of direct empirical experiences, certain properties may be considered to be objective, but these experiences cannot be extrapolated arbitrarily. For instance, at temperatures above 10^{18} K, matter no longer exists in the ordinary sense; there is only radiation. Experience gained under the extreme conditions of high energy physics have led sporadically to the opposite point of view, viz., considering as real only symmetries, conservation laws, and some universal principles of physics, rather than the existence of matter. However, symmetries are relations that do not exist in an absolute sense, but are linked to the existence of matter or radiation. Such a radical conclusion is neither tenable nor necessary, because these experiments demonstrate that even a property like existence cannot be rated as objective in an absolute sense, but is always tied to certain conditions.

While the dependence of existence and observability of a property on external conditions is easily comprehensible, the fact that the status of objectivity also depends on the theoretical context poses a more difficult challenge. A representative example is provided by the question of whether the trajectory of a particle exists and is observable. Depending on the theory, one obtains the result that the trajectory exists and is observable (classical mechanics), that it can be assumed to exist without encountering inconsistencies although it is not observable (kinetic gas theory), or that it is neither existent nor observable (quantum theory). More generally, the status of the dynamical properties reveals itself as basically dissimilar in classical and quantum mechanics. In classical mechanics, they can be assumed as pre-existent attributes belonging to the system irrespective of any measurements, without encountering inconsistencies. Within the framework of quantum mechanics, dynamical properties can be defined consistently only through measurement processes: quantum systems do not possess dynamical properties as pre-existent system attributes. Consistent at least with nonrelativistic quantum mechanics is the assumption that mass, charge, and spin (but not spin orientation) can be assigned the status of objective properties in the sense of being pre-existent system attributes. Consequently, in physics, the term "objective" does not express anything absolute, so one should preferably speak of "objectifiable" properties. A property E is then said to be "strongly objectifiable" with respect to a theory T if its assumption as a pre-existent system attribute does not lead to contradictions within T. Correspondingly, the measurement of E consists merely of the determination of its quantitative value. In contrast, dispositional predicates are referred to as "weakly objectifiable" because they are observable and measurable only by including the environment. In particular, dynamical properties in quantum mechanics must be considered as weakly objectifiable in this respect.

Overall, any conclusion regarding the status of physical properties without referring to the experimental and theoretical context is revealed as meaningless because objective properties do not exist in an absolute, context-independent sense. Accordingly, the strongly objectifiable properties possess an epistemic, rather than ontological function: they guarantee the recognizability and identifiability of objects, but prove the existence neither of a subject-independent reality nor of objective knowledge. Consequently, reference to the experimental and theoretical context-dependence of physical properties must be understood only as a criticism of such beliefs, but in no way does it advance an epistemological relativism, alleging an arbitrary convertibility of reality by theoretical descriptions. Supplementing the tacit assumption (1A), the general definition of a system in classical physics can be given as:

Definition 2.4 A physical system is defined by a set of quantifiable, measurable properties that are called physical quantities and provide, within the framework of the theory, an unequivocal and complete description of its physical state.

This definition leaves it undetermined whether a real, existing system or a non-existent model system is concerned, and whether, in the latter case, there is a material carrier of these properties. It does not advance any ontological claim of validity, in the sense of a proof of existence, but guarantees solely the representativeness, identifiability, and recognizability of systems, which is a necessary precondition for a physical description. Accordingly, this definition is restricted neither to strongly objectifiable properties, nor to the directly accessible field of experience as defined by the ambient conditions. Since the necessary determination of the external conditions, e.g., pressure and temperature, also rests on physical concepts and properties which cannot be defined without reference to a theory, a theory-independent and nevertheless theoretically precise definition of physical systems is unattainable. The aprioristic and theoretically precise conceptual basis, imagined as the starting point for theory formation, is also a fiction, like the aprioristic empirical basis. This conceptual basis consists of physical quantities, instead of terms for idealized objects. The precise and noncircular conceptual definition of physical systems takes place through the theory-oriented setting of properties, so that the system is constituted solely by this definition, but not through reference to real systems. This has been demonstrated explicitly by the concepts of mass point and ideal gas as representative examples. Similarly, an atom must be interpreted as a system constituted by its quantum theoretical description in combination with various experimental studies, but not as a context-independent thing-in-itself, irrespective of the fact that atoms are elements of the subject-independent reality.

In summary, the methods of formation of the concepts which establish the conceptual basis of a theory are essentially dissimilar from those methods that constitute empirical orders. These dissimilarities between empirical and theoretical concept formation become manifest in the following points:

1. Empirical concept formation starts with real systems, transforms them into idealized objects by the selection of measurable properties, and continues with their operational definition and the introduction of generic terms. This may be termed object-oriented idealization. In contrast, theoretical concept construction consists of the definition of non-existent model systems through a set of mathematically well-defined properties defining the physical state. The representativeness of a real system is furnished by the

proof that the physical quantities defining this state have the corresponding values. The basic concepts, e.g., of classical mechanics are position and momentum, and the possible dynamical states are obtained as solutions of an equation of motion. The main body of mechanics deals with general, system-independent structures and properties. Terms for systems such as "mass point", "rigid body", or "oscillator" do not appear at this level of description. This type of abstraction, in the sense of detachment from direct experience, is the essential reason for the abstractness of physical theories, but not the fiction of an uninterpreted formal calculus.

2. The empirical and operational definition of properties is object-oriented in the sense that real objects act as their material carrier, and the properties have the status of pre-existent system attributes. In contrast, the definition of physical quantities which establish the conceptual basis of a theory T is theory-oriented. Firstly, it is oriented toward the conceptual system of a mathematical theory and results in mathematically well-defined quantities that are independent of the existence of a material carrier. This is particularly clear in quantum mechanics where, e.g., position and momentum are represented by self-adjoint operators. Secondly, the selection of the quantities is oriented toward the objective and the context of a theory, and their status is theory-dependent, i.e., a property is said to be strongly objectifiable with respect to T if its assumption as a pre-existent system attribute does not encounter inconsistencies.
3. In empirical object-oriented concept formation, for example, a mass point is paraphrased through mass and negligible extension, while position and momentum are classified as contingent due to their time-dependence. In contrast, theory-oriented concept formation defines the mass point through position and momentum, which thus constitute the necessary basic concepts of the theory. In the framework of theory conception introduced in the next chapter, mass has the status of a contingent property because it is not deducible from the theory, but is added as a system-specific property only at the level of the empirically relevant applications. Only this abstraction from system-specific conditions enables one to formulate generally valid relations.
4. The most important dissimilarity between an empirical object description and the theoretical system definition is the model-theoretical aspect that determines both the scope of the theoretical description and, in an immediately comprehensible manner, its complexity: the smaller the number of defining properties, the more pronounced is the model character of a theory. This model-theoretical aspect has important consequences. The appropriateness of an empirical predicate can be assessed by checking how precisely an object is described and demarcated from similar objects. It can be decided whether it has a certain property or not by comparison with the real object: the statement "S has the property E" permits the meaningful question of whether it is true or false. In contrast, the appropriateness of a theoretical system definition is determined by the theoretical

context: taken as an object-oriented description, the statement "S is a mass point" is always false because the predicate "is a mass point" cannot be assigned to any real body. This statement becomes meaningful only if it is understood in such a way that a system may be considered to be representative of a mass point with respect to the model assumptions of a certain theory. Accordingly, a physical system S can be defined impicitly by complying with a theory, whereupon the statement "S is a mass point" means that the dynamical behaviour of S can be described by the classical mechanics of the mass point, but not that mass points exist as a part of reality. This interpretation not only leads to the unexpected applications mentioned previously, but also demonstrates how physics generates its objects of investigation by construction at the conceptual level; and it also shows in which sense a theory "decides" about the properties assigned to a certain system. Finally, the question of whether a theoretical conceptual description is true or false cannot be asked in the same manner as for empirical descriptions. The same applies to physical laws and theories resting on theoretical concept constructions, but does not permit the general conclusion that the question of whether a physical theory is true or false is meaningless.

2.6 Theoretical Orders

The next constituent of theoretical methodology concerns the structures that are considered as a part of reality, according to the tacit assumption (1B). For this reason alone, their study belongs to the universe of discourse of physics. Furthermore, taking into account the fact that the structuring of knowledge is one of the three basic objectives of physics, structures are the central topic of fundamental physical research. This view is in distinct contrast, for example, with empiricist positions, because the conceptual description of structures is faced with the challenge that they are not directly perceivable as things, but require constructive cognitive effort. In fact, observations and sequences of events are sometimes correlated with structures that do not exist; hence, non-existent objects (phlogiston, caloric, ether) or fictitious structures (constellations of stars) are constructed and believed to be a part of reality. Therefore, the construction of structures bears strongly hypothetical traits and is subject to a distinctly higher probability of error than the conceptual description of objects and phenomena. This is why attempts at a hypothesis-free reconstruction of physics aimed primarily to eliminate assumptions about structures. However, such attempts never get off the ground because the "theoretical orders" obtained through structure construction constitute the non-eliminable kernel of our physical knowledge of nature. In order to understand why these theoretical orders, in spite of certain hypothetical traits, are not at all arbitrary metaphysical constructs, indications of the existence of subject-independent structures will first be presented. Afterwards, criteria for the distinction between fictitious and real structures will be discussed, followed by a description of the construction and properties of theoretical orders.

The question of the existence of structures as a part of reality is answered in physics in the same way as the one for the existence of objects. In general form, it is simply presupposed as a tacit assumption. In specific cases, the application of physical methods may contribute to determining the type and extent of these structures. As exemplified previously by the mass density, the epistemic significance of this property is grounded in the invariance of the ratio M/V with respect to the size and shape of the body, and corresponds, in this sense, to a structure of reality. Analogously, the epistemic significance of theoretical orders is grounded in the fact that qualitative regularities exist in nature as constancies, periodicities, symmetries, and hierarchies, whence structurings compatible with these regularities appear particularly natural.

To what extent symmetries correspond to structures of reality, and how they influence structural thinking has been repeatedly documented (Wigner 1967, Hargittai and Hargittai 1994, Rickles 2016). On the other hand, the existence of hierarchies as structures of reality is less familiar, and will be elucidated with some representative examples. Firstly, the fundamental interactions in physics constitute a natural hierarchy due to the dissimilarities in coupling strength and range. For example, as far as the strong interactions are concerned, the electromagnetic interactions can be treated as small corrections. This hierarchy is grounded on small ratios, like the fine structure constant in this case. Similarly, the different size of the masses of protons and electrons enables the separation of nuclear and electronic degrees of freedom and, in particular, the definition of the concepts of molecular and crystal structure by taking the nuclei as a scaffold for the distribution of the electrons. This can be justified by the distinctly dissimilar time scales of the dynamics of electrons and nuclei, and other characteristic times exist for other natural processes. With reference to such dissimilar time scales, a hierarchy of types of motion can be constructed (see Table 2.1) (Primas and Müller-Herold 1984).

As a characteristic feature, the conceptual description of a higher level, although less detailed, cannot be reduced entirely to that of a lower level, because each level contains terms that are specific to it, but are not necessarily meaningful on a lower level. For instance, the concept of temperature can be defined for a gas, but not for a single gas particle. Properties of higher levels which are not defined on lower ones have been termed emergent, and have been illustrated previously with the hierarchy of approximations for many-particle systems with the stepwise inclusion of interactions. Generalized, the construction of hierarchies is a thought process at the conceptual level that can be accomplished only approximately, and is not necessarily appropriate even when it refers to a hierarchy perceived as natural. Finally, since a notionally constructed structure does not necessarily correspond to a structure of reality, the question arises of criteria that would enable one

Table 2.1 Hierarchy of types of motion

Type of "motion"	Typical time (in s)	
Evolutionary processes	10^{11}–10^{14}	Highest level
Ontogenetic processes	10^{6}–10^{8}	–
Epigenetic processes	10^{4}	–
Metabolic processes	10^{2}	–
Biochemical reactions	10^{-1}	–
Molecular rotations	10^{-12}–10^{-9}	–
Molecular vibrations	10^{-14}–10^{-11}	–
Electronic excitations	10^{-16}–10^{-14}	Lowest level

to distinguish between fictitious and real structures. The following example provides a heuristic to establish such criteria.

Theoretically, the probability P of throwing a number between 1 and 6 with a (homogeneous) die is 1/6. In contrast, determining the distribution of the sum for throws with two such dice, the results are not equally probable. Counting the possible combinations, the following theoretical distribution is obtained:

Total	2	3	4	5	6	7	8	9	10	11	12
$36 \times P$	1	2	3	4	5	6	5	4	3	2	1

Obviously, a correlation exists between the total from the two dice and the probability P, with a maximum at 7. According to the heuristic, referring to the tacit assumptions (3B) and (3C), one may ask whether there is an interaction due to which 7 occurs most frequently. This is not the case, of course, because it is the result of a purely statistical correlation that is not deducible from a physical interaction, but rests on the different combinatorial possibilities for each sum. However, instances exist which have played an important role in the historical development of physics where this has not been so apparent. For this reason, it is helpful to have criteria at one's disposal to distinguish statistical correlations from those that can be deduced from a physical interaction ("dynamical correlations"). Regarding this aspect, the correlation for the two dice possesses the following characteristics:

(S1) The correlation is distance-independent: it does not matter whether the dice are in the same cup, or whether two different people throw one die each, or whether they are sitting at the same table or spatially separated.

(S2) The correlation is time-independent: it does not matter whether both dice are thrown at once, or thrown one after the other.

(S3) The individual events are dynamically independent in the following sense: when a throw with one die yields 5 spots, the probability of throwing a 2 with the other does not increase just because the total

7 has the highest probability when combining the two tosses. Irrespective of the result of one toss, the probability of throwing a given number with the other remains unchanged at 1/6: the probability distribution of statistically correlated events does not exert any influence on the results of the individual events.

Relations between statistically correlated events are thus "non-material", not because they are too weak to be detected, but because they do not correspond to anything physically detectable. Only the individual events are real, while their joint distribution constitutes a constructed reality in the literal sense. Nevertheless, such joint probability distributions have objectifiable properties like reproducibility and intersubjectivity. This results in a figure of thought that may serve as the paradigm of "bad metaphysics". In the first step, due to the existence of objectifiable properties, the distribution is ascribed an existence independent of the individual events. This results, in the second step, to the notion that the distribution itself produces the arrangement of the individual events according to a certain pattern. Finally, one arrives at the conclusion that the probability distribution might influence the output of individual events, in total ignorance of the fact that the individual events alone determine the distribution, and not the other way round. A representative example of this figure of thought are the "pilot-wave" interpretations of quantum theory, which are nothing else than an attempt to interpret dynamically an essentially statistical correlation. Of a similar type are the "realistic" interpretations of the wavefunction, which consider it to correspond to an element of reality. A prominent example is the "many-worlds" interpretation (Everett 1957), according to which it is not just one of the many possible states that is realized in a measurement, but all possible states, so that many, although non-interacting, worlds are generated in each measuring process. Actually, the wavefunction as a generally complex function is merely a mathematical auxiliary quantity without any physical meaning. Such a meaning may be ascribed only to the square of its absolute value, at best.

Altogether, the dynamical interpretation of statistical correlations as propagating spatio-temporal physical interactions results in "nonlocal" interactions, spreading out instantaneously over the entire universe. Beside violating the principle that the speed of light constitutes an upper limit to the propagation of physical interactions, such a result contradicts standard ideas about interactions and will be considered nonsensical. However, it should be admitted that "nonsensical" may only mean unfamiliar: notions about physical interactions could be revised in order to interpret statistical correlations in such a manner. These notions are established from the following properties of dynamical correlations:

(D1) Dynamical correlations are distance-dependent: either their range is finite, or their strength decreases with increasing distance between the correlated systems.

(D2) Dynamically correlated events are time-ordered: the change in the physical state at one system occurs at the other correlated system with a certain time delay (retardation).

(D3) The change in the physical state at one system influences the state of the other one according to the type of interaction. Consequently, knowledge about one of the two systems in combination with the interaction frequently provides information about the correlated properties of the other one.

In particular, (D3) indicates that information transfer is possible between dynamically correlated systems, but not between statistically correlated ones. In addition, a dynamical correlation can be derived from a physical interaction that possesses, e.g., in the form of a radiation field, a carrier as empirical counterpart, which is regarded as really existent as an object or effect. The interaction exists independently of the individual events and determines the distribution in the sense that the latter is deducible from the interaction, in principle. As a consequence, dynamical correlations may be causally interpreted without problems since, due to (D1) and (D2), the interaction is local, and the correlated events are time-ordered. In contrast, the corresponding properties (S1) and (S2) of statistical correlations preclude any causal interpretation. A sufficient criterion for the identification of dynamical correlations is thus the validity of (D1)–(D3). A description of structures, constrained to dynamical correlations, could support the view that, according to the tacit assumptions (3B) and (3C), the explanation by causes belongs to the objectives of theory formation in physics. In contrast, the discussion of Bohr's theory in Sect. 1.3 emphasized that its achievement rested primarily on the fact that the empirical order of spectral lines, constituted, for example, by the series laws and the combination principle, has been deduced from autonomous theoretical assumptions. Accordingly, the objective of theory formation in physics is neither the discovery or the construction of causes, e.g., for the origin of the spectral lines, nor the study of causal connections, but the deductive systematization and resulting unification of knowledge. Moreover, theory formation in physics is not in any way constrained to the description of structures of reality, because ordering schemes are also used that do not correspond to structures of reality, provided that they contribute to structuring and augmenting knowledge, so that they are epistemically significant. In particular, this illustrates the conception of force that constitutes one of the fundamental structural principles of classical physics, and was regarded, for a long time, as the scientific specification of the principle of causality.

The identification of force and cause has a long-lasting tradition going back in principle to Ionic natural philosophy. Newton follows this tradition by tracing changes of motion to the effect of external forces. In this way, he establishes not only the basis for the use of the force concept in classical physics, but also the ideological basis for the causal-mechanistic world view. The conclusion, however, from observed changes of motion to the existence of forces is neither compulsory nor unequivocal, because beside the trajectory, only the kinematic properties velocity and acceleration are observable and measurable. Assuming a constant mass m, the relation between them can be represented, e.g., for one-dimensional motion, by

$$m\frac{\mathrm{d}^2x}{\mathrm{d}t^2} + b\frac{\mathrm{d}x}{\mathrm{d}t} + cx + d = 0\ .$$

By specifying the coefficients and initial conditions, the solution of this differential equation yields all of the measurable properties of the motion. Everything else rests on thought construction, so subsuming the last three terms on the left-hand side of this equation under the concept of force according to Newton's second law is by no means compelled by empirical experience. Indeed, every directly observable motion on earth undergoes friction so that, with respect to empirical considerations, a separate treatment of the friction term $b\,\mathrm{d}x/\mathrm{d}t$ would be more obvious. Consequently, the mere description of motion is feasible without the force concept, and some would deny any autonomous physical importance of this concept, considering the second law only as a nominal definition of force. However, this recognizes neither the physical content, nor the methodological significance of this law and the force concept. The physical content consists of the interpretation of changes of motion as the reaction of a system to influences from its environment, represented symbolically by an external force, rather than an inherent property of the system itself, as was assumed in the impetus theory. The methodological significance consists in achieving a unified representation of classical mechanics, since the laws of motion can be formulated without referring to specific physical systems and processes. For this reason, both Newton's laws and the force concept must not be viewed as isolated, but in combination with the particular force laws. The two together pave the way to measuring certain forces, and to deducing empirically comparable results. The dominant role of this structural schema in classical physics reveals itself in many different ways:

- The very conception of motion itself: the dynamical ground state is no longer the state of rest, but the force-free state, i.e., the state of rectilinear uniform motion.
- The decision about what needs explanation: every deviation from rectilinear uniform motion is in need of explanation, but not this type of motion itself. According to Aristotle, circular motion, as the ideal form of motion, did not require explanation.
- The reinterpretation of already known facts: that a moving body comes to rest is no longer assumed to be a property of the body itself, as in the impetus theory, but is ascribed to the action of an external force, e.g., friction.
- The introduction of fictitious forces to incorporate constraints ("pseudo forces") and to describe motion in non-inertial systems ("inertial forces").

- The transfer of the second law to the rotational motion of rigid bodies by formal analogies.
- The development of continuum theories of matter, e.g., hydrodynamics or the theory of elasticity, according to the principles of Newtonian mechanics: the internal interactions in liquids or solids are transformed into external forces between fictitious material elements.
- The transfer of the concept of force to describe electromagnetic phenomena with, e.g., Coulomb's law in electrostatics, the interpretation of the interaction between current flows in wires as the action of magnetic forces, as well as the extension to the concepts of force field and radiation field.

Although the force concept is no longer considered as a methodological prerequisite for any physical knowledge, as Newton believed, these examples illustrate the efficiency of this concept. Its capability rests primarily on the fact that the force concept has led to a previously unknown unification of knowledge. In contrast, it is not at all related to a causal interpretation of forces, although there may appear to be some reasons for such an interpretation. Firstly, the Newtonian schema reflects pre-scientific and empirical experience according to which forces cause change. Secondly, although the possible trajectories can be deduced from known forces, the opposite is not usually the case, so this suggests that the direction of reading the second law from force to acceleration is favoured. Thirdly, in the deduction of the trajectories, the force occurs among the premises so it fulfills a necessary condition for causes. However, this is not sufficient and does not justify the identification of force with cause, just as any preference in the direction of reading does not correspond to anything in the formal structure of the second law: as a mathematical equation, this law represents a functional relation in which the two sides are equivalent.

The most striking evidence against a causal interpretation of the force concept, and thus of mechanics in its entirety, is provided by the formal structure of the equations of motion, which is not consistent with the postulate that the cause must precede the effect in time. In the second law,

$$\mathbf{F}\big(\mathbf{p}(t), \mathbf{r}(t), t\big) = \frac{\mathrm{d}\mathbf{p}(t)}{\mathrm{d}t} ,$$

the time dependence of the force $\mathbf{F}$ (the "cause") refers to the same point of time t as the change in the momentum (the "effect"). In order to permit a causal interpretation, the equation must read

$$\mathbf{F}\big(\mathbf{p}(t-t_0), \mathbf{r}(t-t_0), t-t_0\big) = \frac{\mathrm{d}\mathbf{p}(t)}{\mathrm{d}t} .$$

Additional arguments against a causal interpretation of mechanics are provided by specifications of the force concept through pseudo and inertial forces that cannot be causally interpreted at all, and finally, the deduction of the possible trajectories from extremal principles within the framework of the

Lagrange or Hamilton schemes, respectively. The latter representation is, under certain conditions, equivalent to the Newtonian one, and does not change anything concerning the deterministic character of mechanics, but suggests a final rather than causal interpretation. Taking all this together, the causal interpretation, being inconsistent with the formal structure anyway, reveals itself as a relic of notions which originate from pre-scientific experience; such notions are physically irrelevant and can be eliminated without difficulty because the structure and content of mechanics remain unaffected. On the other hand, the elimination of the force concept would not only fundamentally change the structure of the theory, but would also reduce its empirical content, as exemplified by the discovery of the planet Neptune. The calculation of the position which led to its discovery relied essentially on the assumption of a gravitational force as perturbing interaction. Consequently, if taken seriously, dropping the concept of force would imply that the observed anomalies in the orbit of Uranus could merely be stated, and that any hypothesis about their origin must be rejected as metaphysical speculation. In contrast, use of the force concept not only expands the empirical content of the theory through the acquisition of new factual knowledge, but also provides for the structuring of knowledge. The resulting deductive systematization and unification extends far beyond any causal description because the force concept is applicable irrespective of the existence of a causal connection. Actually, causality in physics plays a marginal role, at best, as will be argued in detail in Sect. 5.2.

Another example of theoretical order is provided by the description of the various forms of ordered states of matter as the result of the competition between disordered thermal motion and an ordering interaction. The degree of disorder corresponds to the kinetic energy increasing with rising temperature, while the ordering interaction is related to the potential energy. The ground state, as the most stable state of the system, follows from the postulate that the total energy takes a minimum. With increasing temperature, the energy of the thermal motion increases, and when it becomes greater than a given ordering interaction, the ordered state of matter existing at lower temperatures breaks down. Consequently, depending on its strength, each interaction can be roughly ascribed a critical temperature T_{cr} below which it occurs (see Table 2.2). Within the framework of this scheme, most of the phenomena associated with the different forms of matter can be described and interpreted, including the stability of nuclei, atoms, and molecules, but also the solid, liquid, and gaseous aggregate states of matter, up to cosmic galactic clusters, and the transitions between the various states of matter, i.e., phase transitions.

These examples demonstrate that a theoretical order leads to a unification of knowledge that extends far beyond any capacity of empirical orders. Their formation, e.g., the inductive generalization of direct observations to rules and empirical generalizations, the setting up of empirical correlations, or the causal explanation of effects, is oriented toward specific systems and pro-

Table 2.2 Temperature range of interactions

Stable state or interaction	T_{cr} (in K)
Atomic nuclei	10^{10}–10^{12}
Atoms	10^5–10^7
Molecules	10^4–10^5
Liquids	10^3–10^4
Solids	10^2–10^3
Organic life	$\approx 3{\cdot}10^2$
Magnetic order	10^1–10^3
Superconductivity	10^0–10^2
Electronic spin–spin interaction	10^{-2}–10^0
Nuclear spin–spin interaction	10^{-5}–10^{-4}

cesses. In contrast, the structural principles of theoretical orders are system-independent in the sense that they are detached from direct observations and experience, just as model systems are independent of the existence of a material carrier because they are defined exclusively through properties. This system-independence is a prerequisite for constructing a unified conceptual system enabling the structuring of a multiplicity of diverse findings under a uniform aspect. The epistemic significance of these theoretical orders rests on the fact that their structural principles belong to the general principles of physical knowledge of nature, which can neither be gained by inductive methods, nor be reduced to empirical knowledge. For these reasons, they represent theoretically autonomous knowledge and constitute the non-eliminable part of physical knowledge, since otherwise a distinctly dissimilar type of research and comprehension of nature would result. In conclusion, the construction of theoretical orders is, subsequent to theoretical concept formation, the second step of theory formation, and it is fundamentally dissimilar from theorizing at the level of empirical methodology.

In contrast to empirical classifications, the system-independent structural principles defining a theoretical order possess a mathematical representation. Examples are Newton's laws of classical mechanics, Maxwell's equations of electrodynamics, and the state equations of thermodynamics. Due to their system-independence, these equations do not enable the deduction of experimentally comparable results. That necessitates, in a first step, the addition of system-specific assumptions, e.g., the specification of the force function in mechanics. Based on examples such as the gravitational or Coulomb force, such a complement is referred to as a mechanism, but it is not constrained to a dynamical meaning; rather it comprises, in a more universal sense, the specification of an abstract structural principle. In the second step, contingent conditions must be specified and added, e.g., initial conditions and values of the system-specific quantities. This eventually enables the deduction of statements that are directly comparable with experiments. Precisely this separation of the system-independent structural principle from both its system-specific realizations and the contingent conditions is the reason for the comprehensive unification of knowledge which makes theoretical orders considerably more efficient for structuring knowledge than empirical classifications, with their orientation toward specific object domains.

In contrast to empirical orders, the scope of theoretical orders is not the description and classification of factual knowledge, but the structuring of knowledge *about* it. This presupposes that a certain amount of knowledge exists from which a general structural principle can be abstracted. Newton's second law, for example, may be interpreted as abstraction from the equation of motion for harmonic oscillation, central motion, and projectile motion. Conversely, a theoretical order receives empirical content via those realizations of the structural principle which correspond to structures of reality. Similarly to the measurability and the existence of a material carrier, which contribute to the empirical significance of a property, such specifications serve as a type of empirical referendum by imparting an image of the force concept which is independent of theoretical assumptions. Further instances, which contribute both to the physical meaning of the force concept and to establishing a relationship with reality, are realizations in which either the action of forces is directly observable, or the basic equation, in combination with a mechanism, enables the deduction of a measuring instruction and an operational definition. The possibility of directly measuring a force supports the view that it actually exists in nature, but is also an important requirement in establishing experimental preconditions to test theoretically derived statements. In this respect, the physical conception of force and interaction is distinct from its usage in astrology or social sciences, where the "empirical content" merely results from metaphorical, but vacuous analogies with the physical force concept, while the quantitative and operational aspect is totally absent.

Finally, theoretical and empirical orders differ in the role they play in the development of research and science. Although empirical orders may possess a heuristic value for specific research goals, just as the Titius–Bode series initiated the search and discovery of the planetoids, they do not define a research program, in the sense of determining the collective objectives and interests of a whole range of research groups. In order to establish a research program, an order must offer a universe of discourse which is versatile enough for several research groups to be able to work at the same time within that framework without being in permanent direct competition. Maxwell's equations, for example, establish the framework for such diverse fields of application as electro-technics and optics. According to Kuhn (1970), research activity within the framework of a theoretical order may be considered as normal science because the structural principles and basic equations are not subject to refutation. Research in mechanics, for example, never aimed to test Newton's second law, but to construct, within this conceptual frame, an appropriate force function from which the desired type of motion could be deduced. Consequently, Newton's law was not corroborated when this was successful, because no one was looking for anything else than an appropriate force function. In conclusion, acceptance of Newton's idea just involves opting for a particular theoretical view and interpretation of dynamical processes: the objective does not consist in constructing alternative structural principles to describe motion, but in constructing an appropriate force function within the framework of the theoretical order established by Newton. This order has prevailed because it defines an efficient research program for the whole area of mechanics. The same applies to the schema for the structure of matter: research activities consist in filling out the general framework of this theoretical order by the construction of quantitative and system-adapted interactions. Once again, the requisite flexibility of a research program results from the system-independence of the structural principles. The strategy for supplementing the abstract framework with suitable, additional system-specific assumptions enables not only the subsumption and interpretation of new empirical experiences within this framework, but also paves the way for unexpected applications. This openness to new experience gives the impression that theoretical orders, unlike empirical classifications, encompass a broad scope virtually exhaustively, so that research strategies within its framework essentially rest on a belief in its completeness. Although this is unprovable, and has actually been revealed as incorrect, at least for classical theories, it still constitutes a normative ideal.

In summary, the construction of theoretical orders and their status as autonomous knowledge constitute the essential part of the autonomous theoretical component of physical methodology and knowledge. These theoretical orders have the following properties. Firstly, they are based on structural principles that are system-independent since their construction occurs neither in an object-oriented way nor inductively from observations, but is based on considerations which are theoretically autonomous in that they are detached from direct observation and experience. Secondly, a structural principle can have realizations which correspond to structures of reality, although this does not necessarily apply to all of them. It contains, in combination with a mechanism, quantifiable and operational elements enabling the deduction of quantitative, experimentally comparable results. For these reasons, despite its abstractness, it is not a conventionalist construct, and nor does it have a purely instrumentalist function. Thirdly, by subsuming different mechanisms under a unifying structural principle, a theoretical order best realizes the objective of structuring knowledge through unification and deductive systematization. Fourthly, theoretical orders provide a framework for the practice of normal science, typical of physics, because their patterns of structuring and explanation are much more general and comprehensive than any explanations by causes or other notions of empirical origin. Finally, as the central structural patterns of physics and due to their comprehensive, unifying character, they influence our understanding of nature in a much more enduring way than any type of empirical knowledge, and contribute decisively to the formation of our physical world view. Accordingly, they reveal the typical features of world views in that they rest on empirically unprovable assumptions, and determine certain perspectives. The advantages and deficits of a world view based on theoretical orders is exemplified by Newtonian mechanics, which has led to an enormous expansion and unification of knowledge about nature. On the other hand, it has essentially reduced the physical world view due to its causal-mechanistic understanding of nature. However, such shortfallings are accepted in consideration of the compactness and unity of a world view established by theoretical orders, and all the more so as, at the end of the process of familiarization with this view, these shortfallings are no longer perceived, because all thoughts and actions occur within this framework anyway. As a consequence, the formation of alternative fundamental approaches is no longer pursued, and those structures existing already but not conforming to this order are eliminated from accepted knowledge in the course of time. At the conclusion of this development stands the conviction that the possibility of an alternative description does not exist, and the world view eventually acquires the status of seemingly necessary and objective knowledge.

2.7 Modeling

Both theoretical concept formation and the construction of theoretical orders rarely occur according to exclusively formal rules, but are commonly driven by illustrative images and analogies. Denoting these across the board as models, modeling is undisputedly a central auxiliary means to develop concepts and theories in physics. Indeed, model conceptions are used for the conceptual description and interpretation of empirical knowledge, but also for the development, application, and interpretation of theories. The meanings of "model" are correspondingly multifaceted and diffuse, not only in physics, but also colloquially, where it is applied with directly contradictory meanings. On the one hand, it is a prototype for something to be created, as in the fine arts. On the other hand, it is an image of some pre-existing object, as in architecture or technology, where models denote material, reduced images of buildings, ships, or engines. Similar opposites are found among metaphors aiming to illustrate the relationship between reality and its conceptual description. According to Plato, the material world is a model of the world of ideas, while in physics theories are considered as models of reality. Occasionally, models are grasped as mere reflections of reality, and their function is compared with maps serving the purpose of orientation in a predetermined reality. Such metaphors do not even possess any heuristic value, since they impart the erroneous impression that physics is the exploration of a predetermined reality. In contrast, according to the previous discussions and results, the performance of experiments, the manufacturing of nearly ideal systems, and the formation of concepts and theories in physics are largely a construction of reality. This must be accounted for in the discussion of the status and functions of models and the methods of modeling.

A suitable approach for distinguishing and specifying the types of models used in physics classifies them according to whether the object to be modeled ("model substrate") and the model belong to the domain of concepts (C) or the domain of objects (S). This division yields four classes of models:

- $\langle S, S \rangle$: material models of reality,
- $\langle S, C \rangle$: conceptual models of reality,
- $\langle C, S \rangle$: material realizations of abstract conceptions,
- $\langle C, C \rangle$: conceptual realizations of abstract conceptions.

Unlike the usage, e.g., in the fine arts or in some branches of technology where nature serves as a prototype for technical constructs (bionics), $\langle S, S \rangle$-models do not play any significant role in the formation of concepts and theories in physics, as illustrated by the following representative examples:

- a rolling wheel as a model for the epicycles of planetary motion,
- the simulation of gravitational interactions through mechanical links between the Sun and planets in a planetarium,
- water waves as a model for light waves,
- a water flow circuit as a model for an electric circuit,
- illustration of magnetic fields by means of iron filings,
- the planetary system as an analogy for the atom,
- models of molecules, like calotte models or "ball and stick" models.

Although the illustration by such models may serve didactical purposes or may fulfill a heuristic function in solving specific problems, they are irrelevant for physical knowledge, if not even misleading. This is exemplified by the illustration of electromagnetic phenomena through mechanical models during the second half of the nineteenth century. These were advanced, in particular, by Lord Kelvin. Although such mechanical analogies are reflected in the conceptual system of electrodynamics, they are not only irrelevant for physical knowledge, but have even led to erroneous interpretations. Another example of misleading analogies between dissimilar universes of discourse is provided by the illustration of atomic and molecular systems by macroscopic ones, like the planetary model of the atom. In this case, what is ascertainable through the concepts of classical physics is concrete. The characteristic shortcoming and limitation of this analogy rests on the implicit assumption of an essentially identical structure of atomic and macroscopic systems. Numerous difficulties in the interpretation of quantum mechanics date back exactly, in several crucial respects, to the circumstance that this is not the case.

The use of $\langle S, C \rangle$-models as conceptual models for certain aspects of reality is frequently considered to be typical of model formation in physics. Firstly, this comprises models such as mass points, rigid bodies, or the ideal gas referring to real systems. Secondly, there are the "black-box models" constituting, e.g., the hypothetical connection between the input and output of an experiment. Finally, the most comprehensive examples are entire theories corresponding to the notion that every physical theory represents a model of a certain part of reality. With the reservation that a model concept with such a broad spectrum of meanings is necessarily rather unspecific, a physical model in the narrower sense is a conceptual representation that refers to an element of reality, e.g., a real object, a real process, or a real structure. The model substrate corresponding to the relevant part of reality is called the representative of the model, and the assignment of a particular representative to a model as its realization. Such a model may possess none or many realizations. In the first case, it is said to be fictitious. An example for the latter case is the model of the harmonic oscillator, which has numerous dissimilar realizations as a pendulum, an elastic solid, a vibrating molecule, or more generally, any small change of state around a stable equilibrium configuration. The most general version of this model type are the metaphysical models or world views, falling into three representative classes. The first are the cosmological models, e.g., the geocentric world view of the ancient world, or the modern models of the universe based on the space-time structure of the general theory of relativity. The second is the atomistic world view that goes back to Democritus and received a first empirical basis during the nineteenth century. In this context, the atomistic world view means the very general notion that a set of smallest particles exists, irrespective of whether it concerns the atoms of the ancient natural philosophy, the chemical atoms of Dalton, or the quarks and leptons of the standard model of modern physics. The third class comprises, e.g., the mechanistic world view, as well as non-physical models of the world such as mythological and religious world views.

The class of the $\langle C, S\rangle$-models, i.e., the material realizations of a complex of abstract conceptions, constitutes a facet of the relationship between reality and its conceptual description that is complementary to the $\langle S, C\rangle$-models. Examples are Plato's model conception, comprehending reality as a model of ideas, the planetary system as a model for a system of mass points, or the experimental realization of theoretically introduced physical systems, like the ideal crystal, a light ray, or a homogeneous field. Closely related to Plato's view is certainly the experimental demonstration of physical laws if they are interpreted as the ideas of physics. The naive-realistic metaphor considering physical laws as the ideals or norms that govern nature can also be viewed as Platonic. Finally, many of the previously mentioned antithetic uses of the model concept can be traced back to the complementary character of the $\langle C, S\rangle$- and the $\langle S, C\rangle$-models.

The class of models that is most relevant for the formation and reconstruction of theories are the $\langle C, C\rangle$-models, i.e., the conceptual realizations of a complex of abstract conceptions. This class comprises three types that are referred to as mathematical, formally analogous, and theoretical models. The mathematical models are the physical interpretation of results of a mathematical approximation and constitute one of the methods for introducing new physical terms. The starting point for this procedure of concept formation is the insight that, due to the arbitrarily complex reality, a substantial gap exists between an exact formulation of a physical problem and a mathematically solvable one. This gap may be bridged through two complementary methodological approaches. The first one is grounded on highly simplified concepts, in order to transform the corresponding equations of motion, field equations, or state equations into an exactly solvable form. Although this necessitates model systems which are unelaborate with regard to their physical proprties, e.g., mass points, point charges, or the ideal gas, they already produce qualitative insights in many cases. Due to their relatively simple mathematical structure, the issues treated in textbooks are largely of this type, so one may get the impression that this is the standard method. Actually, the opposite is true, because this procedure, beside its strongly reducing aspects, has the disadvantage that it does not enable systematic improvement towards a more realistic description. Therefore, the most common approach in theoretical physics is to start from less reduced concepts, and then apply or develop mathematical methods that also initially yield approximate solutions, but permit the systematic improvement of the outcomes. The results are mathematical representations of physical systems and processes that can gradually be improved and augmented:

1. Linearization is a series expansion of a physical quantity with respect to another one that is terminated after the first non-vanishing term. In some cases, the results of the simplified systems are recovered, but even then this procedure provides additional information. Firstly, through the inclusion of further series members, the results can be systematically improved. Secondly, the scope of the linear approximation can be determined quantitatively. Representative examples are as follows:

   - In order to describe motions or, more generally, changes of state around a stable equilibrium position, the potential energy is expanded in a Taylor series with respect to the displacements, and is terminated after the first nonvanishing term containing these displacements. The linearized result is the harmonic approximation, e.g., Hooke's law for mechanical vibrations. Additionally, the systematic inclusion of higher-order ("anharmonic") terms not only reduces the discrepancy between calculated and measured values, but also extends the scope by permitting the theoretical description of further phenomena, such as the thermal expansion of solids.
   - The phenomenological description of the properties of solids is grounded on the following model. Under external influences, e.g., electromagnetic fields, called inducing quantities, the solid reacts with measurable effects which are ascribed to certain properties of the solid (induced quantities). The induced quantities are expanded in a Taylor series as a function of the inducing ones. If the inducing quantity is an electric field and the

induced quantity an electric current, the linear approximation of the expansion for the current density yields Ohm's law.

- Analogous expansions of the scattered waves in quantum mechanical scattering processes lead to Born's approximation as the simplest case. Similarly, Coulomb's law, valid in its simplest form only for point charges, is obtained as the zero-order approximation of the multipole expansion of the electrostatic interaction between two extended charge distributions.

These models with their potential for systematic improvement, as a characteristic feature of physical knowledge, once again demonstrate that quantitative knowledge can never be considered as complete. In addition, the quantitative improvement of theoretical results by mathematical approximation methods reveals itself as useless without the simultaneous option to improve measured data by experimental methods: quantitative experiments and the application of mathematical methods are mutually dependent. The associated progress of knowledge in physics consists in both the improvement and the *mutual* approach and adjustment of experimental and theoretical results. This progress, however, must not be misinterpreted as an asymptotic approach towards truth.

2. Analogous conclusions apply to approximation procedures exploiting empirically justifiable hierarchies of interactions for dynamical isolation. A representative example are the single-particle approximations of many-body systems, mentioned in Sect. 2.5, because they can be improved stepwise by including additional interactions that may be treated as perturbations. In contrast to the series expansions, this cannot be pursued according to formal rules since there are no universally applicable procedures for generating interactions. This necessitates an educated guess based on physical considerations, but it remains undetermined if other interactions are involved. Furthermore, depending on the type and number of included interactions, different levels of complexity may arise, so different levels of approximation may result in dissimilar conceptual systems. This is exemplified by the Born–Oppenheimer approximation for the separation of nuclear and electronic degrees of freedom in molecules and solids, which provides the conceptual framework for defining the concepts of molecular and crystal structure.
3. Another mathematical approximation procedure is the reduction of spatial dimensions, associated with the conceptual introduction of one- and two-dimensional systems. In spite of the three-dimensionality of real objects, the results of these approximations are not fictitious, since there are many, even technologically important, systems and processes which can be treated as approximately one- or two-dimensional. These methods provide the empirical justification for concepts such as the linear chain, representing a one-dimensional solid, or the two-dimensional electron gas.

The influences due to the omitted dimensions can once again be treated as perturbations.

The second type of $\langle C, C\rangle$-models are the formal analogy models. They can be exemplified by models of light propagation. The empirical fact that the incidence angle equals the reflection angle when a light ray is reflected at a specular surface is described by the reflection law. This law also applies to the reflection of elastic spheres by a smooth plane. Based on the conclusion by analogy that light consists of tiny particles, the application of the mechanics of the mass point to optics enabled Newton to describe successfully several optical phenomena known at that time. Alternatively, Huygens developed the model of light as an undulatory motion, in order to explain diffraction phenomena. He was also guided by a mechanical analogy, viz., the phenomena observed in the propagation of water and sound waves. The drawback for Huygens was that an elaborate mathematical theory of wave propagation did not exist at that time, so he could not exploit the quantitative potential of this analogy. This is precisely the crucial dissimilarity between the formal analogy models and metaphors: insofar as the similarities, e.g., with mechanical systems, are constrained to illustrative analogies, such a model provides nothing more than a metaphorical description without epistemic significance. If, on the other hand, elaborate mathematical theories exist for the relevant mechanical systems, these similarities do not merely possess the status of superficial metaphors, but enable one to derive a number of quantitative analogical relations with considerable heuristic potential. In the case of the wave model, properties of sound waves, e.g., pitch or sound volume, may be identified with corresponding properties of light, viz., colour and brightness, respectively. Since pitch and sound volume constitute, through frequency and intensity, quantifiable properties in the theory of sound waves, the initially solely pictorial analogy becomes a formal one through the interpretation of the corresponding variables of the mathematical theory, viz., frequency and intensity of light. This identification paves the way for the purely mathematical deduction of quantitative, empirically testable effects like the wavelength dependence of patterns of diffraction and interference.

In summary, modeling via formal analogical models rests on the extrapolation from the properties and structures of a well known system, possessing an elaborate theoretical representation M_0, to the hypothetical existence of analogous properties and structures of a less well known system. The advantages of the existence of such a formal analogy are obvious. Since the model already has a realization M_0, the internal consistency of the analogical model M_1 is secured. Additionally, the assumption of maximal congruence between the two systems results in a variety of questions and suggestions for subsequent investigations of the system modelled by M_1. In heuristic respects, a formal analogical relation can thus be exceedingly valuable by acting as a problem generator, whereupon research can be significantly intensified. However, each single analogy, considered as a counterpart concerning a certain property, initially constitutes a hypothesis that requires experimental verification. Similarly, the execution of the relevant experiments exerts an innovative influence on the research.

Consequently, the analogical relation between M_0 and M_1, as far as being relevant for physical knowledge, is not established through illustrative and superficial similarities, but through the assumption that M_0 and M_1 are conceptual realizations of the same formal structure. This provides the basis of the theoretical model as the third type of $\langle C, C\rangle$-model. It is defined as the equivalence class of formal analogical models. Symbolically, the equivalence relation is established through the formal analogy:

$$M = \left\{\mathrm{M}_i | \mathrm{M}_i \sim \mathrm{M}_0\right\} .$$

A representative M_i of a theoretical model is called a realization, representation, or interpretation of M. In the case of the wave model, for example, a directly observable wave-like process, e.g., a vibrating rope or water waves, can be chosen as the model M_0. The mathematical description developed for these examples can then be transferred to any of the representatives M_i by formal analogy. In this way, highly illustrative analogical models are obtained for sound waves, light waves, and electric waves. Further examples of theoretical models are the field model (gravitational field, electrostatic field, magnetic field), and the particle model serving as the basis for most of the applications of mechanics. The theoretical models, with their multifaceted options of realization, provide the most comprehensive theoretical orders with the highest degree of abstraction because they rest solely on formal analogies, entirely detached from illustrative models. Consequently, a theoretical model receives its reference to reality not through correspondence rules between single symbols of M and real objects and observations, but through the various realizations M_i of M, since empirically comparable statements are deducible only within M_i. As will be shown in Chap. 3, this is also a fundamental feature of physical theories, so a theoretical model always establishes a theory. For this reason, due to general linguistic usage in physics, theory and theoretical model are frequently identified. One thus speaks about the wave model or the wave theory of light, without distinguishing between both of them.

A widespread error occurring in the context of theoretical models consists in concluding from the formal analogy between two realizations to the identity of the two object domains, as exemplified again by the wave model of light. Due to the analogy with sound waves, Huygens assumed that light waves were also longitudinal, mechanical waves. This assumption was seemingly verified through the successful deduction of diffraction and interference effects, for which the longitudinal or transverse nature is immaterial. Later on, polarization experiments revealed the transverse nature of light waves. Although this led to corresponding modifications of the model, the identification of light waves with mechanical waves was maintained, despite the considerable resulting challenges. Instead of abandoning the mechanical part of the analogy, the formal analogy was supplemented by the hypothesis of an elastic ether in order to account for light's transversality. Although the attempts to verify its existence acted innovatively in some respects, all in all, the mechanical analogical model of light led to numerous unfruitful speculations and eventually to the conclusion that the process of light propagation is incomprehensible, and all questions regarding its physical nature have to be rated as meaningless. Actually, in this case the problem was not the use of formal analogies, but rather the identification of the physical nature of dissimilar real systems. The formal analogy constructed on the conceptual level was transferred to the material one, and this corresponds structurally with the naive-realistic identification of an object with its conceptual description.

Similar errors occur in the interpretation of mathematical models when properties are ascribed to a physical system which are solely the consequences of mathematical assumptions and approximations, without corresponding to a part of reality. A representative example is the notion that natural processes are inevitably continuous ("nature does not make jumps"). Similarly to the naive-realistic identification of an object with its conceptual description, a $\langle C, C\rangle$-model is identified with a $\langle S, S\rangle$-model. As a consequence, the latter frequently looks like an explanation because, through the ascription to something known, one gets the impression of arriving at a better comprehension. Such "explanations" do not contribute to the enhancement of physical knowledge. Instead, they have been the reason for numerous fallacies occurring during the development of physics, since these explanations were believed to be epistemically significant. An example that is relevant today are all physical interpretations of quantum mechanics which rely on formal analogies with the conceptual system of classical physics.

The alleged advantage of a method based on analogical models for providing direct understanding actually reveals itself to be its crucial drawback, because it is precisely the use of pictorial analogies that introduces a high probability of error. In reaction to such fallacies, the instrumentalist postulate is routinely advanced to remove models entirely from physics, to abstain from any type of interpretation, and to restrict oneself to purely formal descriptions. Such a position, however, ignores the fact that the concept of model is by no means constrained to analogical models, but also the fact that the use of models as auxiliary means for the formation of physical concepts and theories is indispensable. Indeed, together with causal thought processes, thinking by analogy where there are structural similarities constitutes one of our most basic thought patterns: the structuring of new experiences takes place, at least initially, by constructing analogies with something already known. Even the formation of physical theories on the basis of purely formal considerations, prescinding from the physical meaning of the given quantities, is an exception, and is only accepted as long as the full physical interpretation of a formalism is not yet accomplished. The natural quest to understand something, rather than just producing data without rhyme or reason, requires the construction of connections with something known, and this generally occurs through analogy.

As a backlash to this instrumentalist position, the assertion has been advanced that modeling by analogies is the central physical method (Hesse 1963), and that analogies are a non-eliminable part of theories (Campbell 1957). Even more, this should constitute a Copernican revolution in the philosophy of science (Harré 1970):

> The Copernican revolution in the philosophy of science consists in bringing models into the central position as instruments of thought, and relegating deductively organized structures of propositions to a heuristic role only.

Such a position is similarly untenable and contradicts both the historical development and the research practice of physics. Actually, insofar as a formal mathematical framework is elaborated, the use of analogical models is largely removed in favour of modeling associated with mathematical methods, since this is considerably less error-prone due to their deductive structure. Accordingly, a formalism can be applied entirely independently of its physical interpretation, so that the error sources resulting from pictorial analogies are automatically avoided. This is precisely the decisive strength of the theoretical methodology of physics, and to abstain from this advantage does not seem reasonable. In addition, the horizontal analogical relations between formal analogical models are less important than subsumption under a common theoretical model, in accordance with the ideal of deductive systematization. Finally, the analysis of the structure of physical theories in Chap. 3 will show that the physical interpretation of theories does not occur by analogy.

These explications, differentiations, and applications of model conceptions in physics show why any question of the status and function of models cannot reasonably be discussed without specifying the meaning and the particular context. For instance, studying the use of models in the historical development of physics is quite different from the challenge of understanding their relevance to the reconstruction of theories and the relations between theory and experience, or their status within the methodology of physics. Regarding historical studies, each of the types described may play some role, and may also occur with other meanings, e.g., as a preliminary theoretical approach or a theory not yet fully elaborated.

Regarding the assessment of the status and function of models as a part of physical methodology, it should first be emphasized insistently that modeling in physics is neither an objective nor an end in itself: testing of models is not a part of physical methodology. On the other hand, they are important auxiliary means in the formation of concepts and theories, and frequently pursue the goal of facilitating the mathematical treatment of a problem. Accordingly, the $\langle S, S\rangle$-models are irrelevant since they do not accomplish this task. In the empirical methodology which aims to construct empirical orders, the $\langle S, C\rangle$-models are important. In methods of consistent adjustment between theory and experience, all three model types are relevant. This is exemplified by their interrelation during concept formation. In the first step, based on real objects, a class term is defined by the selection of certain properties. This corresponds to the definition of a $\langle S, C\rangle$-model, since a class term represents a model-like description that is reduced in its extension of meaning. This may be considered as a preliminary stage of theoretical concept formation in the sense that, among the selected properties, some of them are used to define a theoretical model system. This definition is associated with a transformation in meaning since the original empirical term is detached from its object orientation, and is exclusively defined through the selected properties. Exemplified by the concept of the mass point, this transformation in meaning from a tiny amount of matter to a physical system defined by its dynamical state, has been described in detail in Sect. 2.5. In the final step, one has to search for realizations of the given model system, corresponding to citing a $\langle C, S\rangle$-model, where a representative of this model may now look considerably different than the objects that served as the starting point of this concept formation. This type of modeling may be represented symbolically by

$$\langle S, C_{\rm e}\rangle \longrightarrow \langle C_{\rm e}, C_{\rm t}\rangle \longrightarrow \langle C_{\rm t}, S\rangle \, .$$

The decisive step determining the physical meaning of the concept is the transformation of meaning with the transition from the empirical-conceptual description $C_{\rm e}$ to the theoretical-conceptual one $C_{\rm t}$. It is solely due to this transformation that the third step citing $\langle C_{\rm t}, S\rangle$-models gets an autonomous and necessary function. The insight that in a certain context a double star or an entire galaxy may be treated as a mass point is unintelligible without this previous transformation of meaning. Although the $\langle S, C_{\rm e}\rangle$- and $\langle C_{\rm t}, S\rangle$-models establish the connection with reality, the view of physical reality is determined by the $\langle C_{\rm e}, C_{\rm t}\rangle$-transition, rather than through such realizations, so the referential aspect plays a subordinate role.

As a part of the theoretical methodology, modeling occurs largely through $\langle C, C\rangle$-models. The mathematical and theoretical models are not eliminable since they establish the central, deductive structures. Accordingly, they are essential components of the theory. The formal analogical models play an important role because, complementary to the deductive structures, they establish additional horizontal structural relations within the theory and frequently produce new insights. In contrast, they do not contribute substantially to the physical interpretation: physical interpretations due to analogies can be eliminated from the theory without losing anything important. The errors occurring within empirical methodology due to the use of pictorial analogical models may occasionally be frustrating, but they do not leave deeper marks in physical knowledge because they can be corrected by theoretical knowledge, like erroneously constructed causal relations. In particular, this is demonstrated by cases in which both errors occur in combination by establishing incorrect causal connections due to superficial analogies between effects. The status of analogies within physical methodology is thus similar to that of causal relations: the objective of theory formation is no more the construction of analogical models as it is the construction and study of causal connections.

References

Böhme G, van den Daele W (1977) Erfahrung als Programm. In: Böhme G, van den Daele W, Krohn W, Experimentelle Philosophie. Suhrkamp, Frankfurt
Bridgman PW (1928) The Logic of Modern Physics. McMillan, New York
Campbell NR (1957) Foundations of Science. Dover, New York
Carnap R (1928) Der logische Aufbau der Welt. Weltkreis, Berlin
Carnap R (1966) Philosophical Foundations of Physics. Basic Books, New York
Chalmers AF (1982) What Is this Thing Called Science? The Open UP, Milton Keynes
Detel W (1986) Wissenschaft. In: Martens E, Schnädelbach H (1986) Philosophie - ein Grundkurs. Rowohlt, Hamburg
Döring W (1965) Einführung in die theoretische Physik I. de Gruyter, Berlin
Duhem P (1908) Aim and Structure of Physical Theories, translated by P Wiener, Princeton UP, Princeton, N.J. 1954
Everett H (1957) "Relative state" formulation of quantum mechanics. Rev Mod Phys **29**:454
Gilbert W (1600) De magnete. English translation: On the Loadstone and Magnetic Bodies, Britannica Great Books of the Western World. Vol 28, Chicago, London, Toronto, Geneva 1952
Grodzicki M (1986) Das Prinzip Erfahrung - ein Mythos physikalischer Methodologie? In: Bammé A, Berger W, Kotzmann E: Anything Goes – Science Everywhere? Profil, München
Hanson NR (1972) Patterns of Discovery. Cambridge UP, Cambridge
Hargittai I, Hargittai M (1994) Symmetry - A Unifying Concept. Shelter Publ., Bolinas, Cal.
Harré R (1970) The Principles of Scientific Thinking. U of Chicago Press, Chicago, Ill
Herschel J (1830) A Preliminary Discourse on the Study of Natural Philosophy. London
Hesse M (1963) Models and Analogies in Science. Sheed and Ward, London
Holton G (1973) Thematic Origins of Scientific Thought. Harvard UP, Cambridge, Mass.
Jung W (1979) Aufsätze zur Didaktik der Physik und Wissenschaftstheorie. Diesterweg, Frankfurt
Kuhn TS (1970) The Structure of Scientific Revolutions, The University of Chicago Press, Chicago, Ill
Losee J (1972) A Historical Introduction to the Philosophy of Science. Oxford UP, New York
Ludwig G (1974) Einfürung in die Grundlagen der theoretischen Physik. Bertelsmann, Düsseldorf

Ludwig G (1978) Die Grundstrukturen einer physikalischen Theorie. Springer, Heidelberg
Mach E (1919) The Science of Mechanics. 4th edn, Open Court, Chicago, Ill
Nagel E (1982) The Structure of Science. Routledge & Kegan Paul, London
Planck M (1983) Vorträge und Erinnerungen. Wiss Buchges, Darmstadt
Primas H, Müller-Herold U (1984) Elementare Quantenchemie. Teubner, Stuttgart
Rickles D (2016) The Philosophy of Physics, Polit Press, Cambridge UK
Ritter J, Gründer K, Gabriel G (eds)(1971-2007) Historisches Wörterbuch der Philosophie. 13 vols. Schwabe AG, Basel-Stuttgart
Wigner E (1967) Symmetries and Reflections. Bloomington
Ziman J (1978) Reliable Knowledge. Cambridge UP, Cambridge, UK

Chapter 3
The Structure of Physical Theories

Three broad classes of theories are distinguished, viz., descriptive, explanatory, and unifying or grand theories. The structure of kinematics as the paradigm of a descriptive theory will be elaborated in detail and is represented in an axiom-like form. Within a theory, there are two conceptual levels with the abstract theory frame and the scientific applications. The theory frame is abstract, in that it is system-independent and has no direct connection with experiments. This connection is provided only through the scientific applications that are obtained by adding system-specific assumptions which have the status of hypotheses. The various types of these hypotheses will be discussed in detail. The relation between descriptive and explanatory will be exemplified by the relation between kinematics and dynamics, and between thermodynamics of the ideal gas and kinetic gas theory, respectively. A number of intertheoretic relations will be discussed in detail, with the conclusion that physics is a polyparadigmatic science. Therefore, a conceptually uniform theoretical description of reality cannot be expected.

3.1 Descriptive Theories

Any analysis of the conception of theory in physics is faced with the challenge that there are theories of very dissimilar types, beginning with the simplest ones with a very limited scope, like Kepler's theory of the planetary motion, up to theories as comprehensive as classical mechanics and quantum mechanics. According to the differentiation of theories in Sect. 1.3, the simplest variant are the descriptive theories which aim at the investigation, structuring, and mathematical representation of relations with a relatively direct connection to observations and experiments. In spite of their comparatively simple theoretical structure, such theories already contain the essential structural characteristics of more complex theories, viz., deductive and unifying elements. In this respect, they are distinctly different from empirical

M. Grodzicki, *Physical Reality – Construction or Discovery?*,
https://doi.org/10.1007/978-3-030-74579-0_3

classifications since they do not merely represent generalized direct experiences. On the other hand, with respect to their physical content, they are simple enough that the structural analysis is not hampered by problems of comprehension. For these reasons, descriptive theories are best suited to elaborate the structure of a physical theory. Using kinematics as an example, the conceptual foundations and constitutive parts of this theory are presented. Starting with the empirical information the theory is based on, the construction of the basic concepts will show in which respects the theory is detached from direct experience. In this context, an important issue will be to elaborate the distinction between those constituents of the theory that are empirically testable, and those that are not amenable to a direct comparison with experiment and are non-empirical in this respect.

Considering planetary motion or a rolling ball, the empirical basis of kinematics which is obtained through repeated and systematic studies consists of the following objects, observations, and measurement results:

(a_1) the moving object in the form of a rolling ball or a planetary disc moving relative to the background of the stars,
(a_2) the spatial regions where the object is observed,
(b_1) the perception that the object is not observed at the same time in distinctly different spatial regions,
(b_2) the perception that for small differences in time the respective positions of the object in space are close to each other,
(c_1) the systematically produced experimental data in the form of the object's positions measured at certain times.

The formulation of a physical theory from this empirical basis may be decomposed into a number of methodological steps. Since physics is the mathematical form of natural research, every physical theory refers in some way to a mathematical theory MT. In this case, this might be, e.g., the theory of the parametrized, continously differentiable curves on a Euclidean vector space. Accordingly, the first step consists in the construction of a conceptual basis matching MT. To this end, one abstracts from the spatial region where the real object is located and replaces it with a mathematical point in space for two reasons. Firstly, the distance between two extended spatial regions is not a well-defined quantity. In everyday life it may be sufficient to report the distance, e.g., between Hamburg and Munich as 800 km, but this is unsuitable for a quantitatively precise, physical description. Secondly, the conceptual construction "space point" is what allows the conceptual system of Euclidean geometry to provide a well-defined concept of distance. These together enforce the replacement of spatial regions with space points.

Another conceptual construct is grounded on the perceptions (b_1) and (b_2). Although the perceived positions do not form a continuum in the mathematical sense, since only more or less closely spaced spatial regions where the moving object is situated are perceivable, one speaks about an *observed* trajectory. Consequently, the perception of a number of single events is not strictly separated from structuring them: the trajectory as a perception emerges only through connecting single perceptive impressions. A fortiori this is the case for the actually measured positions. Although only single points are measured, a trajectory is automatically constructed by interpolation, so it gets the status of an empirical correlation. Due to empirical experience, this appears to be so self-evident that the insight that the concept of a trajectory loses its meaning for quantum systems is incomprehensible, because it contradicts all direct empirical experience. The third concept needed to describe motion is time, which is not a term of Euclidean geometry. By analogy with the concept of space point, the term "time point" is defined. This is a theoretical construct in relation to the empirical term "time", which refers to time intervals, since any measurement requires a finite time. In summary, the conceptual basis of kinematics consists of space points, time points, and trajectories. These constructs demonstrate that the conceptual basis of a physical theory is not the mere verbalized description of elements of reality, because they are constructed with the conceptual system of a mathematical theory in mind.

In the second step, certain variables of MT are replaced with these physical terms. Let $\mathbf{f}$ be a continuously differentiable function of the parameter u on a Euclidean vector space $\mathbf{V}$. The independent variable u is interpreted as a time point, and the function values $\mathbf{f}(u)$ as space points $\mathbf{r}_i = \mathbf{r}(t_i)$. In other words, one defines $u = t$ or "u is a time point" via the predicate $t \in D$, where D is the domain of $\mathbf{f}$ and now possesses the physical meaning of a time interval. Analogously, the range $\mathbf{B}$ of $\mathbf{f}$ is assigned the meaning of a spatial region via $\mathbf{r}_i \in \mathbf{B}$. Formally, these substitutions can be described by adding to MT the assignment rules $a_i \in Q_i$, where the a_i are physical quantities and the Q_i are sets of MT, i.e., the sets Q_i in MT are specified by assigning them a physical interpretation (Ludwig 1978). In the third step, relations must be specified that establish the connection between the results of the theory and the quantitative experimental data. Assuming as the experimentally comparable quantities the positions $\mathbf{r}_i$, calculated for certain instants of time t_i, such postulation may seem redundant due to the direct connection to observations. This is not the case, however. Firstly, within the framework of kinematics, a problem is considered to be solved when the whole trajectory has been determined, while the computation of particular points is initially not a part of the theory. Secondly, a measured value is a real number, but a single point $\mathbf{r}_i$ of the trajectory corresponds to a vector. For these reasons, MT must be supplemented with instructions specifying how to obtain measurable data from the trajectory. This is realized by relations establishing the connection between $\mathbf{r}_i$ and scalar quantities like lengths, distances d, or angles θ, e.g., $d(\mathbf{r}_i, \mathbf{r}_j) = \sqrt{(\mathbf{r}_i - \mathbf{r}_j)^2}$ or $\theta(\mathbf{r}_i, \mathbf{r}_j) = \arccos(\hat{\mathbf{r}}_i \cdot \hat{\mathbf{r}}_j)$, where $\hat{\mathbf{r}}$ denotes the unit vector.

These steps provide not only the precondition for deducing quantitative data, but also the basis for the formation of new physical terms through theory-internal concept construction. Initially, the assignment rules ascribe a physical meaning to those relations, e.g., equations, functions, or inequalities, in MT where all variables are replaced by physical terms. Additionally, there are relations in MT which do not yet possess such an interpretation, so it is not the whole of MT that gets interpreted physically. For example, in the case of kinematics, one can examine whether the mathematical expressions for the slope and curvature of a curve, i.e., the first and second derivatives of the curve $\mathbf{r}$ with respect to time t, correspond to a physical property of the motion. This is indeed the case, and leads to the concepts of instantaneous velocity and acceleration, understood of course as a reconstruction, while historically this process occurred the other way round. Initially, these theoretically introduced terms are not directly related to experience. By means of the theoretical definition in combination with knowledge about the meaning of this term from the theoretical context, a measurement instruction may be definable that makes this theoretically introduced property measurable. In combination with a measuring instruction, the construction of an appropriate measuring device, e.g., a tachometer, yields an operational definition of the relevant property in addition to the mathematical-theoretical definition. Making a property measurable on the basis of theoretical knowledge constitutes an important part of the relation between theory and experience, and it also leads to a remarkable change in the status of the given concept. As a theoretical construct, it had meaning initially only within the corresponding theory. The empirical-operational aspect, received through the measurability, transforms it into a term that acquires meaning independent of the theory, whence it appears as an empirical term.

The partial physical interpretation of MT, in combination with the instruction for computing measurable data, specifies the formal framework of kinematics, but not the physical content. To this end, the objective of the theory must be defined. This does not consist in the computation or prediction of some points on a trajectory, but rather in the classification of the possible types of motion on the basis of a unifying structural principle. Accordingly, that property must be specified that defines the various classes of motion. It is one of the fundamental insights of the seventeenth century that this property is neither the geometry of the trajectory, according to an empirical classification due to direct observation, nor the velocity of the motion, but the acceleration. This constitutes the basic structural principle of kinematics: within the framework of kinematics, an acceleration function defines a class of trajectories (a "type of motion"). Trajectories of the same class may then look very dissimilar regarding their geometry. For instance, the class of central motions comprises circular, elliptic, parabolic, and hyperbolic trajectories that would rarely be assigned to the same type in the frame of an empirical classification exploiting the geometrical shape. Consequently, the quantitative representation of the acceleration function constitutes the fundamental axiom of kinematics:

> The acceleration function $\mathbf{a}$, defining the type of motion in the framework of kinematics, is most generally a function of velocity $\mathbf{v}$, position $\mathbf{r}$ and time:

$$\mathbf{a} = \mathbf{a}(\mathbf{v}, \mathbf{r}, t) = \frac{\mathrm{d}^2\mathbf{r}(t)}{\mathrm{d}t^2} , \qquad (*)$$

where $\mathbf{r}$ and $\mathbf{v} = \mathrm{d}\mathbf{r}/\mathrm{d}t$ are themselves, in general, functions of the time t.

This postulate constitutes the metaphysical basis and the epistemological kernel of kinematics. On the one hand, it defines the general framework of the perspective and interpretation of patterns of motion. On the other hand, as a symbolic representation, the equation $(*)$ does not provide concrete solutions, so it is non-empirical in the sense that it is not amenable to any direct experimental comparison. In detail, $(*)$ corresponds to a system of three equations for the acceleration components, which are themselves functions of seven variables, viz., the three velocity components, the three position components, and the time. In addition, each of these components may itself depend on time. In Cartesian coordinates, $(*)$ has the form:

$$a_x\big(v_x(t), v_y(t), v_z(t); x(t), y(t), z(t); t\big) = \frac{\mathrm{d}^2x(t)}{\mathrm{d}t^2} ,$$

$$a_y\big(v_x(t), v_y(t), v_z(t); x(t), y(t), z(t); t\big) = \frac{\mathrm{d}^2y(t)}{\mathrm{d}t^2} ,$$

$$a_z\big(v_x(t), v_y(t), v_z(t); x(t), y(t), z(t); t\big) = \frac{\mathrm{d}^2z(t)}{\mathrm{d}t^2} ,$$

The deduction of experimentally comparable results occurs in three steps. Firstly, the number of independent variables is reduced in order to obtain explicitly integrable cases, e.g., a = constant, $a = a(t)$, or $a = a(x)$. Such special cases will be referred to as specializations. Their methodological function consists in establishing mathematical methods for the integration of special problems, irrespective of any empirical relevance. These studies comprise that part of theoretical physics commonly called mathematical physics. This alone does not yield experimentally comparable trajectories. In the second step, acceleration functions are constructed which define empirically relevant types of motion. These empirically relevant realizations of a specialization are referred to as scientific applications. Examples are free fall near the Earth's surface, defined by $a = g$ where g is the acceleration due to gravity, or the one-dimensional harmonic vibration defined by $a = -\omega^2 x(t)$, with a system-specific angular frequency ω. A specialization of $(*)$ is thus supplemented with system-specific, *externally* introduced information, e.g., g or ω. This yields the *possible* trajectories, but not the actually observed ones. In the third step, assumptions about "contingent" conditions, viz., numerical values of g and ω and initial or boundary conditions, are added. These must be chosen in consistent adjustment with the experimental design. Among the possible trajectories, this eventually selects the actually observed ones. Overall, this procedure exhibits the types of assumptions that are necessary to move from the non-empirical structural principle $(*)$ to empirically comparable statements and data.

This description of the various steps in the reconstruction of kinematics spells out the constituents that must be included in an account that is not constrained to purely formal-logical aspects, but puts the physical and epistemic content in the central position. Accordingly, a systematic representation of kinematics comprises the following constituents or postulates:

(A1) "Theoretical model". Consider a mathematical theory DG($\mathbb{R}^3$) (differential geometry in three-dimensional Euclidean space) and a domain P containing the physical terms space point $\mathbf{r}$, time point t, and trajectory $\mathbf{r}(t)$. Let $\mathcal{C}^2$ be the set of continuously differentiable curves $\mathbf{f}$ of $\mathbb{R}^3$ with $\mathrm{D}_f \subset \mathbb{R}$ and $\mathbf{B}_f \subset \mathbb{R}^3$. Then,

$$t \in \mathrm{D}_f,\ \mathbf{r} \in \mathbf{B}_f,\ \mathbf{r}(t) \in \mathcal{C}^2 \text{ (physical interpretation)} \qquad (\#)$$

Let DGP denote DG($\mathbb{R}^3$) supplemented with the domain P and the predicates (#).

(A2) "Epistemological kernel". A class of types of motion is defined by an acceleration function $\mathbf{a}(\mathbf{v}, \mathbf{r}, t)$ with the property

$$\mathbf{a} = \mathbf{a}(\mathbf{v}, \mathbf{r}, t) = \frac{\mathrm{d}^2\mathbf{r}(t)}{\mathrm{d}t^2} \qquad (*)$$

where $\mathbf{r}$ and $\mathbf{v} = \mathrm{d}\mathbf{r}/\mathrm{d}t$ are themselves generally functions of the time t.

(A3) "Theory–experiment comparability". The measurable quantities are distances d and angles θ defined by

$$d(\mathbf{r}_i, \mathbf{r}_j) = \sqrt{(\mathbf{r}_i - \mathbf{r}_j)^2}\ , \qquad \theta(\mathbf{r}_i, \mathbf{r}_j) = \arccos(\hat{\mathbf{r}}_i \cdot \hat{\mathbf{r}}_j)\ .$$

(A4) "Empirical relevance". The set of scientific applications of (∗) is not empty.

The postulates (A1)–(A4) represent an axiom-like account of kinematics. The subsequent comments will demonstrate that this type of axiomatization is very different to the concept of an axiomatized theory in mathematics, logic, or philosophy of science.

Via the triple MT, P, and (#), (A1) defines the formal framework MTP (= physically interpreted MT) of the physical theory. Firstly, it is essential to distinguish between the externally given MT and MTP. MT is a syntactic system in that its symbols and relations do not possess any particular meaning. In contrast, MTP as a partially physically interpreted realization of MT is a semantic system. As a consequence, an expression of the type $f(x) = x + x^2$ may be a reasonable relation in MT, but not in MTP, if x is ascribed the meaning of a length, since the sum of a length and an area yields a nonsensical result. Similarly, concerning MT, $\mathbf{r}_i$ and $\mathbf{v}_i$ are both elements of an unspecified three-dimensional Euclidean vector space. With respect to MTP, they belong to different vector spaces, i.e., $\mathbf{r}_i + \mathbf{v}_i$ is a nonsensical expression. Secondly, an axiom or theorem in MT has the typical structure "for all x, one has $A(x)$". In contrast, a physical law in MTP never has this form, since the scope of any physical theory is limited, and in addition, not definable within the conceptual framework of the theory, but only through the comparison with experience. A proposition of the type "for all p, V, T, the gas law $pV = nRT$ is valid" or "for all gases the gas law is valid" is empirically false. A physical law, even when corresponding to a theorem in MT, never contains all-quantifiers. What actually matters is the acquisition of the conceptual system, structures, theorems, and methods of proof of MT. In particular, in the course of elaborating the specializations, other mathematical theories are applied, e.g., the theory of linear differential equations in the case of kinematics. Overall, the use of mathematical methods enables one to gain and to secure theoretical knowledge by largely deductive methods. In particular, the theory-internal introduction of new quantities, and the deduction of physical laws and empirically comparable propositions, concern an aspect of the construction of a conceptual reality that plays an important role in establishing physical reality.

Setting P reveals three possibilities for theory-external assignment of meaning. The meaning of the space point comes from another theory, viz., physical geometry. In the framework of kinematics, the concept of time gets its meaning only via the operational definition of time measurement because a theoretically exact foundation requires considerably more sophisticated theories. Finally, the concept of the trajectory represents an empirical correlation that is constructed as a continuous curve by interpolation between discrete measured points. Hence, none of the terms occurring in P is defined explicitly and in a formally exact way. Actually, this is unattainable without referring to other, more fundamental physical theories, and poses a central challenge to the reconstruction of physical knowledge.

Regarding epistemology it is important to note that both P and MT are externally given. On the one hand, MT contains certain presuppositions, e.g., the continuous differentiability of curves. On the other hand, certain notions are connected with P which go back to empirical experiences. Of course, these enter the theory via MT and P. Although these are not at all dubious or metaphysical accessories to be eliminated, erroneous conclusions will result if one ignores the fact that these presuppositions and notions are externally introduced, rather than consequences of the theory. Therefore, the term trajectory appears in (A1) in order to emphasize that the continuity of curves enters the theory as a presupposition, rather than being a necessary result of the physical description of motion. If, in contrast, this assumption, contained in the theory for mathematical reasons, is transformed into a result of physics and subsequently reinterpreted as a property of reality, this inevitably leads to the erroneous notion that "nature does not make jumps". Such a transformation of an assumption into a result, in combination with the naive-realistic attitude of identifying reality with its conceptual representation, has occasionally led to the notion that nature is mathematical. However, since every physical theory is based on a MT, physics automatically provides a mathematical representation of reality, but this does not mean that reality is itself mathematical. Further erroneous results are generated if parts of MT that cannot be ascribed a physical meaning at the conclusion of theory formation are transformed into elements or structures of reality. It is thus inevitably necessary to carefully distinguish between the variables of MT and the symbols a_i denoting terms of P, as well as between the a_i and their empirical counterparts.

Finally, this distinction is a necessary prerequisite for understanding the status and meaning of the assignment rules. A predicate $a_i \in Q_i$, as occurring in (#), would obviously be nonsensical if a_i were a real object. Only under the presupposition that a_i is a term are the assignment rules structurally compatible with MT. In this case, they supplement MT with elements of the conceptual domain, and such a supplementation cannot lead to logical contradictions since the symbols of MT are not assigned any meaning. Because these substitutions occur in the conceptual-theoretical domain, the assignment rules are theory-*internal* statements, but do not establish relations with the empirical basis. These relations are implicitly inherent in the antecedent concept formation that has led to the terms "space point", "time point", and "trajectory". In addition, as single-digit predicates, the assignment rules represent ascriptions of meaning, so the domain of **f**, for example, gets the meaning of a time interval. Consequently, they are neither syntactic rules, nor conventionalist arrangements, nor mere nominal definitions, but semantic propositions. Therefore, on the one hand, MTP is not a formal calculus without any meaning, in contrast to the theory conception of hypothetico-deductive empiricism. On the other hand, the fact that some symbols of MT are ascribed a physical meaning does not change the abstract nature of MTP, which is actually grounded in the independence of system-specific and contingent conditions.

The structural principle in (A2) constitutes the epistemological kernel of the theory with the mathematical representation $(*)$ in accordance with (A1). This central postulate is frequently reduced to the formal-mathematical aspect, so that the significance of its physical meaning is ignored or explicitly denied, in the sense that the equation $(*)$ is interpreted solely as the definition of the acceleration $\mathbf{a}$. Although formally correct, since $\mathrm{d}^2\mathbf{r}(t)/\mathrm{d}t$ is well-defined in MTP as the second derivative of the curve $\mathbf{r}(t)$ with respect to the parameter t, the crucial property is the methodological and semantic function. Firstly, it constitutes the fundamental structuring principle for the classification of trajectories. Secondly, it establishes the view of reality provided by kinematics as a physical theory. On the one hand, (A2) signifies that the structuring property of motion is neither the geometrical shape of the trajectory nor the velocity, but the rate of change of velocity, i.e., the acceleration. The need for a *variable* velocity follows from the fact that the existence of stable structures and bound systems cannot be deduced by assuming exclusively constant velocities. On the other hand, (A2) contains the implicit assumption that the acceleration is sufficient as the structuring principle, in that a unification with simultaneous differentiation is feasible in a meaningful and efficient fashion. Mathematically, this means that higher derivatives of the trajectory with respect to time are not required. This may be considered as some kind of principle of economy because the inclusion of higher derivatives means that significantly greater effort is required to treat specific tasks. In any case, this assumption involves a particular interpretation of motion and simultaneously implies a conceivable limitation of knowledge. It cannot be excluded that the classification of trajectories with respect to higher derivatives yields another classification of motion. Finally, since a set of seemingly dissimilar trajectories is deducible from a single acceleration function, they are explained, in this respect, but this is obviously not a causal explanation. Instead, it establishes a structural connection comprising both deductive and unifying aspects, in accordance with the objective of kinematics to accomplish the deductive systematization of types of motion according to acceleration as the unifying structural principle.

Another reason to interpret (A2) as epistemological kernel is the ability to gain knowledge. The virtually unlimited number of possible acceleration functions far exceeds any type of knowledge available through empirical methods. Since it is impossible to encompass systematically the full spectrum of conceivable types of motion in the Euclidean space $\mathbb{R}^3$, the elaboration of the theory is constrained to specializations which are either representative for the methods of solution, or are empirically relevant in that they lead to scientific applications. Irrespective of this practice, the multiplicity of possible acceleration functions proves that the theoretical potential of the theory ("theoretical surplus") is considerably greater than any knowledge following from empirical experience of motion. Furthermore, the choice of appropriate additional system-specific assumptions needed to deduce experimentally comparable data within a scientific application paves the way for incorporating novel empirical experience into the theory ("empirical openness"). Hence, the theoretical surplus is the reason for the empirical openness, and both result from the fact that the structural principle is detached from direct observations. Consequently, the theory represents an autonomous form of knowledge, since the mere inductive generalization of empirical experiences never contains this theoretical surplus of virtually unlimited applications. Just because the option of these additional applications is never mentioned in textbooks, although they belong to the potential of the theory, it is usually ignored that they exist and contribute to empirical openness. In combination with the structuring ability, this makes the essential difference to empirical classifications, even when only descriptive theories are concerned. On the other hand, a theory may constrain conceivable future knowledge by establishing theoretical views that might obstruct the pathway to alternative thoughts and insights. These shortcomings and negative implications, mentioned in Sect. 2.6, are well known from the historical development of physics. The designation of (A2) as epistemological kernel is thus justified with all its positive and negative aspects, and in the case of the most comprehensive theories, it may establish a certain world view, e.g., the causal-mechanistic world view of Newtonian mechanics.

Together, (A1) and (A2) define an autonomous theoretical conceptual system and a theoretical order designated as the theory kernel. Together with the specializations, the theory kernel constitutes the theory frame that establishes the foundation for the structuring of knowledge. This theory frame consists of functional relations between physical quantities, entirely independent of the existence of material carriers for the respective properties. The theory frame is thus non-empirical because there is no direct connection with experiments. This connection is established via a number of additional assumptions, with (A3) and (A4) as the general basis. The need for (A3) follows from the circumstance that, on the one hand, every measuring result is given as a real number. On the other hand, the theory frame does not necessarily contain directly measurable scalar quantities α_j. Therefore, it must be supplemented by relations that may be represented symbolically as $\mathrm{R}(a_i;\alpha_j) \subset$ MTP $\times$ **R***, with $a_i \in$ MTP and $\alpha_j \in$ **R*** (Ludwig 1978). This is symbolical because neither MTP nor the domain **R*** of scalar quantities are sets, in the sense of set theory. The fact that the construction of these relations is not at all self-evident and may play a central role is demonstrated by Bohr's theory containing as an essential part the relation between the theoretically calculated energies of the electronic states and the measurable frequencies ν of the spectral lines. In this case, the relation R corresponds to the postulate $\nu = (E_\mathrm{f} - E_\mathrm{i})/h$. The need for such relations becomes evident, in particular, when the space of the theoretical description is not identical with the Euclidean $\mathbb{R}^3$, but an abstract space like the Hilbert space in quantum mechanics. Their function is then to establish the connection between elements of the underlying space of the theory and scalar quantities, i.e., how the measurable quantities are connected with the physical quantities of the theory. In addition, the relations R provide the theoretical basis for the operational definition of theoretically introduced quantities. Finally, they do not contain elements of the empirical basis, e.g., experimental data or effects, and nor are they generally relations within MT or MTP. For example, equations describing the behaviour of quantities under coordinate transformations do not belong to (A3), since they do not establish a connection with scalar quantities.

The postulate (A4) asserts that the scientific applications are indispensable constituents of the theory because it is only within these applications that the deduction of experimentally comparable results is possible. Accordingly, they establish the connection between the theory frame and experience, and taken together these comprise the potential empirical content of the theory. Therefore, the elaboration of scientific applications belongs to fundamental research. It is precisely this that should express the predicate "scientific", in contrast to mere technical applications. Although a distinction between the two types may not always be clear-cut and may sometimes rest on convention, scientific applications are primarily method-oriented, while technical ones are usually object-oriented in the sense that practically relevant systems are studied by physical methods that are already available. As a pure existence postulate, (A4) constitutes a norm that must be fulfilled by any physical theory: a theory without such applications cannot be a part of physics, since it does not contain any quantitative, experimentally comparable results. A representative example of this case is string theory, which does not yet contain such applications. The demand for empirical relevance sets additional boundary conditions regarding the structure of the theory. Firstly, the mathematical representation of the structural principle in (A2) must permit solutions. Secondly, (A4) implicitly contains the condition that the theoretical considerations must be adapted to the experimental capabilities, so that the theoretically deduced results are actually experimentally provable. Finally, this demand implies that the elaboration of the theory frame is largely constrained to specializations that lead to scientific applications. Concerning the structure of kinematics, this means that, in the first step, the properties of (A2) are studied relative to purely mathematical aspects, i.e., with regard to the mathematical structure and methods of solution. The first distinctions are made between rectilinear and curvilinear motion, as well as between one-, two-, and three-dimensional motion, depending on the number of nonvanishing components of $\mathbf{a}$. Rectilinear motion can always be treated as one-dimensional, since a coordinate system can be chosen in such a way that $\mathbf{a}$ has a single non-vanishing component. Regarding three-dimensional curvilinear motion, it can be shown that there is a coordinate system relative to which $\mathbf{a}$ has only two non-vanishing components. This is defined by the Frenet frame consisting of tangent, principal normal, and binormal, where the component of $\mathbf{a}$ in the direction of the binormal vanishes. Subsequent studies are reduced to two classes of motion, viz., one-dimensional rectilinear and two-dimensional curvilinear motions. Among the one-dimensional rectilinear motions, the specializations $a = 0$, $a = \text{const}$, $a = a(t)$, $a = a(x)$, and $a = a(v)$, are discussed explicitly as prototypes in order to explore how the respective equations of motion are integrated, i.e., how classes of trajectories can be derived from these acceleration functions, irrespective of any particular systems and initial conditions. Two-dimensional curvilinear motion is discussed from a somewhat different point of view, viz., to see how, by choosing an appropriate coordinate system, $\mathbf{a}$ can be decomposed in such a way that the components yield one of the one-dimensional prototypes. All these studies are entirely system-independent and unrelated to experiment, notwithstanding the fact that empirically relevant, qualitative insights can already be obtained at this stage.

Once these theoretical studies have made available the solvable problems with solution methods, concrete types of motion are treated according to these patterns. While the specializations are obtained by constraining only the general structure of the acceleration function, the study of actually occurring motions requires one to preset a specific acceleration function with assumptions resting on empirical, and thus *externally* introduced information, e.g., the experimentally determined value of the acceleration due to gravity g in the case of free fall near the Earth's surface. Additionally, the deduction of quantitative, experimentally comparable data requires one to specify initial conditions to solve the equation of motion, and if necessary, the values of system-specific parameters, e.g., masses or force constants. In this way, one proceeds stepwise from the abstract structural principle to the specializations through constraints, and afterwards to the scientific applications by adding theory-external assumptions to the specializations. Consequently, experimentally comparable results can be deduced only within the framework of a scientific application. Examples are free fall, harmonic, damped, and forced vibrations, and Kepler's laws of planetary motion. Each scientific application arises as an empirically relevant, physical interpretation of the theory frame. The diversity of specializations of a theory is a measure of its complexity, while the full set of successful scientific applications represents the empirical content.

The success of an application cannot be determined theory-internally, but solely through comparison with experience. This connection with experience is not established by single correspondence rules (Hempel 1952, Nagel 1982), but through an extensive network of relations between experience and theory. It is essential for the comprehension of both the theory concept in physics and the comparison between theory and experience that these relations are not part of the theory. The implications of this important fact are exemplified by Kepler's theory of the planetary motion in the heliocentric system, which is a scientific application of central motion. It is based on Kepler's three laws:

1. The planets describe elliptical orbits around the Sun, which is located at one focus of the orbit.
2. The radius vector from the Sun to a planet sweeps out equal areas in equal time intervals.
3. The square of the period is proportional to the cube of the mean distance from the Sun.

The first law is a statement about the geometry of the orbits, while the second is about the velocity of the motion, and both together describe the properties of the central motion for elliptical orbits. The third law establishes an additional structural relation between different orbits around the same centre. First note that Kepler's theory does not provide a description of direct perceptions since the directly observed planetary trajectories are not ellipses, so there is no direct connection between theory and experience. The shape of the directly observed trajectories thus demands an explanation. In contrast to a geocentric description, one must take into account the fact that the observed planetary motions are deduced as the combination of the simultaneous motion of planet and Earth around the Sun, and their mutual relative positions. This requires some computational efforts, so Kepler needed several years to calculate the observed positions of the planets. In view of these efforts and in combination with the fact that the theory can neither be empirically tested nor applied to practical situations without these additional calculations, one might consider including these conversions in the theory, e.g., as a part of

(A3). That this is not actually done has important consequences regarding theory conception in physics:

- The subject matter of the theory is not the directly perceived reality, but the structures of reality: theoretical physics is not the calculation of particular positions of the planets, but the insight that all planetary orbits can be described by a few common structural elements.
- The unification accomplished by the theory is not an inductive generalization of observations, but rests on structural principles arising from autonomous theoretical considerations. Consequently, this type of generalization is not an inductive inference from the particular to the general.
- The principle of economy in physics concerns theoretical, not practical economy: although Kepler's theory is more complicated for practical purposes than Ptolemy's theory of epicycles, its theoretical structure is simpler because the orbits of all planets are reduced to a few uniform structural elements, rather than considering each trajectory as an isolated phenomenon.
- A theory is already regarded as empirically relevant due to the knowledge that the required connection with experience can be established, in principle. The actual execution of these calculations is not assigned to the theory because this is considered to be a mere technical problem that does not produce new insights and is insignificant for the general structure of the theory.

The last point demonstrates once again that the relations (A3) are theory-internal relations, rather than relations between theory and experience. Therefore, they do not correlate with the correspondence rules of empiricist theory models since the experimental data do not belong to the theory, and neither do the methods that are used in the comparison with experiments. Consequently, a problem that is deemed to be solved within the theory does not necessarily provide a result which is directly comparable with observation or experiment. Although closing this gap may require considerable efforts in complex theories, this part is not attributed to the theory, but to the relations between theory and experience. This emphasizes the fact that the objective of theory formation is to structure knowledge, rather than to provide predictions and explanations of particular events. Similarly, operational definitions of the basic quantities are not part of the theory. Firstly, they are not necessarily scalar quantities so they are not directly operationally definable, but only in combination with the relations (A3). Secondly, although the operational definition of a quantity contributes to its physical meaning, it does not possess the status of a precise theoretical definition. Including operational definitions in the theory could then give the erroneous impression that one could have at one's disposal an explicitly defined conceptual basis upon which the theory could be built as a strictly deductive schema. Finally, the essential feature of operational definitions is not the conceptual, but the experimental, action-oriented component. Their ontological status is thus similar to that of instruments and experimental data, so they should not appear as part of the theory in the name of a consistent theory conception. This view is in accordance with all common representations of theoretical physics which do not contain operational definitions.

In summary, as developed in this section, theory conception comprises four structural elements, viz., the theoretical model, the epistemological kernel, theory–experiment comparability, and scientific applications. Central importance is assigned to the strict separation of the physical theory from both the mathematical theory and the empirical domain. On the one hand, in spite of its abstractness, as a semantic system, the theory frame is not structurally identical to a mathematical theory. On the other hand, the empirical propositions and orders do not belong to the theory, because they do not presuppose the conceptual system of the theory. In particular, the empirical basis as the representative of reality is not a part of the theory, which constitutes a system of knowledge *about* reality but does not contain elements of that reality. The connection with empirical experience is established, not via single correspondence rules, but through a comprehensive network of relations between experience and theory which is likewise not a part of the theory. The consequences for theory conception and for the connection with experience is exemplified by Kepler's theory.

Within the theory, two conceptual levels are distinguished, viz., the abstract theory frame and scientific applications. The theory frame is independent of any system-specific assumptions and contingent conditions. Consequently, it is empirically underdetermined and without direct connection with experiments, so its axioms and theorems are not testable experimentally. The comparison with experiments necessitates additional system-specific assumptions which supplement the theory frame with external, empirical information and lead to scientific applications dealing with special classes of model systems. These retroact on the theory frame in that the motivation for elaborating a specialization is frequently the existence of a scientific application. This retroaction changes the abstract character of the theory frame just as little as does the fact that it has a partial physical meaning. Accordingly, a physical theory is not a propositional system, in the sense of propositional logic ("statement view"), but must be considered, due to the predicates (#), basically and essentially as a model theory. However, the model constituted by (A1) must not be misunderstood as an iconic, object-oriented, or ostensively imaginable model. Even when theories are frequently associated with such models, going back mainly to the developmental phase, they are not a part of the theory. Epistemic significance is assigned to the theory not through such models but via the structural principle in combination with the claim that at least some of the axioms and theorems correspond to structures of reality. Accordingly, in spite of its abstractness, even the theory frame is far from representing a fictitious or purely instrumentalist construct. This defines a distinct antithetic position against notions that reduce the function of theories to solely instrumentalist or descriptive aspects. In addition, it accounts for the fact, easily verifiable historically, that at least the most comprehensive theories of physics have established world views with all of their positive and negative consequences.

3.2 Formation of Hypotheses

Another important contribution to understanding the structure of physical theories is provided by the connection between the abstract theory frame and the scientific applications which may be represented as the supplementation of a specialization with a number of system-specific assumptions having the status of hypotheses. In a first step, the intended meaning of the term "hypothesis" must be specified since, like the terms "model" and "theory", it has a wide variety of interpretations. This began in ancient Greek philosophy, where a hypothesis could mean either an unproven proposition (in mathematics), or a tentative assumption (in dialogues), or a claim resting on unproven premises, and it extends to modern philosophy of science where a hypothesis can refer to anything from a single claim up to an entire theory. Due to this spectrum of usage, all claims advanced in this context become arbitrarily vague and empty (Ritter et al. 1971ff.):

> The question becomes considerably bedeviled by the fact that many authors speak in generalized fashion about hypotheses without explaining sufficiently or not at all the meaning of this term as assumed by them, frequently for the reason that this meaning is not clear to themselves.

According to a definition by Pascal, a hypothesis is any unproven conjecture that raises a claim of validity and is not an axiom. Transferred to physical statements and propositions, this definition results in the following conclusion. Firstly, with the exception of conventionalist or instrumentalist views, physical axioms are associated with a claim of validity, so they must be considered to be hypotheses, in contrast to the axioms in mathematics. Secondly, physical laws representing statements assumed to be strictly valid are ultimately neither empirically nor theoretically provable. Empirical proof would require an infinite number of observations or experiments, and every theoretical proof contains other hypotheses as premises that are not ultimately provable. Consequently, no statement deduced within the framework of a physical theory would be provable, so that all physical laws and theories, and eventually the whole of physical knowledge must be considered as inevitably hypothetical, and never able to attain the status of secured knowledge. The term "hypothesis" thus degenerates to an empty all-embracing concept which has no value for the representation of theories, nor for the description of theory formation and the research practice of physics.

That such an attitude, occasionally called hypothetical realism, contradicts the research practice of physics, is demonstrated by a trivial example. Due to previous tests it may be established that a wire of a certain material and certain thickness can be loaded with masses up to 2 kg. Next, a mass of 1.8 kg is attached to this wire, and against any expectation it begins to give. Taking hypothetical realism seriously, everything must be put in question, i.e., from the law of gravitation down to the details of the fabrication of the wire, because in order to search for the error, one must systematically take into account the possibility of error in every relevant area of physical knowledge. Although some philosophers may consider this type of skepticism as a sign of outstanding scientific rationality, such an attitude would actually be exceedingly irrational. Indeed, those who advocate hypothetical realism do not behave in such a manner, so their attitude reveals this to be an irrelevant background metaphysics, without influence on scientific practice. In contrast, the strategy pursued in practice aims to localize, in the first step, the contradiction between the expected and the actual result, in order to remove it afterwards. Localization involves distinguishing between those statements of the theory which are not subject to revision and those assumptions that are. The least hypothetical part is the theory frame, which is not subject to modification in the case of discrepancies between theory and experience since it has no direct connection with experiment. Accordingly, a feasible definition of "hypothetical" must consider first of all that, in the case of discrepancies between theory and experiment, it is not the entire theory that should come under scrutiny. A physical statement is then hypothetical if it is subject to modification, revision or refutation in the comparison between experiment and theory. Secondly, a hypothesis in physics is commonly understood as an assertion about an unknown or incompletely known element of reality that is not provable by purely theoretical methods. Consequently, the formulation of a hypothesis is always associated with the idea of arriving in some way at an extension of knowledge either by postulating something new ("Beyond Uranus there is another planet") or by deriving new knowledge from the hypothesis. Thirdly, every hypothesis is associated with a claim of validity: an assertion without such a claim is a mere statement or a fact, a convention, or a proposition, like a mathematical axiom. A definition of hypothesis appropriate for physics thus reads:

Definition 3.1 A hypothesis in physics is an assumption that postulates new knowledge, raises a claim of validity, and is subject to refutation in the comparison between theory and experience.

Notably, direct experimental testability is not a necessary property of physical hypotheses. Even this more precise definition possesses a sufficiently broad spectrum of meanings to cover the many and varied types of hypotheses in physics. In particular, most assumptions at the empirical level are hypothetical because they attempt to arrive at generally valid and epistemically significant results on the basis of limited experience. Since securing their claims of validity can only be attained through deduction from a theory, they are of subordinate importance as long as they are not connected to the theoretical level. Finally, this definition enables the demarcation of prognoses. A prognosis is a statement that is already considered to be valid due to its deduction from a theory, so that it is neither hypothetical nor associated with new knowledge, because it is merely the statement of an undisputed fact, although not yet empirically proven. That men are sent to the moon by rockets on the basis of precalculated trajectories is grounded on the safe conviction that they will arrive there provided that no technical malfunction occurs. Thus, it is assumed that a prognosis fulfills theoretical expectations, while this must be proven in the case of a hypothesis.

In contrast to empirical hypotheses, the assumptions hypothesized within theories are of crucial importance for gaining physical knowledge and also for its status. Firstly, they are the proof that there is an autonomous theoretical component of physical knowledge. Otherwise, theoretical considerations would not yield any new knowledge, and there would be no real hypotheses at the theoretical level. Secondly, this formation of hypotheses is entirely dissimilar to the corresponding empirical methods. In addition to the formation of concepts and models, the formation of hypotheses reflects the general difference between empirical and theoretical methodology. Indeed, it is only via hypotheses at the theoretical level that the transition can be made from the theory frame to experimentally comparable deductions, and it is only due to this process that theoretical knowledge becomes hypothetical. This is shown by analyzing potential hypothetical elements of the theory frame. First note that (A4) is not concerned: as a pure existence postulate it is a norm that cannot be questioned, since the theory cannot otherwise be a part of physics. In principle, in (A1)–(A3), the choice of MT, the assignment rules (#), the mathematical representation of the structural principle, and the particular form of the relations in (A3) may be hypothetical. At least (A2), as the epistemological kernel, represents novel knowledge and advances a claim of validity, irrespective of what this may look like. In this context, one must distinguish between the historical development of a theory and its reconstruction. As long as theory formation is not completed, a theory frame does not exist and there are therefore no postulates of the type (A1)–(A3). In this case, the formation of hypotheses is structurally similar to what happens on the empirical level. When theory formation is completed, the postulates (A1)–(A3) are not hypothetical, since they are not directly connected with experiments and cannot therefore be subject to revision in the comparison between theory and experiment. Finally, the definition of specializations does not require any hypotheses since only constraints of the general form of (A2) are concerned, and these exhibit the same status as the propositions in a mathematical theory.

In contrast, the transition from a specialization S to a scientific application ScA is realized through the addition of assumptions which are not contained in the theory frame and which cannot be interpreted as mere conventions or statements of facts. According to the above definition, they exhibit the status of hypotheses and comprise three structurally dissimilar classes. The first comprises the system-adapted realizations of the postulates (A1)–(A3), which occur at the transition from S to ScA:

(H1) Assumptions regarding the physical nature of the model system under study.

(H2) Assumptions concerning a mechanism for the system-specific dynamics.

(H3) The system-adapted formulation of the relations in (A3).

These three hypotheses establish ScA and may be summarized as model assumptions $\langle M \rangle$. The concept of scientific application may then be assigned a precise meaning:

Definition 3.2 The scientific application of a physical theory is a specialization of the theory kernel supplemented with hypotheses of the type $\langle (H1),(H2),(H3) \rangle = \langle M \rangle$.

The assumptions (H1) may refer to models, e.g., the planetary model of the atom in Bohr's theory, or the mass point model of the kinetic theory of gases. However, as shown in Sect. 2.5, physical systems are not defined by idealized objects in the form of $\langle S, C \rangle$-models. Instead, the specification of the model system occurs through the definition of system-specific measurable quantities, such as masses, force constants, moments of inertia, or charge distributions. In contrast to the object-oriented idealizations, these abstract models of physical systems are theoretically well-defined.

The hypotheses (H2) define the dynamics of the particular system. In Newtonian mechanics, these assumptions comprise the specific force laws. The specialization of (A2) for position-dependent forces may be replaced by specific force functions, e.g., $F = -fx$ (Hooke's law) or $F = Gm_1m_2/r^2$ (law of gravitation). If these particular force laws are interpreted as mere definitions or conventions, they are not hypothetical. Indeed, the hypothetical element is buried in the claim that these assumptions are related in some way to reality rather than being restricted to merely enabling the deduction of experimentally comparable data. This is supported by numerous examples which demonstrate that the physical content of these assumptions is key to the correct comprehension of the physical nature of real systems and processes. In Bohr's theory, it is the correct assumption that the detectable radiation is generated by an electron jump. In Drude's theory of electrical conductivity, it is the erroneous assumption that electrical resistance is caused by the scattering of conduction electrons on ionic cores. The discussion of the problem of truth in Sect. 5.4 will show that the assessment of physical knowledge as true or false is crucially associated with this claim of validity, and that an eschewal of this claim is neither meaningful nor necessary.

The assumptions (H3) may, on the one hand, supplement the general relations of (A3) with the definition of system-adapted reference frames or coordinate systems. Regarding kinematic studies in a laboratory, this may be chosen as the reference frame, and the trajectories of planets or comets in celestial mechanics are described by specifying their orbital elements. On the other hand, specifications may be necessary to provide a basis to correlate the new quantities introduced via (H1) and (H2) with scalar quantities or real numbers. Due to these interrelations, the model assumptions $\langle M \rangle$ must be seen as an interconnected entity which defines the physical model of the system under study. The scientific applications, i.e., the empirically relevant realizations of the theory frame of a physical theory, constitute a special type of $\langle C, C \rangle$-models, but they are not models as used in mathematics. Although a model in mathematics is also an interpretation of the system of axioms, it does not contain additional hypotheses, because it does not need to meet a criterion of reality in the sense of having empirical relevance. The specializations of the theory kernel are in this respect mathematical models. In contrast, the key feature of scientific applications is the supplementation with hypothetical model assumptions $\langle M \rangle$ resting on external system-specific assumptions and knowledge that are not contained in the theory frame. For these reasons, the formulation and elaboration of scientific applications constitute an indispensable part of fundamental research in physics, and of the extension of our knowledge about nature.

The model assumptions establish the ScA, but they do not yet enable one to deduce quantitative results that are directly comparable with experimental data. To this end, within ScA, a second class of assumptions is needed, comprising hypotheses of the type (H4)–(H8):

(H4) The assignment of numerical values to the system-specific quantities defined in (H1).
(H5) Assumptions about initial or boundary conditions as the necessary prerequisite to compute particular solutions of equation $(*)$.
(H6) The physical interpretation of symbols and theorems of the corresponding specialization.
(H7) Simplifying mathematical assumptions facilitating or enabling the quantitative solution of the problem.
(H8) Other ad-hoc assumptions, e.g., existence hypotheses (Neptune, Maxwell's displacement current, neutrino).

The necessity of (H4) and (H5) is immediately clear and does not require any explanation. (H6) corresponds to the introduction of a new physical quantity. It is not a hypothesis if interpreted as a merely nominal definition, but such a reduced view does not well represent the character of physical concepts. Indeed, the hypothetical element is the demand for empirical significance, i.e., that the new term is not an arbitrary theoretical construct, but provides additional physical insights as previously exemplified by the mass density. In particular, this becomes obvious when such a hypothesis implies the existence of a material carrier of a property. Through hypotheses of the type (H7), additional mathematical simplifications may be introduced if necessary for the deduction of quantitative data. An ad-hoc hypothesis (H8) postulates supplementary, experimentally provable information in which an ScA is underdetermined. Newton's theory of gravitation, e.g., applied to the Solar System, may be suppplemented with the existence hypothesis for a new planet, since the theory does not contain a statement concerning their number. Another type of ad-hoc hypothesis involves empirical modifications of (H1) or (H2) that are not deducible theoretically, but result in improved agreement with experiments. The empirical modification of the ideal gas law by adding terms for volume and interaction of the gas particles may be viewed as such an ad-hoc assumption.

The third class of hypotheses comprises just a single type, namely:

(H9) Each claim about an observable fact deducible within the scientific application.

As deductions in ScA, hypotheses of this type seem to be neither hypothetical nor to provide new information. This is not the case, however. They are hypothetical because the premises are hypothetical, and they lead to new knowledge in various ways. Firstly, due to the addition of the model assumptions ⟨M⟩ in combination with an extension of the physical interpretation, a mathematical theorem of the specialization S may be transformed into a physical law in ScA, i.e., into a thesis associated with the assertion to correspond to a structure of reality. Secondly, new infomation can be obtained through the calculation of additional empirically testable data. For example, within the framework of Kepler's theory, new points of a trajectory can be calculated from known ones, and can be compared with measured planetary positions. Formally, this situation may be expressed in such a way that this type of hypothesis is a statement in ScA that has been supplemented with symbols for virtual facts, objects, or properties (Ludwig 1978):

Definition 3.3 A hypothesis of type (H9) is a proposition in an ScA of the form

(H9) For $a_i \in Q$, there exists an $h \in Q$ with $\mathrm{R}(\ldots, a_i, \ldots, h; \alpha_h)$.

Here α_h is a scalar quantity derived from h in combination with a related (hypothetical) numerical value that is directly comparable with an experimental datum.

This is exactly the formal representation of the example of a calculation of a new point h of a trajectory from some known a_i. This can be extended without difficulty to several h and Q. According to their formal structure, these are hypotheses that are directly falsifiable or verifiable. However, this alone is without significance for securing the knowledge contained in the model assumptions. The correctness of the premises does not follow in the case of a verification. Conversely, if the deduction does not agree with experiment, it remains undetermined, in the first instance, what has been shown to be incorrect.

As a first result, one may say that, on the one hand, the hypothetical elements of physical theories come into play through the scientific applications. On the other hand, experimentally comparable results can be deduced only by means of these additional hypothetical assumptions. Einstein expressed this situation as follows: "When a statement is safe, it does not state anything about reality, and when it states something about reality, it is not safe." As a consequence, the question arises whether theoretical knowledge can ever achieve the status of secured knowledge, or whether it inevitably remains hypothetical, as alleged in the hypothetico-deductive empiricism. The remainder of this section deals with methods for securing the knowledge contained in the various types of hypotheses. Knowing the strategies used to accomplish this objective constitutes another important step towards both the comprehension of physical methodology and the assessment of the status of physical knowledge.

This task is relatively easy to handle for hypotheses of the type (H4)–(H9), which can be largely controlled by experimental methods. In physics this pertains primarily to (H4) and (H5). One reason for performing experiments, instead of restricting oneself to mere observations, is precisely the experimental control of the contingent conditions, i.e., system parameters and initial conditions. The relevant hypotheses are thereby transformed into sets that are fixed by the experimental design, whence they are no longer hypothetical. This constitutes the crucial difference with sciences based on pure observation, where these conditions are not manipulable and usually not known. The speculative feature, e.g., of cosmological or paleontological theories rests at least partially on the fact that these initial or boundary conditions are unknown. (H7) becomes experimentally controllable by designing the experiment in accordance with the simplifying mathematical assumptions. Examples are the experimental realization of systems as ideal as possible (ideal gas, homogeneous field) or the systematic experimental selection of variables. Hypotheses of type (H6) are empirically controllable by manufacturing an object experimentally, or by defining a measuring instruction for a newly introduced property. The introduction of quantities that are not directly measurable (potentials, wave functions) necessitates particular justification and additional conditions, e.g., gauge invariance.

Regarding (H8), it should be noted that the introduction of ad-hoc hypotheses may appear dubious because they are frequently associated with some arbitrariness. This may apply to assumptions at the empirical level, but not to ad-hoc hypotheses within an ScA. These are usually advanced in combination with the deduction of experimentally testable data. Concerning the existence hypotheses for Neptune or the neutrino, this is obvious, because they were transformed into secured knowledge after the discovery of these objects. In contrast, phenomenological corrections of the model assumptions due to empirical knowledge appear more arbitrary. The transformation of such hypotheses into secured knowledge is commonly achieved through theory extension, so that the hypothesis appears either as an axiom or as a deducible theorem of the extended theory. As an important criterion, ad-hoc hypotheses must be adapted to the current state of instrumental development because otherwise there is no chance of empirical control. For this reason, Prout's hypothesis was originally rejected, although later on it was revealed to be qualitatively correct. The same applied to the hypothesis that assigns a well-defined geometrical structure to molecules before experimental methods of structure determination were developed. Ad-hoc hypotheses adapted to the current state of theoretical and experimental development may be accepted as admissible. In conclusion, the claim sometimes put forward that ad-hoc hypotheses should be eliminated from science as being nonscientific is unjustified, both for methodological reasons and also due to their numerous successes. Finally, the hypothetical part of (H9) is only the premises. If the deduction is assumed to be formally correct, (H9) provides itself with an option of control due to the direct connection with experiment, since it constitutes a necessary criterion for securing the knowledge postulated by the other hypotheses.

As a result, hypotheses of the type (H4)–(H9) which occur within an ScA can be considered as relatively unproblematic in terms of their hypothetical character, since they are largely empirically controllable. Regarding the model assumptions (H1)–(H3), this is not an option. In this case, theoretical methods of control exist because the addition of hypotheses to a specialization is subject to a number of constraints and criteria. In order to demonstrate this, some general considerations are required that will also provide the basis for analyzing the relations between theory and experience in Chap. 4. The necessary concepts have been taken in part from Ludwig (1978), who has already described this addition of hypotheses at the theoretical level. Consider an ScA, i.e., a specialization S of the theory frame that has been supplemented with model assumptions $\langle M \rangle$. The minimum condition for adding hypotheses is that this must not lead to logical contradictions:

Definition 3.4 A hypothesis H is said to be permitted with respect to ScA if ScA&H (= ScA supplemented with H) is logically consistent. Otherwise, H is said to be false (with respect to ScA).

This definition can be extended to several hypotheses $H_1, \ldots, H_n$ and may be elucidated with some simple examples. Concerning the classical mechanics of the mass point, an assumption about the initial velocity v of a particle of the type $v > c$ (= speed of light) is permitted. If ScA is an application of the special theory of relativity, such a hypothesis is false, since WA&H becomes contradictory. Again concerning classical mechanics, initial conditions for the position and velocity of a particle at the same time can be given, but in the framework of quantum mechanics, such a hypothesis is false. Considering Bohr's theory as a scientific application of classical electrodynamics, the postulate that electrons might remain without radiating on certain orbits is false. In contrast, assuming that electrodynamics is not applicable to atomic processes, one is not concerned with any of its applications, and such a hypothesis is permitted. These examples demonstrate that H is defined as false only relative to ScA, so that the statement "H false" does not mean that H is physically impossible. For instance, if the comparison with experiment shows that a hypothesis, being false in this sense, is realized, one of the model assumptions of ScA might be incorrect or the theory may not be applicable because no specialization exists that can be supplemented to yield a suitable ScA. In conclusion, one cannot conclude from a theory to the non-existence of objects or processes. Conversely, the statement "H is not false" does not indicate that H is true.

The transition from a specialization S to one of its ScA via the addition of model assumptions $\langle M \rangle$ will lead, in general, to an extended physical interpretation of S, since further symbols of S are assigned a physical meaning.

Every theorem already deducible in S, whose symbols are replaced by the addition of ⟨M⟩, is then transformed into an empirically testable hypothesis in ScA, at least in principle. When, e.g., S is the mechanics of the mass point for position-dependent forces, the law of energy conservation is deducible in S. An ScA of S is the theory of harmonic oscillations realized, e.g., by a mass m on a spring. The force is thus substituted with $-f \cdot x(t)$, where f is the spring constant, and the system is specified by f and m. Thus, the law of energy conservation as deducible in S becomes an empirically testable hypothesis. This leads to the definition (Ludwig 1978):

Definition 3.5 A hypothesis is said to be theoretically existent in ScA if it is already deducible in S as a theorem.

Theoretically existent hypotheses can therefore be attributed to type (H9) and are obviously always permitted, since otherwise S itself would be logically inconsistent, and this is assumed to be excluded. On the other hand, ScA may contain hypotheses without there being a corresponding theorem in S, so these are not theoretically existent, like the previously mentioned hypothesis "Beyond Uranus there is another planet". This is a permitted, but not theoretically existent hypothesis within Newton's theory of gravitation applied to the Solar System, because this theory does not contain any statement concerning the number of planets. Conversely, the negative hypothesis "Beyond Uranus there is no other planet" is also permitted with respect to this theory, notwithstanding that, since 1846, it has been proven to be empirically wrong. This results in the conclusion, appearing initially as paradoxical, that together with a permitted hypothesis, the "negative" hypothesis

$$(\neg\text{H}) \quad \text{For } h \in Q \text{ the relation } \neg\text{R}(\ldots, a_i, \ldots, h) \text{ is valid}$$

is also permitted and may be added to ScA without WA&(¬H) becoming inconsistent. Otherwise, H must be a theorem in S, i.e., theoretically existent. Indeed, both S and ScA are underdetermined regarding H, i.e., they do not contain any statement concerning either H or ¬H, so ScA can be supplemented either by H or ¬H without running into logical inconsistencies. The decision as to whether H or ¬H is realized must be made either through comparison with experience or by switching to another theory where one of the two hypotheses becomes theoretically existent. Even a theoretically existent hypothesis does not necessarily provide an unequivocal statement about an empirical fact. Particularly important for the comparison between experiment and theory are those hypotheses that meet this condition (Ludwig 1978):

Definition 3.6 A permitted or theoretically existent hypothesis is said to be determinate if the proposition advanced by the hypothesis (H) is unequivocal.

A simple example is provided by the theory of central elastic collisions, constituting a specialization of the classical mechanics of many-particle systems. The final velocities $\mathbf{v}_{1f}$ and $\mathbf{v}_{2f}$ of two central elastically colliding balls can be derived from the given initial velocities $\mathbf{v}_{1i}$ and $\mathbf{v}_{2i}$ by means of the conservation laws of energy and momentum:

$$\text{momentum law:} \quad m_1\mathbf{v}_{1i} + m_2\mathbf{v}_{2i} = m_1\mathbf{v}_{1f} + m_2\mathbf{v}_{2f}\,,$$

$$\text{energy law:} \quad m_1\mathbf{v}_{1i}^2 + m_2\mathbf{v}_{2i}^2 \;=\; m_1\mathbf{v}_{1f}^2 + m_2\mathbf{v}_{2f}^2\,.$$

These laws allow two pairs of solutions:

$$\text{(L1)} \quad \begin{array}{l} \mathbf{v}_{1f} = [(m_1 - m_2)\mathbf{v}_{1i} + 2m_2\mathbf{v}_{2i}]/[m_1 + m_2] \\ \mathbf{v}_{2f} = [(m_2 - m_1)\mathbf{v}_{2i} + 2m_1\mathbf{v}_{1i}]/[m_1 + m_2] \end{array} \quad \Bigg| \quad \text{(L2)} \quad \begin{array}{l} \mathbf{v}_{1f} = \mathbf{v}_{1i} \\ \mathbf{v}_{2f} = \mathbf{v}_{2i} \end{array}$$

Since both hypotheses are theoretically existent, only the comparison with experience can show which one is realized. Without going into details, three possibilities are conceivable for a non-determinate hypothesis:

- In the simplest case, one of the two options is realized, as in this example. For (L2) correponds to a process where both balls keep their velocities, i.e., would penetrate each other unhindered, which contradicts all notions of material bodies. In this case, a thought experiment or plausibility argument suffices to obtain a determinate hypothesis. This is commonly expressed by saying that a solution must be excluded for "physical reasons", i.e., without actually carrying out any empirical test. Due to this argument solutions of equations are usually rejected as unphysical when they do not fulfill certain conditions. As Dirac's equation demonstrates, however, solutions may initially be eliminated that are actually realized, against common expectation.
- Both cases may be realized. This raises the question of whether particular conditions exist due to which one or the other possibility occurs. If this is not the case, one may investigate whether the behaviour of the system can be encompassed by a statistical law or whether it occurs purely accidentally.
- Methodologically, the most interesting case is the one in which none of the possible hypotheses are realized, especially when both are theoretically existent, because the question arises as to what has been falsified. Although formally this is, in the first instance, only the deduced hypothesis, it is obvious that something must be wrong when a theoretically existent hypothesis cannot be realized experimentally. Knowing the strategies for localizing and eventually removing such contradictions provides important insights into the methods of securing knowledge. These strategies will be discussed more generally in Chap. 4, while here only the theoretical part will be elucidated.

According to empiricism the strategies for securing knowledge are essentially, if not exclusively, of an empirical kind, in that they consist in scrutinizing the theoretically deduced hypotheses by empirical methods. This immediately poses several questions because at best only hypotheses of type (H4)–(H9) are directly empirically provable. In contrast, the model assumptions ⟨M⟩

establishing an ScA are not amenable to a direct comparison with experiment without additional hypotheses. Moreover, the model assumptions enter the empirically comparable hypotheses as premises, so that a discrepancy occurring in the comparison of experiment and theory cannot be localized initially. This emphasizes the general difficulty of any empirical examination of theoretical claims: the concrete empirical test is local in the sense that it refers to real systems and processes, while the theoretical claim possesses numerous connections within ScA and beyond, and is therefore of global type. This severely restricts the empirical options for checking the model assumptions $\langle M \rangle$, as is easily exemplified. An ScA may work well locally, i.e., it may produce results in agreement with experiment, but it may nevertheless be globally false because it rests on erroneous assumptions. The crucial speculative elements of physics enter through the model assumptions $\langle M \rangle$. For this reason, repeated attempts have been made to develop a physics without this type of hypotheses or at least to reconstruct physical theories without them.

One option to eliminate the speculative content of the model assumptions relies on the fact that their hypothetical character primarily results from the claim that they correspond to structures of reality. For this reason, a "hypotheses-free physics" would be conceivable by retaining the assumptions, but by denying any ontological or epistemic relevance. Instead they are considered as mere auxiliary means to deduce empirically comparable facts. Such an instrumentalist view concerning the status of physical knowledge may appear justified in the light of the conventionalist elements inherent in all quantitative theoretical knowledge. Eventually this would mean, however, that physics abstains from any claim to provide knowledge about reality. Taken seriously with this consequence, instrumentalism is demonstrably not advanced in physics, since in this case any physical interpretation of experiments and effects would be worthless for gaining insight, and would have to be considered as an intellectual game played by a few cranks straying along metaphysical paths. Accordingly, advanced in this way, instrumentalism is highly unscientific unless science is to be understood as a mere collection of factual knowledge. But when instrumentalism is just preached verbally, it only expresses a fear of errors or an implicit confession of lacking understanding without willing to admit it. The alternative to such antirealistic attitudes consists in searching for efficient, non-empirical options to control the model assumptions. That such options actually exist is largely ignored in empiricist philosophies. On the other hand, it must be conceded that such strategies for securing the model assumptions are considerably more intricate than any empirical examination since they necessitate substantial expert knowledge. For this reason, some methods will be exemplified here.

Many scientific applications of electrodynamics were established, shortly after Maxwell's development of the theory frame, through model assumptions which were in close analogy to mechanics. Accordingly, the hypotheses (H1) specifying electrodynamic systems consisted largely of mechanical analogical models which were later revealed to be misleading. This excessive use of mechanical models was the main reason for the demand around the end of the nineteenth century to stop using any kind of models and to restrict to the description of formal relations and properties. This demand is to be obeyed in the sense that the model systems introduced via (H1) must be specified through precisely defined theoretical properties, instead of using qualitative analogical models. The secureness of the assumptions in (H1) can thus be increased by abandoning pictorial models and analogies which are dispensable anyway when establishing scientific applications and should be eliminated for this reason.

In a similar vein, many fallacies regarding the assumptions (H2) for the mechanisms are based on incorrect analogies, as exemplified by the theories of Bohr and Drude. Both postulate mechanisms resting on apparently plausible but incorrect classical analogies. As a theory of atomic constitution, Bohr's theory attributes the stability of atoms to the balance between the electrostatic nucleus–electron attraction and the centrifugal force. In contrast, in quantum mechanics, stability is deduced from the uncertainty relations. In Drude's theory, the origin of the electrical resistance in metals is attributed to the scattering of the conduction electrons on the ionic cores of the lattice atoms. According to quantum mechanics, the scattering of electrons on the periodic lattice does not lead to energy loss. Instead, the electric resistance is due to the scattering at deviations from the ideal periodic structure, viz., lattice vibrations, defects, or surfaces. In both cases, the classical analogies yield a misconception of the relevant physics, although they enable the deduction of some empirically correct results. In these cases, (H2) is checked by comparison with another theory. (H3) can be checked because a theory frame has various scientific applications that are not isolated from each other, but are cross-linked through numerous interconnections. Hence, there are a number of consistency conditions between the hypotheses (H3) of different scientific applications of the same theory frame which may be utilized to check them.

Due to the horizontal and vertical interconnections within a theory, a number of non-empirical methods exists to check and increase the secureness of physical knowledge. Due to these interconnections which each hypothesis possesses at the theoretical level, both empirical and theoretical checking strategies are based on the fact that the examination never concerns an isolated hypothesis. Even when a concrete empirical test refers to a single hypothesis, a group of interconnected hypotheses, if not an entire ScA, is actually concerned. The ScA itself is obtained from a specialization through supplementation, not by a single hypothesis, but by a complex sentence $\langle M \rangle$ containing interconnected hypotheses, so that once again a number of consistency conditions persist between them. Accordingly, one defines:

Definition 3.7 Two or more permitted hypotheses are said to be compatible if their combination yields a permitted hypothesis.

This is a necessary criterion concerning, in particular, the model assumptions, for the follwing reason. Just as the addition of the predicates in (A1) to a mathematical theory is always possible, each single hypothesis of the type (H1)–(H3) is trivially permitted when added to S, because S does not contain experimentally comparable statements. Accordingly, one defines:

Definition 3.8 A scientific application ScA is said to be internally consistent if (H1)–(H3) are compatible, i.e., if the combination ⟨M⟩ of the hypotheses establishing ScA represents a permitted hypothesis.

It may happen that a theoretically existent hypothesis becomes empirically false due to the supplementation of S by the model assumptions ⟨M⟩. This opens the most direct option for their empirical examination. If all the other hypotheses, especially (H4) and (H5), that are required for the empirical comparison can be checked experimentally, i.e., can be transformed into fixed settings by appropriate experimental design, discrepancies to experiment can be localized within the model assumptions. Furthermore, the various scientific applications of the same theory frame are not isolated from each other, but exhibit interconnections that must not lead to inconsistencies. This yields the definitions:

Definition 3.9 An ScA is said to be empirically consistent if it does not contain empirically false hypotheses that are determinate and theoretically existent.

Definition 3.10 An ScA that is compatible with the other scientific applications of the same theory frame is said to be theoretically externally consistent.

Two applications of a theory are compatible if their constituent model assumptions are compatible. Analogous interconnections may exist between scientific applications of dissimilar theories. Their compatibility is then a supporting instance for the ScA in question, but not a necessary condition, as will be argued in more detail in Sect. 3.5. Finally, a condition can be defined that enables one to consider theoretical knowledge as secured:

Definition 3.11 An ScA is said to be secure if it is internally, empirically, and theoretically externally consistent. The model assumptions of a secure ScA are not subject to refutation in a comparison between experiment and theory, so that in this regard they are no longer hypothetical.

Most important in this context are the external checking methods because, exclusively through experiments and investigations within the ScA, the model assumptions can never be transformed into secure knowledge. Discrepancies with experiments in a secure ScA must be sought in the hypotheses (H4)–(H8). This emphasizes the practical relevance of this definition. It is one of the strategies for localizing inconsistencies and contradictions, but must not be misunderstood as saying that the secureness of the knowledge provided by an ScA is guaranteed in an absolute sense. Furthermore, one should not believe that, when theory formation is completed, the theory contains exclusively secure scientific applications. This goes back to the historical development. If an approach devised before the completion of theory formation works, in the sense that it reproduces experimental data satisfactorily and in a simple manner, it is still being used, even when it rests on model assumptions that have revealed themselves to be non-compatible with other parts of the theory. Since precisely this provides the interesting instances for the assessment of physical knowledge, some of them will be discussed in Sect. 5.4 in the context of the problem of truth. For this reason, the concepts needed to describe this situation will be introduced briefly. Firstly, the term "secure", as defined for scientific applications, can be defined in modified form for hypotheses:

Definition 3.12 A hypothesis is referred to as uncertain with regard to an ScA if it is incompatible with a hypothesis permitted in the ScA, otherwise secure.

From this definition, it immediately follows that every theoretically existent hypothesis is secure in a secure ScA. Secondly, the terms "hypothetical theory" and "model theory" can be assigned a more precise meaning:

Definition 3.13 An ScA is hypothetical if it contains an uncertain hypothesis, i.e., if it is internally inconsistent.

Definition 3.14 An ScA is called a "model theory in the narrower sense" if it is theoretically externally non-consistent.

The term "model theory in the narrower sense" is used because, in principle, any physical theory may be considered to be a model theory. Although such a model theory works for the deduction of a number of empirical data, it is fictitious because its model assumptions do not claim to correspond to structures of reality. Accordingly, such a model theory has a similar status to an algorithm, which is easier to handle or produces more precise results than the epistemically significant physical laws, while its functional form is not associated with any ontological claim. A representative example is the crystal-field theory, which enables the description of a number of optical and magnetic properties of transition metal compounds, although it is based on false model assumptions (Lebernegg et al. 2008).

In summary, it has been shown that the hypothetical elements of theoretical knowledge occur at the transition from the specializations of the theory frame to the scientific applications ScA. Accordingly, decisive dissimilarities exist between the theory frame and the ScA concerning both status and claims to validity. The theory frame is non-hypothetical since it is not directly connected to experiments, so it cannot be subject to refutation in the comparison between experiment and theory. This connection is established via the various ScA. Their hypothetical character is primarily due to the model assumptions ⟨M⟩, while hypotheses of the type (H4)–(H8) can usually be checked experimentally. In purely observational sciences, as far as they are based on physical knowledge, the critical hypotheses are mainly (H4) and (H5), because (H1)–(H3) are accepted as secured knowledge, so they are not subject to refutation. Both the options of securing and the secureness of the knowledge inherent in the model assumptions rest crucially on the circumstance that a physical theory is not an accumulation of isolated facts and statements, but constitutes a network of interconnected knowledge. The transformation of hypothetical into secure knowledge occurs according to the following strategies. The securing of empirical hypotheses is carried out by theoretical methods, i.e., through incorporation in the theory or by deduction, but not by experimental methods. The securing of theoretically deduced knowledge takes place primarily by empirical methods, although not exclusively, because the knowledge inherent in the model assumptions cannot be secured without taking into account the interconnections at the theoretical level. Put simply, one may say that the methods for forming and for justifying hypotheses in physics are largely complementary. Consequently, the strict separation between context of formation and context of justification is obsolete, even for the reconstruction of physical knowledge, because it is precisely this complementary structure of the methods for gaining and securing knowledge which essentially contributes to the reliability of physical knowledge.

3.3 Explanatory Theories

The central feature of explanatory theories is the deduction of knowledge contained in descriptive theories from theoretical assumptions possessing the status of accepted patterns of explanation. Actually, even descriptive theories are not purely descriptive, because they provide explanations corresponding to deductive structures. In kinematics, for instance, the trajectories are deduced from an acceleration function, and in Bohr's theory the Balmer formula and Rydberg constant are deduced. From this point of view, the distinction between descriptive and explanatory theories appears to be somewhat arbitrary. At the outset, possible differences between the two types of theory are elaborated by exploring the relation between kinematics and dynamics.

A representative example of the relationship between a kinematic and a dynamic theory is the one between Kepler's theory and Newton's theory of gravitation. Firstly, Kepler's theory does not constitute a description of direct observations because the observed trajectories of the planets are not ellipses. The theory is explanatory in the sense that the shape of the directly observed trajectories can be deduced from the theoretical assumption that they result from the simultaneous elliptical motions of both the Earth and the planet around the Sun. Secondly, the theory provides a unifying aspect since analogous structural elements are assigned to the various planetary orbits, instead of considering each orbit as an isolated phenomenon. However, it cannot justify its own assumptions. Why the planets revolve around the Sun and why the orbits are elliptical is not explained. On the one hand, the theory thus defines an idealized standard by setting the possible trajectories as ellipses: the planetary orbits are assessed with respect to their deviation from the elliptical shape. On the other hand, deviations from this theoretical standard can only be stated, but not substantiated, so any conjecture concerning their causes necessarily remains speculative.

According to standard textbooks, Newton's theory of gravitation provides an explanation of Kepler's laws in the sense of a theoretical deduction. This suggests a tentative idea of how to define an explanatory theory: a theory T explains a theory T_0 if the axioms of T_0 are deducible as theorems within T. Conversely, T_0 is said to be reducible to T. Although this definition is correct in that T does not in the first instance explain observations or empirical facts, but theoretical assumptions, the relation between the two theories is actually considerably more complex because, among other things, the conceptual systems of the two theories are not consistent with each other, as will be shown in detail in Sect. 3.5. In spite of these difficulties, one may ask whether the general relationship between kinematics and dynamics can be reconstructed in such a way that dynamics appears as an explanatory theory of kinematics. Early reconstructions of this relation by Mach (1919) and Hertz (1965), however, aimed rather to reduce dynamics to kinematics in order to eliminate the concept of force from dynamics. To this end, the concept of mass was either explicitly defined through the ratio of velocities (Hermes 1964) or accelerations (Mach 1919), so that it became a kinematic term, or it was introduced axiomatically (Hamel 1949) or operationally as a new term. The concept of force was then totally eliminated or else defined through Newton's second law, which thus gained the status of a nominal definition of force rather than representing a physical law. The subsequent analysis will show, on the one hand, that such reconstructions not only ignore the physical content of dynamics, but also unnecessarily conceal the relationship between classical mechanics and both relativity and quantum theory. On the other hand, according to the previous definition, the explanation of kinematics by dynamics is incomplete in the strict sense.

The conceptual foundation of the kinematic description of motion comprises the terms "position", "velocity", "trajectory", and "acceleration". The definition of the kinematic velocity by the differential quotient, i.e., $\mathbf{v}_{\mathrm{kin}} = \mathrm{d}\mathbf{r}/\mathrm{d}t$, is directly associated with the trajectory of the moving entity whose physical nature remains undetermined, so that kinematics does not contain any constraints in this respect. As usual, material bodies may be concerned, but non-material, geometric points are not excluded, e.g., the motion of the intersection point of two crossed, moving bars (Falk and Ruppel 1972) or the motion of a laser point on a whiteboard. In contrast, the concept of motion in dynamics is constrained to motions in which certain physical properties are transported. This is usually the mass, but dynamical transport is not linked to mass, as illustrated by the propagation of light or of waves on a water surface. Consequently, restricting the dynamical concept of motion to mass, as found in many textbook presentations of mechanics, introduces a constraint that does not correspond to the general character of dynamical processes. The decisive quantities transported in a dynamical process are rather energy and momentum. They are not independent of each other. The relation between them is specific to the transport process, and that is completely described by the way in which the energy depends on the position and momentum. That the energy may depend explicitly on time is irrelevant in this context, since time is a parameter in classical mechanics, not a dynamical variable. One of the central statements of mechanics is then that the mechanical properties of a system are completely described by functions of the independent dynamical variables position and momentum, which specify the dynamical state. The definition of the mass point in Sect. 2.5, via its dynamical state rather than its mass, is based precisely on this circumstance.

For these reaons, the kinematic, trajectory-linked concept of motion must be distinguished from the dynamic one that is characterized by the dynamic velocity $\mathbf{v}_{\text{dyn}} = \nabla_p E$ of the transport of energy and momentum (Falk and Ruppel 1972). In the case of the one-dimensional motion of a system in a position-dependent potential, it is defined through $v_{\text{dyn}} = \mathrm{d}E/\mathrm{d}p$. With $E = p^2/2m + V(r)$, this becomes $v_{\text{dyn}} = p/m$. If, in addition, the position can be described by a position vector, assumed to be a continously differentiable function of time, the kinematic velocity is definable, and $\mathbf{v}_{\text{kin}} = \mathbf{v}_{\text{dyn}}$. In order to ascribe a kinematic interpretation to the momentum, the system must be localizable according to the mathematical condition of the existence of $\mathbf{r}$, and the motion must be describable by a trajectory, i.e., $\mathrm{d}\mathbf{r}/\mathrm{d}t$ must exist. These assumptions impose two geometrical constraints on the energy–momentum transport which a general dynamical description need not satisfy, as illustrated by the quantum theoretical concept of "motion" and the propagation of waves. The kinematic interpretation of the motion of quantum systems is impossible because $\mathrm{d}\mathbf{r}/\mathrm{d}t$, i.e., the concept of a trajectory, cannot be defined, and for the propagation of waves the requirement of localizability is not fulfilled.

In summary, the dynamical view of motion does not rest on the concept of trajectory, but on the general representation of the time evolution of the dynamical state. Accordingly, the velocity $\mathbf{v}_{\text{kin}}$ is less fundamental than $\mathbf{v}_{\text{dyn}}$, because the former is linked to certain geometrical conditions while the latter constitutes a concept of velocity abstracted from these notions. Any reconstruction of dynamics starting from kinematics necessarily ignores this abstract character, because it reduces the extent of meaning to this particular aspect. In contrast, the conceptual system of a general dynamics constitutes a considerably higher degree of abstractness, leading not only to greater difficulties of understanding, but also rendering comprehensible the long historical evolution towards dynamical concepts like momentum, energy, or angular momentum. These difficulties result in part from the fact that their full extent of meaning becomes comprehensible only within the theory as a whole. The theory kernel of the classical mechanics of the mass point may finally be constituted through the following postulates:

(A1) Let be given two parametrized, continuously differentiable vector functions $\mathbf{r}(t)$ and $\mathbf{p}(t)$ in a Euclidean vector space, referred to as position and momentum. A mechanical system whose dynamical state is completely defined by position and momentum, irrespective of size and shape, is called a mass point.

(A2) The dynamics of the mass point is defined as follows:

(a) The dynamical ground state is defined by $\mathbf{p} = \text{const}$.

(b) The dynamical velocity is proportional to the momentum and may be identified with the kinematic velocity:

$$\mathbf{v}_{\text{dyn}} = \mathbf{v}_{\text{kin}} \iff \mathbf{p}/m = \frac{\mathrm{d}\mathbf{r}}{\mathrm{d}t}\,.$$

The proportionality factor m ($\neq 0$) is called the mass.

(c) Any change in the dynamical state occurs exclusively due to external influences, denoted symbolically as forces $\mathbf{F}_{\text{ext}}$, according to the equation of motion

$$\mathbf{F}_{\text{ext}}(\mathbf{p}, \mathbf{r}, t) = \frac{\mathrm{d}\mathbf{p}}{\mathrm{d}t}\,.$$

(A3) The reference system is an inertial system defined by

$$\mathbf{p} = \text{const} \iff \mathbf{F}_{\text{ext}} = 0\,,$$

and is described mathematically by a Euclidean vector space.

Comparing with kinematics, the velocity has been replaced by momentum and the trajectory by the dynamical state. The position and momentum are defined in (A1) solely through their mathematical properties. Apart from the loss of generality, the complete and explicit definition of their physical meaning is neither possible nor a part of the theory frame. All operational definitions, e.g., of the momentum are based implicitly or explicitly on the law of momentum conservation which is itself a result of dynamics. On the other hand, the property of momentum that it is a conserved quantity which is merely exchanged in physical processes, and not created or destroyed, constitutes only one of several things required to explain it fully. Others are provided first by the kinematically interpretable specialization of mass transport, then Newton's second law $\mathbf{F}_{\text{ext}} = \mathbf{dp}/dt$ ascribing any change of momentum to the influence of the environment, and finally the scientific applications of dynamics. Knowledge from the whole of physics is actually required to understand its full physical significance, as exemplified by the fact that the need to distinguish between dynamic and kinematic quantities becomes apparent only via the relationship between classical dynamics and both relativity and quantum theory. Consequently, any explicit definition of position and momentum which comprises not just certain partial aspects is unattainable without referring at least to dynamics. All attempts to define these basic concepts independently, in order to establish a theory-independent conceptual basis of dynamics, are necessarily doomed to failure.

This means that, through the terms position, momentum, mechanical system, and dynamical state, (A1) defines the general theoretical model of the dynamics of systems without internal degrees of freedom, entirely independently of any particular conditions. The postulate (A2a) corresponds to Newton's first law and defines uniform rectilinear motion as the dynamical ground state which exhibits within dynamics the status of a natural motion, in the sense that it does not require an explanation. Equation (A2b) is not the nominal definition of the dynamical velocity, but a postulate characterizing the special case of classical mass transport, where the two velocities can be identified with each other. This necessitates as a further condition a non-vanishing mass that is in this representation not a basic concept of dynamics. It is precisely this that enables a seamless transition to the special theory of relativity, which allows transport with vanishing rest mass (Falk and Ruppel 1972). Since the form of the time evolution in (A2c) is independent of the mass of the system, it is also valid in this form in the special theory of relativity. The necessity of (A3) rests on the fact that, in non-inertial systems, accelerations occur that are not associated with external forces.

In the special case of constant mass, the relationship between kinematics and the theory of classical mass transport, as a specialization of the general dynamics, may be extended beyond (A2b). In this case, (A2c) can be transformed into $\mathbf{F}_{\text{ext}} = m\mathbf{a}$, and the force is directly proportional to the acceleration as the structure-determining quantity of kinematics. Therefore, this structure is largely transferable to the dynamics of the mass point. It was precisely this structural similarity of this specialization that was the reason for attempts to reconstruct dynamics from kinematics by adding the term "mass". Apart from the previously mentioned reduction of the physical meaning of dynamics, the spirit of such attempts is rarely comprehensible because conceptual systems of classical mechanics exist with the long-standing formalisms of Lagrange and Hamilton which explicitly emphasize the general character of dynamics and its relationship with other physical theories. The dynamical state of a mechanical system is described, e.g., by the Hamilton function, which is a function of the generalized coordinates and momenta. Geometrically, the dynamical state of a system may be interpreted as a point in the phase space, i.e., the abstract product space of the generalized coordinates and momenta. The time evolution of the dynamical state is described by the Hamilton equations and is displayed as a trajectory in the phase space. This trajectory, however, is by no means related to the one which a mass point traverses in position space, so this representation does not correspond to a reduction to kinematics. A reconstruction oriented toward this conceptual system is not pursued since it does not contribute further to the analysis of the relations and dissimilarities between kinematics and dynamics. The results of this analysis can be summarized as follows:

1. Regarding the formal structure, abstracted from the physical content, there are no substantial structural differences between the two theories. In the framework of the theory model developed in Sect. 3.1, both can be reconstructed with axioms of the type (A1)–(A4). This becomes notably apparent in the case of the dynamics of systems with finite, constant mass, so that (A2c) can be written as $\mathbf{F} = m\mathbf{a}$. The specific design of (A1)–(A3)

may vary, of course, but such differences are irrelevant to formal studies and comparisons.

2. The two theories do not differ regarding their relationship with the empirical level. Either both are classified as descriptive because, in both cases, the deduction of empirically comparable statements rests on unexplained premises, or both are explanatory in that observations and experiments are explained by subsumption under the theory.
3. The relationship between kinematics and dynamics does not conform to the previous tentative definition of an explanatory theory because neither of the theories is reducible to the other. On the one hand, dynamic motions exist without a kinematic counterpart, in the sense that they are independent of the concept of trajectory. On the other hand, kinematic motions exist that do not permit a dynamic interpretation as energy–momentum transport. Accordingly, axiom (A2b) constitutes a horizontal structural relation between the two theories by equating two quantities that are defined differently within the framework of each theory. A reconstruction of dynamics, based on kinematics, which properly accounts for the physical meaning of both theories cannot succeed, and neither can the reduction of kinematics in its entirety to dynamics in such a way that it appears, e.g., as a partial model of dynamics (Sneed 1971).
4. There are dissimilarities between the two theories, of course, concerning the physical meaning of the conceptual systems. The kinematic concept of motion rests on terms like "velocity" and "trajectory" that are familiar from direct experience and describe the speed and shape of motion. In contrast, the conceptual system of dynamics evolved in a process of abstraction that lasted more than two centuries and is detached from direct experience. The terms "dynamical state" and "dynamical change of state" are considerably more abstract, but more universal and comprehensive than the kinematic terms "trajectory" and "motion".
5. Differences also exist in the objectives of the two theories. The connection of kinematics to direct experience becomes evident through the view that motion, in general, is considered to be in need of explanation. In this respect, kinematics follows the tradition of Aristotle and the impetus theory. According to Newton's first law, only motion with a change in momentum is in need of explanation with respect to dynamics. Consequently, while motion with $a = 0$ is one among several types in kinematics, in dynamics it occupies a special position because it defines the dynamical ground state. Finally, the objective of a general dynamics is a description of state changes of physical systems entirely detached from the direct view of motion.

However, these differences in meaning and objectives hardly provide sufficient justification to say that kinematics is descriptive and dynamics is explanatory. Since the relationship between the two theories is not an explanatory one, and at the same time forces but not accelerations are viewed as the patterns of explanation of motion, one might conjecture that this differentiation is due to reasons that are not related to the formal structure. This is also demonstrated by the relationship between mechanics and thermodynamics as the second representative example of a putative reduction relation (Nagel 1982). Consider this relationship in the case of gases, since the connection between the measurable state variables pressure P, temperature T, and volume V is relatively simple. In this case, the descriptive theory is the thermodynamics of the ideal gas. Kepler's laws correspond to the ideal gas law $PV = nRT$ with the amount n of matter and the molar gas constant R. Both theories may be described as ontologically undetermined or neutral in the follwing sense:

- Just as kinematics does not contain any assumption regarding the physical nature of the moving entity, thermodynamics does not contain any assumption regarding the internal structure of the systems under study.
- Just as Kepler's theory does not explain the elliptical shape of the planetary orbits, thermodynamics cannot provide the reasons for the universality of the gas law, i.e., the reason why the various real gases behave approximately similarly, while other dissimilar properties do not play a role.

Indeed, this universality cannot be understood without model assumptions regarding the internal structure. These are necessarily of non-empirical type and must be viewed in the context of the historical background, wherein mechanics was taken as the universal theory until approximately the end of the nineteenth century. Accordingly, assumptions must be made in such a way that the behaviour of gases can be described in the conceptual system of mechanics, in order to establish the resulting "microscopic" theory of the ideal gas, referred to as the kinetic gas theory, as a scientific application of the classical mechanics of systems of mass points. Afterwards, one must examine to what extent mechanics provides an explanation of the thermodynamics of the ideal gas, in that, in particular, the ideal gas law is obtained as a deducible theorem of the kinetic gas theory.

The attempt to realize this program is, in the first instance, faced with the challenge that the temperature, occurring in the ideal gas law, is not a term of mechanics. As a matter of principle, no law containing the temperature is deducible from mechanics. Such a deduction may be expected, at best, for gas laws like Boyle's law, that contain only terms that can be assigned a meaning in mechanics. Even then, there remains the challenge that the model assumptions which ought to describe the system "ideal gas in thermal equilibrium", and ought to establish the kinetic gas theory as an application of mechanics, cannot be formulated without concepts and knowledge of thermodynamics. In order to demonstrate this clearly, and to establish the basis for an analysis of the general relationship between mechanics and thermodynamics, all assumptions needed to integrate Boyle's law into the conceptual system of mechanics will be itemized explicitly. Accordingly, what is to be deduced is the claim of Boyle's law: PV is constant for the ideal gas at constant temperature.

The first prerequisite to perform the deduction are model assumptions (S), corresponding to hypotheses of type (H1), that specify the system and establish the qualitative connection between the thermodynamic concept of "ideal gas in thermal equilibrium" and the vocabulary of mechanics:

(S1) The gas consists of N particles of the same mass in a container. The particles are described dynamically by the three degrees of freedom of their translational motion.
(S2) The total volume of the particles is small compared with the volume of the container.
(S3) The particles interact solely via elastic collisions.
(S4) The particles are distributed homogeneously over the volume of the container.
(S5) The distribution of the particle momenta is isotropic.

Assumption (S1) indicates that internal degrees of freedom of the particles are neglected. The "dilution assumption" (S2) implies that the average distance between two particles is large compared with their size or that the average time between two collisions is large compared with the collision time. These two assumptions together are usually summarized by postulating that the particles are mass points. Actually, they are independent of each other since the deduction of the specific heat, for example, requires one at least to take into consideration the rotational degrees of freedom, irrespective of (S2). Assumption (S3) about interaction solely via collisions means that long-range intermolecular interactions (gravitational, electromagnetic) should be negligible. This is fulfilled, in general, in gases of electrically neutral particles sufficiently far from the condensation point. That the collisions should occur elastically appears less plausible, especially in the light of the contemporary knowledge of molecular structure. Actually, the particles do not interact like macroscopic balls through direct contact, but via short-range repulsive interactions. While the first three assumptions provide the mechanical description of the system, the last two are of statistical type and define thermal equilibrium. They cannot be reconstructed conceptually as part of classical mechanics, but are necessary because interactions solely via elastic collisions do not lead to thermal equilibrium. Since due to (S1) all particles have identical mass, (S5) is equivalent to the assumption that the velocity distribution of the particles is isotropic.

The second group of assumptions, corresponding to hypotheses of type (H2), specify the dynamics, i.e., the interactions and mechanisms required for the deduction:

(D1) Classical mechanics of mass points applies to the particles.
(D2) The particles interact with the container walls only via collisions.
(D3) Collisions of the particles with the container walls occur elastically.

All of these assumption are of a theoretical kind because they cannot be proven for the individual particles. From (D1) and (D3), it follows in particular that the conservation laws of energy and momentum apply to the motions of the particles. (D2) implicitly postulates the mechanism for the thermodynamic pressure, analogously to the electron jump for the origin of the spectral lines in Bohr's theory.

The third group of assumptions establishes the connection between the microscopic mechanical and the macroscopic thermodynamic quantities. Similarly to the act of equating $\mathbf{v}_{\text{kin}}$ and $\mathbf{v}_{\text{dyn}}$ in the relationship between kinematics and dynamics, they have the function of correspondence principles between the two conceptual levels:

(C1) The "average of the squared velocities" of the N particles is

$$\langle v^2 \rangle = N^{-1} \sum v_i^2 \,.$$

(C2) The "pressure of a particle" on a wall with area A is

$$P_i := \frac{F_i}{A} \,.$$

(C3) The total pressure and mass are

$$P = \sum P_i \text{ and } M = Nm \,,$$

respectively.

F_i, P_i, v_i, and m are the microscopic quantities and P, A, and M the measurable, macroscopic ones. The crucial assumption is (C1) introducing a term, viz., the average of the squared velocities $\langle v^2 \rangle$, that occurs neither in thermodynamics nor in mechanics. It is in fact a statistical assumption, implicitly defining a concept of probability. The average of the velocity must be defined by $\langle v^2 \rangle$, instead of $\langle v \rangle$, because $\langle v \rangle$ vanishes due to (S5). The mechanical particle velocity v_i is time-dependent since it changes after each collision, while the average velocity is constant in time in thermal equilibrium. Finally, the term "pressure of a particle" is not a genuinely mechanical term, but it is a formally permitted concept analogous to the mechanical pressure, establishing the connection with (D2).

Prior to starting the actual deduction, two simplifications (V) may be made which are not assumptions since they can be derived theoretically:

(V1) The collisions of the particles with the walls occur perpendicularly to them.
(V2) The container may be taken to be a cube with edge length L.

The first simplification is proven by the fact that only the velocity component perpendicular to the wall contributes to the pressure. This implies the empirically significant fact that the pressure does not depend on the shape of the container, and this proves the permissibility of (V2). Since (S5) implies that all walls are physically equivalent due to symmetry, only a single wall need be considered, and without loss of generality, this can be chosen perpendicular to the x-axis.

The subsequent deduction consists of two parts. In the first step, the microscopic problem of the interaction of a single particle with the wall is solved within the framework of mechanics. Afterwards, the connection with the macroscopic, measurable quantities will be established by means of the correspondence principles. In this context, a problem in mechanics is considered to be solved when the prevailing forces are known. The equation of motion $F = \mathrm{d}p/\mathrm{d}t$ with momentum p is approximated by $F = \delta p/\delta t$ because, for processes in the microscopic regime, differential coefficients may be replaced with difference quotients. Accordingly, δp and δt must be calculated for a single particle in the container, so the particle index i can be omitted in this first step.

(T1) From (D1), (D3), and (V1), it follows that the momentum transferred to the wall by a particle is

$$\delta p_x = m(v_{x1} - v_{x2}) = 2mv_x \ , \ \text{ with } v_x = v_{x1} = -v_{x2} \ , \ \ v_{xn} = v_x(t_n) \ .$$

This is the conservation law of momentum for the perpendicular collision of a particle with the wall when moving in the x-direction, where t_1 and t_2 are times before and after the collision, respectively.

(T2) The time interval δt between two collisions is obtained from the following model consideration. The particle has the velocity v_x, and hits the wall perpendicularly, according to (V1). The time needed from one wall to the opposite one is $t = L/v_x$, and for the time interval δt between two collisions with the same wall it follows that $\delta t = 2L/v_x$. If δt is the time in seconds between two hits, then $1/\delta t$ corresponds to the number of hits per second.

(T3) From (T1), (T2), and Newton's second law, applicable according to (D1), the force exerted by the particle on the wall is

$$F_x = \frac{\mathrm{d}p_x}{\mathrm{d}t} = \frac{\delta p_x}{\delta t} = 2mv_x \frac{v_x}{2L} = \frac{mv_x^2}{L} \ .$$

Since $v_x^2 = v_y^2 = v_z^2 = v^2/3$ due to (S5), the pressure P_i of particle i on the wall is obtained from (C2) as

$$P_i = F_i/A = mv_i^2/3V \ .$$

The "microscopic" part of the problem is thus solved. In the second step, the connection with the macroscopic quantities is established by means of the correspondence principles:

(T4) According to (C3), the pressure is

$$P = \sum P_i = \frac{m}{3V} \sum v_i^2 \ .$$

(T5) With (C1) and (T4), it follows that

$$P = \frac{N}{3V} m \langle v^2 \rangle \ .$$

(T6) With (T5), the product of the pressure and volume of the gas can be cast in the desired form:

$$PV = \frac{2}{3} N \frac{m}{2} \langle v^2 \rangle = \frac{2}{3} N \langle E_{\text{kin}} \rangle = \text{const.}$$

The crucial step is (T5), because averaging over v_i eliminates the dependence of the macroscopic quantities on the dynamical properties of the individual particles, and in particular, the time dependence of v_i. In this context, one must assume that $\langle v^2 \rangle$ depends solely on the temperature, and not on other state variables, e.g., the pressure. Since this is fulfilled with sufficient accuracy, Boyle's law has been explained in the sense of a deductive argument. This alone, however, is not the decisive reason why this deduction is usually considered to be the paradigm of a successful physical explanation. Rather, it has a number of epistemic and empirically significant consequences in that it explains further facts that are in need of explanation within thermodynamics, and contributes to the enlargement of physical knowledge:

- Since the total energy of a system of non-interacting mass points is identical with the kinetic energy, (T6) provides a consistent physical interpretation of the internal energy U, which remains undefined within thermodynamics, viz., $U = N \langle E_{\text{kin}} \rangle$.
- The system-specific model assumptions (S1)–(S5) provide possible reasons for deviations from Boyle's law that can be reduced by phenomenological corrections, e.g., for intermolecular interactions, for the particle volume, or for additional degrees of freedom of the particles. Due to such a heuristic potential, the model assumptions themselves appear empirically significant.
- The result of the deduction explains why all real gases behave similarly, so that the gas laws are independent of the physical nature of the various gases.
- Finally, from the interpretation of the gas pressure as an effect of molecular motions, several consequences ensue that are, in the spirit of physical methodology, directly provable by experiment. In particular, the average velocity, which appears initially to be fictitious, can be assigned a physical meaning:
  - From (T5), it follows immediately that $\langle v^2 \rangle = 3P/\rho$. Since the pressure P and the density ρ are measurable, numerical values are obtained for $\langle v^2 \rangle$ that can be checked regarding their plausibility and consistency.
  - If two gases with particles of different masses $m_1 < m_2$ are enclosed in a container with permeable walls, the lighter gas will leak faster because

its particles have larger velocity due to $v_1^2/v_2^2 = m_2/m_1$. This fact is exploited in gas centrifuges.

In short, the deduction of Boyle's law can be rated as successful because:

- facts and concepts are explained that are in need of explanation in thermodynamics,
- it contributes to the enlargement of physical knowledge about nature,
- it is grounded on assumptions that are no longer considered to be in need of explanation.

These successes suggest attempting analogous derivations of further laws of thermodynamics, and possibly reconstructing the entire theory in this way. Such an enterprise, however, is faced with two fundamental problems:

- Thermodynamics contains concepts like temperature and entropy which cannot be assigned a physical meaning within the framework of mechanics.
- The connection between thermodynamics and mechanics rests crucially on statistical concepts which do not belong to either of these theories.

The first problem already comes into play when one attempts to derive the ideal gas law $PV = nRT$ by analogy with Boyle's law. This features the temperature, which is not a term of mechanics and has no mechanical analog, in contrast to the gas pressure or the internal energy of a gas. Therefore, this law is not deducible from mechanics, and this theory does not provide any insights into the temperature dependence of the properties of gases. However, rewriting the gas law in the form $PV = Nk_{\mathrm{B}}T$, the comparison with (T6) yields

$$\langle E_{\mathrm{kin}} \rangle = \frac{3}{2} k_{\mathrm{B}} T \; ,$$

where k_{B} is the Boltzmann constant. By analogy with the act of equating $\mathbf{v}_{\mathrm{kin}}$ and $\mathbf{v}_{\mathrm{dyn}}$ in dynamics, the mechanical interpretation of both the temperature and the gas law does not rest on a deductive argument, but on a correspondence argument. This correspondence is highly significant from an epistemic point of view. Firstly, the connection between the kinetic energy and temperature links the molecular mechanical conceptual domain with the macroscopic phenomenological one of thermodynamics. Secondly, it establishes the qualitative framework for the theoretical order of the forms of appearance of matter described in Sect. 2.6. Thirdly, the existence of a lower bound on the temperature of a physical system follows because the kinetic energy cannot become negative. This limit is attained when all particles are at rest.

The proportionality between $\langle E_{\mathrm{kin}} \rangle$ and the temperature indicates, at least in principle, how the conceptual connection between mechanics and thermodynamics may be established, although it operates by correspondence or analogy rather than by deduction. In a strict sense, analogies do not explain, as illustrated by the "explanation" of electric phenomena by mechanical analogy models. Therefore, other reasons must exist for accepting this structural relation as an explanation, all the more so as $\langle E_{\text{kin}} \rangle$ only seems to be a mechanical term, while it is actually a statistical one. This is the explicit expression of the fact that the behaviour of a single particle is irrelevant for the macroscopic properties, and constitutes the second fundamental challenge: the connection between the mechanical properties of the individual systems and the thermodynamic properties of the total system necessitates statistical concepts. Although their introduction in the deduction of Boyle's law occurred more in passing and appeared relatively unproblematic, this nevertheless represents the central challenge. Beside the historical development, this is demonstrated by the fact that the corresponding questions have not yet been clarified in a formally satisfying manner. The reasons should be indicated, at least briefly.

The first attempts to reduce the thermal properties of gases to the laws of mechanics go back to Bernoulli, but they turned out to be so difficult that they were not pursued further. It was not until more than hundred years later that this program was actually carried out, primarily by Clausius, Maxwell, and Boltzmann, based on the insight that the description of systems consisting of a very large number of subsystems whose dynamical states cannot be completely specified necessitates statistical methods associated with the introduction of a concept of probability. In this context, it remains undetermined at the outset whether statistical methods are necessary due to the large number of subsystems, due to not knowing the initial conditions, or due to other reasons. The theory of physical systems with a very large number of degrees of freedom that was developed on this basis during the second half of the nineteenth century is called statistical mechanics and constitutes, beyond any doubt, a self-contained theory, rather than a mere specialization of the classical mechanics of many-particle systems. Understanding this self-reliance will help to understand why thermodynamics cannot be reduced to mechanics, as will now be argued.

Since statistical mechanics treats physical systems with a large number of degrees of freedom, e.g., a gas with about 10^{23} particles per cm^3, the need for statistical methods is obvious for pragmatic reasons. The question is, however, whether their application is required solely for those pragmatic reasons or as a matter of principle. The commonly recognized self-reliance of statistical mechanics supports the second view, but the first view seems to be put forward through statements in numerous textbooks which "apologize" for the introduction of statistical methods because, due to the large number of particles, not all the initial conditions needed for a strictly deterministic description can be determined. According to this view, statistical mechanics is not a self-contained theory, but a specialization of mechanics for incompletely specified many-particle systems. Indeed, the use of probabilities for systems with incompletely specified initial conditions is definitely compatible with the deterministic laws of mechanics. Since initial conditions are not part of the theory frame and nor do they belong to the model assumptions of its scientific applications, such probabilities do not refer to the laws which then only transform probabilities for one set of initial conditions into another set of probabilities occurring for a certain event. The determinism characterizing mechanics is not concerned, because the laws describing this transformation remain deterministic. Consequently, the statistics involve only the results, but not the laws. Such a statistical mechanics would indeed be a defective mechanics because the description of the systems is considered to be deficient compared with a completely specified and strictly deterministic one. Behind this attitude stands the desire to abstain from statistical methods if only the dynamical state of the total system could be completely known at the microscopic level. Although compatible with mechanics, probabilities are ultimately undesirable and considered to be a weird element: they come into play solely via initial conditions but not through the laws or model assumptions.

The structure of statistical mechanics is basically at variance with such a view. This is already demonstrated by the way the pressure is expressed through microscopic terms in the deduction of Boyle's law. According to the correspondence principles (C2) and (C3) and theorem (T3), the macroscopic thermodynamic pressure which the gas exerts on the container walls and which is defined as the force per area unit is related to the momentum transfer of a particle on the wall. If this momentum transfer could be identified directly with the pressure, nothing new would be added to mechanics. This is not the case, however. The momentum transfer varies with the time, the particle, and the point on the wall. In contrast, the macroscopic pressure as a thermodynamic state variable is constant in thermal equilibrium. The pressure would not change, e.g., if just two particles hitting the wall were interchanged while everything else remained unchanged. Consequently, the thermodynamic observables are not directly correlated with the dynamical properties of the individual particles, so a transformation among the microscopic states does not necessarily imply a change in the macroscopic state of the system. In other words, the same macroscopic state can be realized by different microscopic states. In the deduction of Boyle's law, this incompatibility is removed by the assumption (C1) replacing the particle- and time-dependent velocity $v_i(t)$ by an average velocity which does not depend on particle or time. Consequently, in order to arrive at a constant pressure when starting from the momentum transfer varying with position, time, and particle, some averaging process is mandatory. In doing so, it is not unequivocally determined which of the fluctuating quantities must be averaged over to obtain an experimentally comparable, macroscopic pressure. The momentum transfer may be averaged over many particles, or over the entire wall at a certain point of time, or over a time interval of the order of magnitude of a pressure measurement. Each of these possibilities is conceivable and cannot be excluded a priori.

In statistical mechanics these qualitative considerations are formally expressed via the definition of canonical ensembles specifying how a macroscopic state is realized by a distribution of microscopic states. The functions describing such distributions are thus called distribution functions. The thermodynamic quantities A_{td} are obtained as expectation values of the mechanical quantities A_{m}, weighted with a distribution function σ:

$$\text{(M1)} \qquad A_{\mathrm{td}}(\tau) = \int \sigma(p,q,\tau) A_{\mathrm{m}}(p,q) \mathrm{d}^N p \, \mathrm{d}^N q$$

Here, p and q are the mechanical, dynamical variables momentum and position, respectively, τ is a parameter, e.g., the temperature, and the integral extends over a volume of the phase space. Exempt from this definition are all quantities, of course, which do not possess a mechanical analog, like the entropy, for which a relation of its own is postulated:

$$\text{(M2)} \qquad S(\tau) = -k_{\mathrm{B}} \int \sigma(p,q,\tau) \ell n\{\sigma(p,q,\tau)\} \mathrm{d}^N p \, \mathrm{d}^N q$$

Taking σ as the distribution function of the microcanonical ensemble yields the Boltzmann relation$S = k_{\mathrm{B}} \ln W$with the statistical weight W. This emphasizes the special status of the entropy, because W is not a mechanical quantity.

The connection between the two conceptual systems is thus established by interpreting the dynamical variables of the subsystems as random variables whose distribution is a function of some parameter, e.g., the temperature. The average values of the microscopic observables depend only on these parameters and are identified with the macroscopic properties of the total system. Since these are directly measurable, one can test whether the calculated average values are in accord with the actually measured data. The theoretical rationale of this correspondence rests on the law of large numbers, so that thermodynamics appears formally as the asymptotic limit of statistical mechanics when the particle number N, energy E, and volume V all become infinite, while the ratios N/V and E/V remain finite. Solely for this asymptotic limit do the laws of statistical mechanics turn into statements that are no longer of a statistical nature, so that they may be identified with the corresponding deterministic laws of thermodynamics. Indeed, thermodynamics is already applicable in practice for finite N, as long as the relative fluctuations $\delta A/A$ of the relevant observables are negligible. Such fluctuations are a function of $1/N$, so they become unimportant for sufficiently large particle numbers, although they vanish only when N becomes infinite. In this context, it does not matter whether the fluctuations are observable or not: thermodynamics is a theory where these do not occur, regardless of whether they exist in reality.

Statistical mechanics, on the other hand, is applicable to systems with an arbitrary particle number and describes such fluctuations. When these are observable, e.g., in Brownian motion or in the density fluctuations of a gas, the theory exhibits direct connections with reality while otherwise these are mediated through thermodynamics in combination with the correspondence principles (M1) and (M2). Conceptually, fluctuations occur neither in thermodynamics nor in mechanics, but necessitate a theoretical description as observable phenomena. As a consequence, the introduction of statistical concepts is mandatory for empirical reasons, and is associated merely parenthetically with large particle numbers and not at all with incompletely specified initial conditions. There is another reason why statistical mechanics is inadequately characterized as a theory of incompletely specified systems. The statistical-mechanical description of the properties of macroscopic systems presumes that the time scale of the macroscopic time evolution is large compared with the time scale of the motion of the subsystems. Only under this condition is the averaging meaningful, even if not carried out explicitly over the time. Regarding the statistics of incompletely specified mechanical systems, the time scales are irrelevant and need not fulfill such a condition. Actually, the interpretation of statistical mechanics as a theory of incompletely specified systems rests on the reduction of the function of theories to simply providing predictions. With respect to such a misconception, it is irrelevant whether the statistics comes into play via the initial conditions or the laws. In contrast, both the structure of statistical mechanics and its status as a physical theory are primarily determined by the objective of establishing the connection between the macroscopic and the microscopic conceptual systems. This is accomplished by relating the thermodynamical properties of a many-particle system, which are constant or slowly variable on the microscopic time scale, to the mechanical properties of the subsystems. Therefore, the goals pursued by the theory must be taken into account to assess its status within physics, while the property of yielding predictions is insignificant in this respect.

The self-reliance of statistical mechanics becomes even more apparent when going from systems in thermal equilibrium to irreversible processes. Irreversibility is an everyday experience in which the notion of time goes from past, to present, to future. It appears so self-evident that it hardly seems to need an explanation why, e.g., birth, life, and death always happen in this order, never the opposite. Analogously, irreversibility in the framework of thermodynamics is stated as an empirical fact, but is not explained. The situation becomes fundamentally different when thermodynamics is related to mechanics. Now the question arises as to how the irreversible thermodynamic processes can be derived from laws of mechanics that are invariant under time reversal. In this case, the particle number is indeed of crucial importance, as illustrated by a simple computer experiment. To this end, imagine a box divided in two parts by a wall. The left half may contain a certain number of particles while the right half is empty. After removing the dividing wall, the particles will enter the empty half due to their thermal motion, and one may ask how high the probability will be that the initial state "left half full, right half empty" is eventually recovered, i.e., that the whole process is reversible. Corresponding model computations show that for a small number of particles this probability is sufficiently high to consider the entire process as reversible. However, even with just forty particles, it is already so small that the initial state will only be recovered after a time exceeding the estimated age of the universe. Although the time evolution of each particle is reversible, the time evolution of the entire system becomes irreversible, for all practical purposes, if the number of particles is sufficiently large. Accordingly, in relation to mechanics, irreversibility is an emergent concept coming into play only with increasing particle number, while strict irreversibility is revealed formally as an abstraction of thermodynamics.

In relation to classical mechanics, the self-reliance of statistical mechanics rests on the fact that statistical concepts with a probability to be defined axiomatically, the distribution function, and the correspondence principles (M1) and (M2) play the key role in establishing the theory frame. (M2) does not contain mechanical terms, (M1) establishes a horizontal correspondence relation with thermodynamic terms which do not occur in any specialization of mechanics, and the statistical concepts are the constitutive part of the theory frame. Accordingly, the laws are statistical propositions which differ structurally from the deterministic laws of mechanics. All this precludes its integration into classical mechanics as a specialization of the mechanics of many-particle systems. Likewise, the reasons for the self-reliance of thermodynamics in relation to statistical mechanics are the structure of the correspondence principles and the essentially dissimilar structure of the laws. The thermodynamic laws are deterministic and strictly valid, while the statistical-mechanical ones are valid only in the mean. The statement that the thermodynamic laws are valid only in the mean, rests on the erroneous identification with their statistical analogues. Although thermodynamics may symbolically be interpreted as the asymptotic limit of statistical mechanics for $N \to \infty$, it is not one of its specializations. Finally, the two theories differ in the characterization of their model systems. Thermodynamics does not contain any assumptions concerning their internal structure and is in this respect less specific, but more universal than statistical mechanics. The model systems of statistical mechanics consist of such a large number of subunits, that the application of statistical methods is justified. On the other hand, the theory does not contain any assumptions concerning the internal structure of the subunits, so it is not a "microscopic" theory in the sense that the subsystems ought to be atoms or molecules.

All in all, thermodynamics cannot be reduced to statistical mechanics, nor a fortiori to classical mechanics, and statistical mechanics cannot be reconstructed as a specialization of mechanics which provides a deterministic and reversible description of the subsystems. Thermodynamics provides a deterministic and, with regard to the second law, irreversible description of the total system, in general. Statistical mechanics establishes the connection between the two conceptual levels and is statistical concerning both the subsystems and the total system. The properties of the subsystems are considered to be random variables over which one averages. The macroscopic observables fluctuate around the mean values, so the statistical analogues of the thermodynamic laws are valid only in the mean. For these reasons, none of these theories is more fundamental than any of the others.

In summary, the relationship between thermodynamics and statistical mechanics exhibits a similar structure as the one between kinematics and dynamics. On the one hand, reducibility in the spirit of a strictly deductive relation does not exist. Although the respective intertheoretic relations contain vertically deductive structural elements, the horizontal ones are decisive for understanding this relationship. In dynamics this came from equating the kinematic and dynamic velocities axiomatically, in statistical mechanics it came from the correspondence between thermodynamical and mechanical observables that was established by averaging with a distribution function. On the other hand, in terms of both the formal structure and the relation to the empirical level, there are no significant dissimilarities between descriptive and explanatory theories. This seems to confirm the view that physical theories provide mere descriptions of reality, but no explanations, and it raises the question according to which feature the two examples are considered as representative of physical explanations. Formal and cognitive reasons cannot be decisive:

- Returning to Newton's goal, "to explore the forces of nature from the phenomena of motion and to explain afterwards further phenomena by these forces", the deduction of trajectories from forces in Newtonian mechanics is frequently considered as the paradigm of a causal explanation, and seems to constitute a program for searching for causal explanations. Such a view is untenable for several reasons. Firstly, the formal structure of mechanics is inconsistent with a causal interpretation, as shown in Sect. 2.6. Secondly, it is not comprehensible that the deduction of trajectories from a force function ought to be an explanation, but not the formally identical one from an acceleration function. Thirdly, in the Hamiltonian schema, the trajectories are deduced from a variational principle that may be interpreted as if a system selects that trajectory along which the action becomes a minimum. Such a picture implies a final pattern of explanation and demonstrates that a general dynamics is in no way related to causality. As a consequence, the externally imposed causal interpretation of Newtonian mechanics does not belong to this physical theory, and searching for causal explanations is not an objective of physics.
- Formal arguments cannot justify the idea that the deduction of Boyle's law is an explanation. Due to the large number of empirically non-testable assumptions, nothing is explained, since the reduction of an explanandum to unexplained premises does not explain anything. When the deduction of this law is nevertheless accepted as an explanation, the reason must be that the premises are not considered to be unexplained assumptions, but rather the specification, not in need of explanation, of a "reality behind the phenomena", in accordance with an atomistic world view.

In formal and cognitive respects, both patterns of explanation only incompletely fulfill their function as the premises of a strictly deductive argument, or indeed not at all. Consequently, their explanatory merits and eventually the distinction between descriptive and explanatory theories rests on entirely different reasons:

1. Both the deduction of trajectories from forces and the deduction of Boyle's law rest on two fundamental patterns of explanation of occidental natural research, with roots going back to ancient Greek natural philosophy:
   - the explanation of observable changes, especially the perceivable motions of objects, through the action of forces as causes (causal-mechanistic world view),
   - the explanation of observable phenomena, especially the perceivable macroscopic properties of matter, through reduction to the motion of tiny particles or to model assumptions about the microscopic structure of matter (atomistic world view).

   These two patterns of explanation are the most widely accepted because they appear to be the most compatible with the physical conception of nature, but also because physics is frequently considered even to be their empirical foundation.
2. The reduction of observable phenomena to not directly perceivable entities or structures is felt throughout to be an explanation and not a mere description, even when nothing is explained regarding the formal structure. It is not clear, for example, whether the kinetic gas theory would be considered as an explanation if the gas particles were directly perceivable like billard balls. In the case of dynamics, this might be a reason not to regard the more directly accessible accelerations as an explanation, in contrast to the forces. However, the notion of an "underlying reality", hidden behind the phenomena, that is established by these patterns of explanation does not conform to the structure and status of theoretical knowledge. At least, the non-direct perceptibility and thus speculative feature of the two patterns of explanation makes it comprehensible that the empiricist program for a setup of physics that would be "hypothesis-free" was primarily directed against the force concept and the atomic hypothesis. A program which is consistently empiricist in this respect and constrained to bare descriptions eventually corresponds to the reduction to geometric and kinematic descriptions which have turned out to be less rich in empirical content, besides having other weak points.
3. An empirically unjustifiable, and in this sense metaphysical, aspect of the explanatory patterns is some kind of claim to validity associated with the tacit assumption (1B). When reality is structured, these structures must be properly represented in the theoretical description. Precisely this function is fulfilled by the patterns of explanation of explanatory theories. Although themselves not directly perceptible, they are deemed to be an essential part of our knowledge about nature, rather than mere thought constructs without epistemic relevance. Dynamics and the kinetic gas theory provide interpretations of motion and heat, respectively, that are not exhausted in the deduction of experimentally comparable data, but are as-

sociated with the ontological claim that atoms, molecules, electromagnetic fields, and interactions are likewise a part of reality and exist as undeniably as directly perceptible objects. In contrast, as descriptive theories, kinematics and thermodynamics leave undetermined the physical nature of motion and heat, and permit several interpretations. Accordingly, descriptive theories are ontologically undetermined or neutral, as physics in its entirety would be if understood as a purely descriptive science. As a result, a qualitative difference persists between the two types of theory: the explanatory theories constitute the physical reality and, eventually, the physical world view due to these ontological claims of validity. On the one hand, this implies assumptions about the physical nature of the systems and interactions under study that lead to more thorough insights, but simultaneously to a finalization excluding alternative interpretations that are conceivable in the framework of descriptive theories. On the other hand, if this ontological commitment is absent from the patterns of explanation, or if it is denied as a matter of principle, theoretical knowledge degenerates to a mere auxiliary means for the deduction of factual knowledge. In this case, differences between descriptive and explanatory theories no longer persist because the patterns of explanation, rated as ontologically and epistemically irrelevant premises, do not explain anything. Opponents of the atomic hypothesis, for example, considered the assumptions of the kinetic gas theory as a mere auxiliary means for deducing empirical facts. Note also that Osiander's preface to Copernicus' *De revolutionibus orbium coelestium* exemplifies the retreat to the instrumentalist interpretation of a pattern of explanation that has not yet been accepted as related to a structure of reality.

4. The existence of ontological claims of validity implies another difference between the two types of theory, although these claims become effective only on the level of the scientific applications. The specification of the general structure principle of an explanatory theory as a mechanism within an ScA is not a mere realization, as understood in model theory, but is associated with the hypothetical claim that it corresponds to a part of reality. On the basis of a theory conception that is reduced to the formal structure, this claim is eliminated. An important goal of the elaboration of the ScA of an explanatory theory is rather the assessment of this ontological commitment. In contrast, its transfer from the ScA to the abstract structure principle of the theory frame is naive realism. For example, the concept of force as the structuring principle of mechanics has realizations like the inertial forces that do not correspond to a part of reality.
5. Another property of explanatory theories is some type of research progress with attributes which descriptive theories do not possess:

   - Only the detachment from direct experience, as achieved by the structure principles of explanatory theories, provides the prerequisite for a comprehensive unification of knowledge that leads to new insights and interrelations. Kepler's three laws, for example, are independent of each other in Kepler's theory, but can be obtained from a single law within the framework of Newton's theory of gravitation.

- The explanatory theory is richer in empirical content because it enables the derivation of additional, empirically testable information, like the calculation of the position of the planet Neptune, or the calculation and physical interpretation of empirical parameters occurring in descriptive theories.
- The explanatory theory answers questions that are in need of explanation at the level of the corresponding descriptive theory, but it does not explain everything. In particular, it does not explain its own patterns of explanation and why phenomena and structures of reality are amenable to a law-like description. There is no more proof that the gravitational force is inversely proportional to the square of the distance than there is that, in statistical mechanics, the correspondence between mechanical and thermodynamical quantities should take the form (M1).
- In the simplified methodological conception with the three steps "observing $\rightarrow$ describing $\rightarrow$ explaining", the explanatory theories represent the last link in this development and are actually considered as progress in this respect. In addition, the explanatory theories constitute a theoretically autonomous form of knowledge because they are grounded on assumptions that are not justifiable by purely empirical means.

In short, a theory is considered to be explanatory because its structuring principles are consistent with commonly accepted patterns of explanation. A theory is explanatory with regard to the conceptual level of a descriptive theory, so that explanation in this respect constitutes a relation between theories, rather than a relation with the empirical level or reality. This relation does not exhibit a strictly deductive structure, but explanatory merit and progress in knowledge rest decisively on correspondence principles as horizontal structural relations. Crucial for justifying the differentiation into explanatory and descriptive theories are the ontological claims of validity of explanatory theories which establish the physical world view, while descriptive theories are ontologically neutral. Without this ontological commitment, significant differences would not exist between the two types of theories, and an explanatory theory would not represent progress in knowledge, because its patterns of explanation are considered to be irrelevant for knowledge about nature.

3.4 Axiomatization

Axiomatic methods are a genuine part of the methodology of mathematics and logic, so attempts to axiomatize physical theories are largely oriented towards it. For this reason, it is helpful to analyze both the achievements and the goals of axiomatization in mathematics. The first theory that anyone attempted to axiomatize was geometry by Euclid. According to him, an axiomatic system is a class of propositions that can be deduced from a finite subclass, viz., the axioms. Not only is the entire theoretical knowledge and potential of the axiomatic system implicitly contained in the axioms, but in addition, questions concerning the epistemic and ontological status of theories can be reduced to the corresponding questions about the axioms. The various meanings of axiomatization differ essentially by the status of the axioms (Stegmüller 1973). According to Euclid, it is determined by the postulate of evidence: the basic terms of the theory possess a meaning that is comprehensible independently of the theory, and the axioms, as relations between these terms, are immediately evident and do not need any additional justification. A similar conception forms the basis of Newton's representation of mechanics, where the evidence of the axioms is justified by their direct relation to empirical experience.

It was only around the middle of the nineteenth century, after the construction of non-Euclidean geometries, that the status of the axioms was critically reviewed and eventually revised by Hilbert (1899), who aimed to free geometry from all features that had erroneously been believed to be a priori geometrical notions, but actually went back to their empirical origin. He accomplished this objective via the strict separation of the formal structure from the semantic content. The axioms are instead assumptions about relations between basic terms which are themselves only defined by the condition that they satisfy the axioms. As a consequence, they acquire meaning exclusively through formal, theory-internal relations, without possessing any additional meaning. In contrast to the concrete Euclidean axiomatization, the one by Hilbert may be called an abstract axiomatization. As a criticism of this type of axiomatization, it has been objected that the definition of the basic terms is circular. However, requiring their explicit definition in a way that is independent of the axioms means nothing else than assigning a meaning to them that necessarily comes from notions external to the theory. In this way, not only the basic terms, but eventually the entire theory gets an externally imposed meaning which the abstract axiomatization was trying to eliminate. Accordingly, the call for an explicit definition of the basic terms is at variance with the basic idea of the abstract axiomatization to restrict oneself to the reconstruction of formal structural relations, and to deliberately leave any meaning beyond that undetermined. Talk of undefined basic terms is thus inappropriate. If the reconstruction is constrained to the study of structural relations, entirely abstracted from any specific meaning, only these relations can be used for the definition. This is precisely what abstract axiomatization is all about.

A subsequent development beyond abstract axiomatization is formal axiomatization (Stegmüller 1973). While the axioms in both the concrete and the abstract axiomatization are formulated colloquially, they are transformed in the first step into a formal language. Axiomatization must then be preceded by the construction of a formal language providing a system S defined exclusively by syntactic rules. A model of S is an interpretation transforming the syntactic system S into a semantic one. Since physical theories do not use formal languages, this type of axiomatization is irrelevant for physics and will not be considered. Instead, the merits and objectives of an abstract axiomatization will be analyzed. In order to serve as a formally irreproachable basis of an axiomatized theory, the system of axioms should fulfill three conditions:

- Consistency: logically contradictory statements must not be deducible from the axioms.
- Independence: none of the axioms is deducible from the others.
- Completeness: all theorems of the theory must be deducible from the axioms.

The frustrating result of fundamental mathematical research regarding the provability of these properties may be summarized as follows: no deductive-axiomatic system can be proven to be non-contradictory, independent, and complete by means of the system itself, i.e., a formally irreproachable self-justification of an axiomatic system is impossible.

Since an abstract axiomatization is unable to provide the proof of these properties, this cannot be the goal. They represent instead a kind of normative ideal to which an axiomatic system should come as close as possible. Accordingly, the axiomatization of geometry by Hilbert aimed primarily to make explicit all of Euclid's implicit assumptions which go back to the empirical roots of geometry, in order to axiomatize or eliminate them. In this way, he was able to demonstrate that a geometry can be established without any empirical knowledge of geometrical relations, and show how it can be separated from concrete notions due to its empirical origins. The concrete content is thus irrelevant for the geometry as a mathematical theory which is restricted to abstract structural relations.

Due to these achievements of an abstract axiomatization for fundamental research in mathematics, attempts to axiomatize physical theories are guided by the expectation of similar successes through an analogous axiomatization. Accordingly, these attempts implicitly rest on the assumption that there is an essentially identical structure in mathematical and physical theories. This assumption is untenable for a number of reasons.

1. The central feature of an abstract axiomatization is, beyond any doubt, the separation of form and content, which is not end in itself, but has a well-defined function. Regarding geometry, its purpose was the proof that a priori geometric axioms with the status of necessities of thought do not exist, but that the geometric views rather rest on a conglomerate of empirical experience and common-sense notions. Hilbert managed to eliminate these non-mathematical elements from geometry through the strict separation of form and content by elaborating these empirical origins, by transforming into axioms those parts that were necessary for the theory,

and by eliminating the remainder. In this case, the separation of form and content had the function of removing those parts from the theory that were not specific to mathematics and could be designated in this respect as "undesirable metaphysics". For this reason, an abstract axiomatization in mathematics was able to contribute substantially to clarifying fundamental problems. In contrast to mathematics, physics claims to describe realms of reality by its theories. A separation of form and content which aims to confine itself to formal structural relations by eliminating physical meaning and connections with experience would remove, not undesirable metaphysics, but the empirical content of the theory.

2. The epistemic status of both the basic terms and the axioms of physical theories is basically dissimilar from that of mathematical axioms. Firstly, physical axioms are associated with both epistemic content and a claim to validity, irrespective of what it may look like. Secondly, the basic terms occurring in the physical axioms are not defined exclusively by formal relations within the theory. The theory frame contains, via the assignment rules in (A1), theory-external connections to experience which must be made explicit in an axiomatization because they codetermine the epistemic content of the theory. Consequently, the challenges associated with the constituents and claims of validity which enter the theory through external empirical experiences must not be ignored by elimination. Interpretations in which physical axioms degenerate into defining equations of the physical quantities occurring therein can serve as a warning example.
3. A mathematical theory constitutes a system of formal structural relations and the objective of mathematical research consists in the construction and study of such formal structures. In contrast, the crucial feature of physical theories is the relationship with experience, even from an instrumentalist point of view. As long as a theory does not possess empirically relevant applications, it cannot be presumed to be physical. The associated structural features must be recovered in an axiomatization. Accordingly, an axiomatic reconstruction cannot be restricted to formal considerations, but must show in particular how the theory is organized with respect to its relationship with experience. This is unattainable if one abstracts from the meaning associated with the content.
4. The final difference concerns the status, significance, and function of the scientific applications ScA postulated in (A4). In mathematical theories, the realizations through the interpretation of the abstract symbols serve solely as illustrations, but are not a part of the theory. Consequently, they do not belong to an axiomatic representation. In contrast, the ScA of physical theories are a constitutive part because they provide the contribution of the theory to the empirical content. Additionally, the transition from a specialization of the theory frame to an ScA occurs by the addition of system-specific model assumptions which themselves exhibit the status of axioms. The empirical information contained in these assumptions cannot be a part of the theory frame, which is entirely independent of

system-specific and contingent conditions, but rather concerns those areas in which the theory frame is underdetermined. Hence, the entire knowledge of the theory is not already contained in the axioms of the theory frame. It is precisely this that is in distinct contrast to the basic idea of the abstract axiomatization of mathematical theories, where all theorems of the theory should be deducible from the axioms of the formal system. Accordingly, the axiomatization of physical theories cannot be restricted to the theory frame, but must comprise the scientific applications, as well as their mutual relations. Although their selection is not definitive and may vary, a comparison of different textbooks shows that there is a broad consensus over which of them are considered as representative. Their diversity and their mutual support via horizontal interconnections contribute significantly to the secureness of physical knowledge by securing the consistent connectedness of theoretical knowledge. In mathematics, such relations do not play any role since there is no problem of secureness for mathematical knowledge, beyond the logical correctness of theory-internal deductions.

For these reasons, a physical theory cannot be assumed to be structurally identical to mathematical theories. Although with the theory frame it contains a network of formal structural relations, there are additional non-formal constituents of equal importance. As the distinct feature of physical theories, these constituents are external relative to the theory frame and must be included in an axiomatic reconstruction. The elimination of these external parts would lead to such a distorted representation of the structure and status of physical knowledge that an axiomatization by analogy with mathematics which is restricted to the formal parts of the theory frame would not contribute to an improved understanding of the structure of physical theories and knowledge. The fundamental challenges are entirely different, viz., regarding content and conceptual problems that cannot even be formulated, let alone solved, by an analysis oriented toward an abstract axiomatization of mathematical theories. The belief that one can elaborate the epistemological structure of physical theories by an abstract axiomatization rests rather on the conviction that all problems regarding content could be reduced to formal ones. Although abstract axiomatization is appropriate for mathematical theories due to their restriction to formal structures, and can contribute to the solution of fundamental problems, an axiomatization of physical theories which is grounded on the total separation of formal structure and physical content will be counterproductive. Seminal results of an axiomatization of physical theories can be expected, at best, if it does not just imitate the axiomatization of mathematical theories.

Actually, there have been some attempts to axiomatize physical theories which account for their empirical character. The most prominent example is, of course, Newton's representation of mechanics in combination with the assertion that the axioms have been derived inductively from experience. Therefore, the axioms claim to have the status of empirical truths, which definitely proves their claim to validity. Other empirically oriented axiomatizations are classical mechanics according to Hertz (1965) and thermodynamics according to Caratheodory (1909). These attempts correspond to the positivistic program that the basis of a theory should contain exclusively directly observable or measurable quantities. Although the thermodynamics of Caratheodory may be considered as the most successful product of this program, it exhibits, beside some shortcomings regarding content (Born 1921), two main deficiencies:

1. The operational definition of measurable quantities is too vague to prove theorems of the theory on this basis, so the associated terms are actually used in their precise theoretical sense. The resulting commingling of operational and theoretical definitions constitutes one of the roots of the naive realistic conception of reality, as is characteristic of empiricism, although internally consistent. Since the existence of an autonomous theoretical conceptual level is denied, there is no compelling reason to distinguish either between operational and theoretical definitions or between an empirical and a theoretical conceptual level.
2. The representation is grounded on the empiricist dogma that perceptions and direct observations constitute a safe basis for a theory, while any theoretical knowledge is suspect and must be either eliminated or reduced to direct experience. A theory reconstructed in this way gives the impression of constituting absolutely safe knowledge, and does indeed pretend to do so, on the grounds that it rests solely on direct experience. In addition, the program of abstract or formal axiomatization of empirical-scientific theories in combination with the simultaneous elimination of the theoretical terms, as is also pursued in the analytic philosophy of science, reveals a remarkable inconsistency. On the one hand, this program of elimination is based on an empiricist position that is molded by a belief in the reliability of direct experience and distrust of any type of theoretical knowledge. On the other hand, it presents an abstract axiomatization with the function of controlling the unreliability, narrowness, and limitation of direct perception and experience. Taking the empiricist position seriously, any type of abstract axiomatization must be declined in favour of a concrete one, because it is this one that is tied to direct experience, something believed to be safe and reliable.

Similarly, axiomatizations reconstructing the relationship between the theory frame and its applications by analogy with the formal model theory do not comply with the structure of physical knowledge. Firstly, the theory frame itself is already a semantic system and cannot be treated as a formal calculus containing only syntactic relations. Secondly, the introduction of supplementing axiom-like assumptions, which constitute the scientific applications and enable the deduction of experimentally comparable results, is at variance with the central idea of the formal model theory, viz., that after axiomatization of the formal calculus the theorems of the realizations or models of the calculus are also deducible without additional axioms. The model-theoretical axiomatization of classical mechanics by a set-theoretical predicate (Sneed 1971) is faced with precisely these objections. According to this view, the statement that c is a classical mechanical system may be expressed symbolically through $c \in CM$, which means that c represents an interpretation of the formal framework of classical mechanics CM. This is directly analogous to the situation in mathematics, where, e.g., $g \in GT$ means that g is a group iff g fulfills the axioms of group theory GT. Again, such a representation is meaningful only if the whole body of theorems is deducible solely from the axioms, i.e., without supplementing axiom-like assumptions, which is not the case for physical theories. Accordingly, the axiomatization of a physical theory through a set-theoretical predicate again rests on the assumption of the structural identity of mathematical and physical theories, and this becomes manifest in particular in the circumstance that the relationship with experience and experiments remains unclear.

In summary, a reasonable axiomatization of physical theories can be accomplished neither on the basis of a conceptual system of operational definitions and the empiricist reduction of knowledge to direct experience, nor by analogy with an axiomatization of mathematical theories. Accordingly, the corresponding reconstructions of physical theories did not win recognition because they did not lead to substantial new insights into the structure and status of physical theories and knowledge due to their serious deficiencies. More generally, one must ask how far, if at all, axiomatization in physics can contribute to the three basic objectives of gaining, securing, and structuring knowledge:

- Axiomatization does not contribute to any gain in physical knowledge. While Hilbert's axiomatization has established geometry as a genuine mathematical theory in the modern sense, axiomatization in physics does not constitute a theory conception. On the one hand, the insight into the essential constituents and attributes of a physical theory, and on the other hand, what is not attributed to it, is a necessary precondition for, but not the result of axiomatization. Moreover, it does not resolve any challenges regarding content and in particular foundational problems, it does not provide a conceptual analysis of physical concepts, and it does not contribute to theory development. On the contrary, the latter must be advanced far enough for a sufficent number of applications to be elaborated to provide a network of theory-internal structures. A premature axiomatization that does not meet this requirement will act rather to inhibit theory development. Therefore, axiomatization may be at best the reconstruction of an intact, elaborated theory. It constitutes primarily a meta-theoretic analysis that does not impart any retroaction on the contents of the theory, or answer questions that are crucial to understanding physical knowledge, e.g., concerning the status of Newton's laws.
- Axiomatization does not directly contribute to securing physical knowledge. The correctness of a theory or proposition is not increased by axiomatization. Knowledge is secured rather through the relationship between

theory and experience, the application of proof-theoretical methods of the underlying mathematical theory, and the mutual support due to the horizontal interconnections betweeen scientific applications, although these may be made more explicit by an appropriate axiomatization. Indirectly, axiomatization may contribute to the securing of knowledge because the identification and localization of "undesirable metaphysics", i.e., of those parts alien to physics, may be facilitated, so that they can be eliminated.

The question remains regarding the contribution to the structuring of knowledge. Provided that structures are made explicit that are not so pronounced in a non-axiomatized representation, this would be the case. In this context, the requirement of a representation of the theory that is appropriate to the status and the specific features of physical knowledge has priority. This means that, firstly, the axiom-like reconstruction should occur in the language of physics. Secondly, it must not lead to a deterioration of the theory by reducing or eliminating its empirical content and by distorting or eliminating established theory-internal structures. Only under these conditions can an intersubjective acceptance of the axiomatized theory be expected that does not fall back behind the non-axiomatized one, and does not degenerate into a linguistic exercise or formal baublery. Taking into consideration the structural differences between mathematical and physical theories, the theory conception developed in the preceding sections meets these criteria. Firstly, it is grounded on the fact that theories provide statements about reality but are not a part of it. Consequently, the relationship with experience and a fortiori the empirical basis does not belong to the theory, so it is not subject to axiomatization. This conforms with the common representations of physical theories which also do not contain these. The separation between formal structure and semantic content, as is characteristic of an abstract axiomatization, corresponds here to the differentiation into the three conceptual levels of the mathematical theory MT, the theory frame, and the scientific appplications. In MT, which is externally given relative to the physical theory, the separation between form and content can be presumed, so that by definition MT does not contain any constituents relevant for physical knowledge of nature. On the level of the scientific applications, the separation between form and semantic content is impossible for obvious reasons: this would reduce or even eliminate the contribution of the theory to the empirical content. It remains to discuss whether such a separation is attainable and meaningful within the theory frame, where (A4) as a pure existence postulate is obviously not concerned.

In (A1) a physical meaning is ascribed to some variables of MT, so that MTP appears as an interpretation of MT in the sense of formal model theory. A reduction on this formal aspect would degrade the assignment rules to mere nominal definitions. This is inadmissible because this assignment defines physical quantities that are empirically significant in the following respect. Although these quantities do not themselves necessarily possess a direct operational meaning, since they need not be scalar quantities and their definition is frequently oriented toward the conceptual system of MT, they are operationally realizable, at least in thought, by deriving from them, in combination with (A3), quantities permitting an operational definition. In contrast to empiricist theoretical models, this condition of operational realizability does not have the function of substituting theoretical definitions, all the more so in that the associated operational definitions are not themselves part of the theory, but instead guarantee the empirical significance of the theoretically defined terms. This aims to prevent the intrusion of undesirable metaphysics, so that operationally non-realizable, and thus empirically meaningless concepts like absolute space or time do not appear in the conceptual basis of the theory. Since for this reason the physical meaning of the basic terms cannot be eliminated, the alternative remains to shift them from the theory frame into the scientific applications, where the separation between form and content is not feasible anyway. Against this, it must be objected that the physical meaning of the basic terms, as ascribed via (A1), is common to all scientific applications. An important step in the process of abstraction during theory development is precisely the elaboration of the common elements of dissimilar special theories in order to integrate them into a unified conceptual system. For this reason, shifting (A1) and (A3) into the scientific applications runs contrary to the goal of unification of knowledge that has the highest priority in the process of structuring physical knowledge. Overall, both the elimination and the shift of the physical meaning introduced through (A1) and (A3) lead to a deterioration of the theory, because either the control options for the intrusion of undesirable metaphysics or the degree of unification is reduced.

The status of (A2) differs from the status of mathematical axioms because the corresponding propositions not only represent the epistemological content, but also advance a claim of validity, although the opinions about its kind are widely dissimilar. According to naive realism, (A2) is a part of nature and is discovered by direct experience. Consequently, this axiom should be accessible to a direct empirical comparison, and also the absolutely necessary, sole possible and true type of description of nature. In contrast, the instrumentalist position reduces the validity claim of (A2) to the function of a merely auxiliary means for deducing empirically comparable statements without ascribing to this property any ontological or epistemic relevance, whence (A2) itself need not be related to a structure of reality. This view is ontologically and epistemically neutral and without any doubt possesses certain advantages, because it admits the option of alternative interpretations and does not give any impression of the necessity of physical knowledge. At least, this position does not adequately account for the epistemic importance of (A2), because rejection of the structural principle established by (A2), which corresponds to the transition to another theory, will usually lead to a change in the understanding of reality. In this respect, (A2) decisively determines the view about reality. This conforms to the conviction that the successful empirically relevant realizations of (A2) in the various scientific applications demonstrate that a structural principle abstracted from them is not an epistemically irrelevant construct or pure convention. For instance, the interpretation of changes in motion as an effect of external forces, instead of being an inherent property of the system, has the status of knowledge about nature, instead of being a purely auxiliary means for deducing experimentally comparable data. In contrast to naive realism, such a position advances an epistemic rather than ontological claim of validity regarding (A2), and accounts explicitly for its importance as the central constituent of physical knowledge about reality. Accordingly, the structural principle established by (A2) is regarded neither as an epistemically irrelevant, theoretical construct nor as a part of nature nor as the true, absolutely necessary type of description of nature. Finally, the physical meaning and epistemic significance of the theory frame is exclusively determined through (A1)–(A3). In contrast, both the premises entering via MT and the constraints leading to the specializations are epistemically irrelevant. Ignoring this fact constitutes one of the roots of undesirable metaphysics.

For these reasons, the separation between form and content is impossible within the theory frame, even in the case of an instrumentalist theory conception eliminating only the epistemic claims of validity. This separation being constitutive for the axiomatization of mathematical theories is here replaced with the differentiation into the abstract theory frame and the concrete, system-specific scientific applications. The abstract character of the theory frame does not rest on abstracting from both the physical meaning of the basic terms and the epistemic content and validity claim of (A2), but on the independence of system-specific and contingent assumptions. In addition, the axiomatization of physical theories is not restricted to the theory frame, which does not contribute much to the secureness of physical knowledge due to its indirect relationship with experience. As a crucial part, it comprises the reconstruction of the relations between the theory frame and the scientific applications, in combination with the horizontal structural connections between them. The reconstruction of the network between the various applications cannot be mastered by restricting to formal considerations because this requires detailed expert knowledge of the physical content of the theory.

All things considered, the main focus of axiomatization in physical theories is basically different from what it is in mathematical theories. Beside the contribution to structuring the secured knowledge, the most important objective is the localization and elimination of undesirable metaphysics in the sense of notions that are either false or alien to physics. These may enter the theory frame at points where it possesses external connections:

- analogies with other physical theories,
- connections with the mathematical theory MT,
- the empirical connection of the basic terms in (A1),
- the relationship with scientific applications.

It is precisely at these points that the historically detectable roots of parts alien to physics are localized. From the analysis of representative examples, one can derive strategies for largely avoiding the permeation of undesirable metaphysics, although it cannot be entirely excluded. The first item indicates that iconic or formal analogies between dissimilar theories are epistemologically irrelevant. The fact that acoustic and electromagnetic waves are described by mathematical equations of identical structure or that electric currents may be illustrated by hydrodynamic analogical models does not permit the conclusion that their physical natures are identical.

Through the connection with MT, erroneous interpretations may enter the theory frame in the following ways. Either a physical meaning is assigned to physically non-interpretable parts of MT or certain premises, which are inherent in MT and enter, of course, the physical theory, are reinterpreted as results of physics. If this is not recognized, fictitious properties or structures of reality are constructed and erroneously assumed to be real. The representative example is determinism as a characteristic feature of the mathematics of classical mechanics. This determinism is not a property of reality, but can be traced back to the fact that the initially investigated types of motion belong to solutions of linear differential equations which, due to their linearity, are stable against small variations of the initial conditions. In classical mechanics such types of motion already exhibit non-deterministic behaviour ("chaos") which are derived from nonlinear differential equations where the nonlinearity is not a small perturbation. Similarly, the continuity of motions dealt with in mechanics is actually a requirement on the solutions of the underlying differential equations, but not a general property of reality deducible from physics. The strategy for avoiding such errors is, firstly, to make explicit the mathematical premises entering the physical theory, and secondly, not to conclude from structures of the mathematical theory to structures of reality. The mathematical theory itself has neither ontological nor epistemic significance for physical knowledge. Although mathematical results may be taken as a heuristic clue for scrutinizing certain properties or structures, nature does not possess these just because they are "postulated" by some mathematical description. The most general example of undesirable metaphysics of this type is the notion that nature itself is mathematical. It is only the physical description of nature that rests on the language of mathematics. Such an inference is not only circular, but also naive-realistic by identifying an object with its conceptual description.

The errors due to the last two external connections can largely be traced back to the historical development. This development commonly begins with concrete questions about specific systems and leads, in the first step, to relatively limited theoretical approaches that may later arise as scientific applications of a comprehensive theory, if they do not completely disappear. Consequently, such a theoretical approach necessarily contains terms and model conceptions that derive from visual or figurative considerations or from analogies to similar, already known systems that are of limited generalizability. A theory frame abstracted from such considerations may become the source of undesirable metaphysics when models and claims of such concrete approaches, where they may be justified, are transferred to the theory frame. This concerns, firstly, the definition of the theoretical model in (A1). The physical meaning of terms like "mass point", "ideal gas", or "atom" is defined by quantifiable and measurable properties rather than through externally introduced, object-oriented models. Models of physical systems derived from empirical or theoretical analogies are indeed generally eliminable, and do not occur as a part of (A1). Secondly, the empirical significance of the defining properties is guaranteed by their operational realizability, but not through figurative descriptions. This guarantees, at least in principle, that figurative common-sense notions are not erroneously considered as empirically well-founded. Through the mathematical theory in combination with the interpretive predicates, (A1) provides the formal conceptual framework of the theory without any ontological and epistemic claims.

In contrast, the epistemically relevant constituents of the theory frame are exclusively contained in (A2) and (A3), while any ontological claims that may arise from realizations of the structural principle in scientific applications are out of place in the theory frame and should be eliminated. This exemplifies the connection between the physical force concept and the philosophical terms "cause" and "causality". Since some mechanisms can be directly experienced as the cause of perceivable effects, the structural principle "force" abstracted from these mechanisms is universally interpreted as cause. On this basis, non-existent causal connections are constructed and result eventually in the conclusion that nature is causal and that the aim of physics ought to be the study of causal relations. Erroneous conclusions of this type are presumably inescapable as long as the theory development is not concluded. Consequently, an important objective of axiomatization is the elimination of all ontological claims and system-specific notions from the theory frame, either by restricting them to the model assumptions of a scientific application or by eliminating them from the theory. In fact, an instrumentalist theory conception is not needed in order to avoid or eliminate undesirable metaphysics: the epistemic claims of validity of the epistemological kernel as advanced by a constructive realism do not at all represent undesirable metaphysics. Correspondingly, numerous erroneous conclusions can be avoided if presuppositions are not identified with results and the theory frame is kept free of ontological claims and figurative analogical models.

Due to the structural differences between mathematical and physical theories, the norms of consistency, independence, and completeness of an axiomatic system exhibit a different significance. Central importance is ascribed rather to the challenge of determining in which respect and under which conditions physical knowledge may be considered as secured and definitive. This results in the conception of a closed theory (Heisenberg 1948), which aims to describe the historical experience that the theoretical knowledge of a certain domain is completed or definitive, in the sense that "the theory cannot be improved by minor changes." This is not changed by the fact that, due to its restricted scope, the theory may produce empirically wrong conclusions and results. Later on, Heisenberg (1972) attempted to cast this intuitive notion of closure of a theory in a more precise form by including further properties. On the basis of the theory conception developed in this chapter, they correspond roughly to the following criteria:

1. Internal consistency: the theory does not contain internal contradictions. An investigation of the theory frame for logical contradictions is unnecessary, because it is presumed that the underlying mathematical theory is consistent. The problem of consistency is then localized in the region of the transition from the specializations to the scientific applications and in the interconnections between them. In Sect. 3.2, it was shown how consistency can in principle be assessed and secured by the construction of compatible hypotheses. More generally, the consistency between scientific applications must be understood as a norm: consistency must be proven and established for new applications.
2. Compactness: corresponding to the normative ideal of independence of the axioms, the theory frame contains a very small number of assumptions relative to the scope of the theory, i.e., the number of its scientific applications. In contrast, notions such as economy or aesthetic criteria, like simplicity and elegance, are only marginally related to compactness.
3. Connectivity: the various scientific applications of the theory frame are not isolated from each other, but through their interconnections establish a coherent structural network of theoretical knowledge. For this reason, discrepancies with experiments cannot be corrected locally, i.e., they cannot be removed by minor modifications of the theory.
4. Definitiveness or theoretical closure: theory development is conclusive in the sense that modifications of the theory kernel either lead to equivalent representations of the theory or to an entirely new theory. Definitiveness does not mean that the elaboration of specializations and the addition of new scientific applications is excluded. Regarding research practice, this means that the theory kernel of a closed theory is not subject to refutation, i.e., it is non-hypothetical and represents definitive knowledge. The physical laws of the theory are valid wherever the conceptual system of the theory is applicable, and the conceptual system forms the basis of future research (Heisenberg 1948):

   > The closed theory belongs to the presuppositions of further research; we can express the result of an experiment only using terms from previous closed theories.

   The closed theories remain valid irrespective of future developments.
5. Empirical completeness: all observations, experiments, and effects falling within the scope of the closed theory can be deduced and in this respect explained within the framework of the theory. This characterization, however, is vacuous or circular as long as the scope of the theory is unknown. Although the entirety of the successful and failed applications may provide indications of its scope, a definitive determination is possible only by

means of other theories. The scope of classical mechanics, for example, could be determined only after its limitations were made apparent by the theories of relativity and quantum theory.
6. Empirical accuracy: within the scope of the theory, the experimentally comparable results can in principle be improved quantitatively by means of the theory, up to the relevant experimental accuracy. Therefore, the improvement of accuracy is not a fundamental problem, but a merely technical one. If this kind of improvement turns out to be impossible, the limit of applicability of the theory is affected.
7. Empirical consistency and openness: there are no contradictory experiences within the scope of the theory, and novel experiences can be consistently incorporated and interpreted in the conceptual and structural schema as defined by the theory ("world view"). That this should be possible is unprovable and must be rated as a conjecture, even though it decisively determines the strategies for removing discrepancies between experiment and theory.
8. Intersubjectivity: the closed theory is intersubjectively accredited within the entire scientific community. Accordingly, it possesses the highest rank that can be attained by human knowledge in this context. To repudiate the closed theories of a science means to query this science itself. The knowledge condensed in a closed theory may be characterized as secured in the sense that it is not subject to refutation. However, it is neither a matter of objective nor of true knowledge, i.e., even definiteness and intersubjective recognition cannot be used to assert absolute validity or truth of the given knowledge.

Characteristic of this more precise conception of a closed theory is the combination of theoretical closure and empirical openness: only that part of the theory which is independent of specific empirical experience is definitive. The feature of invariability, implicitly contained in the term "closure", concerns only the theory kernel, but does not exclude the elaboration of further specializations, because this just means exhausting the theoretical surplus and occurs within the underlying conceptual system. Nevertheless, the theory remains open to new empirical experiences due to the separation of the abstract theory frame from its concrete scientific applications. The options to supplement the theory frame with permitted model assumptions oriented toward empirical knowledge, and to arrive in this way at novel or even unexpected scientific applications, are so large in comprehensive theories that they cannot be exhausted. Due to this separation alone, the conceptions of closed theory and empirical completeness in connection with simultaneous openness acquire a definite meaning. In contrast, a science with theory formation remaining on the level of concrete applications would not exhibit these properties. The property of empirical openness thus provides the strongest argument against all theoretical models assuming that all knowledge is already contained in a formal calculus and that concrete applications of the theory do not provide new knowledge. Accordingly, empirical completeness does not mean that new empirical experiences are impossible nor that they are "forbidden" by the theory. In contrast, a modification or extension of the theory kernel is not admissible. If it should turn out to be necessary, this will involve the transition to a new theory. The generalization of Newton's first law, e.g., to non-Euclidean spaces, leads to the general theory of relativity.

By analogy with consistency, independence, and completeness of the axioms of mathematical theories, the properties of closure are unprovable because no necessary and sufficient criteria exist which would allow one to decide whether these properties are satisfied. Accordingly, considering a theory as closed is a matter of conviction or conjecture, resting primarily on the number and diversity of its scientific applications, but remains a matter of belief in the end. Actually, the significance of the conception of a closed theory does not rely on such cognitive aspects, but on the normative consequences for research practice, in particular, concerning theory formation and the strategies for comparison between theory and experience. Since in a closed theory the theory kernel is not subject to refutation, theory formation consists either in the elaboration of specializations or in the formulation of new scientific applications. As a consequence, the existence of a closed theory for a certain universe of discourse by no means implies that the research activities in this field are completed, but rather the opposite. Due to the existence of a secure theoretical basis, research can be entirely concentrated on empirically relevant problems. This increases the efficiency of the research, instead of losing itself in unproductive grass-roots discussions. In this respect, the research conducted within the framework of a closed theory corresponds precisely to Kuhn's normal science (Kuhn 1970).

Furthermore, regarding the relationship with experience, it is highly significant if a theory is closed, and a scientific application, where the deduction of experimentally comparable data occurs, is rated as secure. How the comparison between theory and experience is dictated by estimates of whether an application is secure or not will be illustrated in Sect. 4.3 by the historical discussion of the specific heat of gases as treated within the kinetic gas theory. In a secure application the underlying model assumptions are not subject to refutation either, and the theoretically deduced results have the same if not a higher status of secureness than the experimental data. Accordingly, the experimental corroboration of data deduced in a secure application is not a research goal. In contrast, research in a field without a closed theory is dominated by substantially dissimilar norms and objectives, because the theoretical approaches themselves are subject to modification or refutation, whereupon greater changes may occur in the theoretical representation. The primary objective at this stage of scientific development is the construction of a closed theory. This type of research is dominated by methods that are, according to Kuhn (1970), characteristic of "revolutionary science". However, scientific revolutions in the sense of Kuhn, i.e., characterized by the incompatibility of the theoretical conceptual systems and by the elimination of the preceding theory, do not occur in physics. Relations exist rather between the dissimilar conceptual systems that may be called generalized correspondence principles (Fadner 1985). For instance, numerous terms of classical physics occur in quantum mechanics, but the interpretation of the shifts in meaning as incompatibilities is inappropriate. In contrast, terms like "impetus" or "caloric" have vanished from the language of physics since they date from a stage of development prior to the constitution of a closed theory for the corresponding universe of discourse. Closed theories, on the other hand, remain a firm part of physical knowledge, even when they turn out to be of limited applicability. Similarly, the related conceptual systems remain an integral part of the language of physics. As a characteristic trait of continuity of the historical development in physics, closed theories survive each theory change. Therefore, the notion of a development in science dominated by scientific revolutions must be modified, in that it corresponds to the transition from one closed theory to another one. All things considered, the existence of non-cumulative periods of scientific development and the distinction of two structurally dissimilar types of research enables one to retain two important aspects of Kuhn's approach without being subject to the objections raised against the cogency of such a distinction (Toulmin 1974).

In short, the axiom-like reconstruction of physical theories may yield meaningful results if the theory is closed and the axiomatization is adapted to the specific structure and properties of physical knowledge, instead of being oriented toward the axiomatization of mathematical theories. Under these premises it may contribute to structuring and securing knowledge, by disclosing how parts alien to physics may be localized and eliminated. Furthermore, a conception of a closed theory can be defined that is not only relevant for actual research practice, but may also serve as a basis for a representation of the development of physics that provides a more reasonable description of its historical development than the purely cumulative models of empiricist philosophies of science or the notion of scientific revolutions.

3.5 Unification

The goal of summarizing the whole of human knowledge in a concise, manageable form (the philosopher's stone, or theory of everything) is as old as epistemology itself, and stands as a more or less acknowledged normative ideal behind any activity producing knowledge. It is thus not surprising to find such objectives in physics, as expressed, e.g., by Planck (Planck 1915):

> As long as physical science has existed, its ultimate, desirable goal has been the solution of the problem of merging all observable and still to be observed natural phenomena under a single fundamental principle that can be used to calculate both past and especially future events from the present ones [...] The overriding target of any science has always been and remains the merging of all theories that have grown within it into a single theory within which all the problems of that science find their unequivocal place and unambiguous solution.

In such an unspecific form this objective must be assessed as a heuristic principle and finds its expression in the tacit assumption (3F). According to this general idea, attempts at a unification of physical knowledge were guided in the nineteenth century by the program of the reduction of all physical theories to mechanics, and since the second half of the twentieth century by the reduction of the fundamental interactions to a unified theory, occasionally referred to as a "theory of everything". In spite of these dissimilarities, the attitude underlying such ideas identifies unification with a reduction in the spirit of a relation as strictly deductive as possible, i.e., that the entire knowledge should be deducible from a small number of fundamental postulates. Apparently, only hierarchically structured systems are perceived as ordered, in contrast to horizontally structured ones.

In concrete form, reduction involves a particular type of intertheoretic relations. In this context, the question arises as to how the various theories of physics are related to each other. More generally, the study of the relationship between two theories involves comparison on three different levels:

(R1) Formal level: the relation between the axioms and theorems of the two theories.
(R2) Semantic level: relations between the conceptual systems, especially with regard to shifts in the physical meaning.
(R3) Empirical level: relations of both theories to experience, i.e., dissimilarities in scope (range of applicability and validity) and accuracy of the experimentally comparable results of the two theories.

In the case of the relationship between a theory $T_<$ (T-minor, or theory to be reduced) and a superordinate reducing theory $T_>$ (T-major), these three aspects lead to the following criteria:

$(R1)_{rd}$ The axioms of $T_<$ appear as deducible theorems in $T_>$.
$(R2)_{rd}$ All terms of $T_<$ are also contained in $T_>$ and share the same meaning. This postulate may be weakened in that some concepts of $T_>$ may acquire a more general or broader meaning. In any case, $T_<$ must not contain terms that are not defined in $T_>$.
$(R3)_{rd}$ $T_>$ is more comprehensive than $T_<$ in that the extension of the statements and the accuracy of the experimentally comparable results of $T_>$ are broader and better, respectively.

Since a reduction process is usually carried out by transforming $T_>$ in such a way that it resembles $T_<$ as closely as possible, the common way of speaking in physics is that $T_>$ is being reduced, while in the philosophy of science $T_<$ is reduced in the sense that it becomes a part of $T_>$.

A simple example of a reduction relation is the connection between the thermodynamics of the ideal gas and the version for real gases which rests on the van der Waals equation:

$$P = \frac{nRT}{V - b'} - \frac{a'}{V^2} = \frac{nRT}{V(1 - b\varrho)} - a\varrho^2 ,$$

with system-specific parameters a, b and "particle density" $\varrho = N/V$. The state variables P, V, T have the same meaning in both theories, and the van der Waals equation yields an improved agreement with experiments over larger (P, T)-regions, with the additional result that a gas condenses to a liquid below a certain temperature. When $\varrho \to 0$, the state equation $P = nRT/V$ of the ideal gas is obtained, so that the universality of the behaviour of highly diluted matter becomes comprehensible. According to Nagel (1982), this relation is called homogeneous reduction and is considered to be unproblematic. A closer inspection of the meaning of the limiting process reveals that this is not the case. Due to $N > 0$, the condition $\varrho \to 0$ is equivalent to $V \to \infty$. On the other hand, every thermodynamic system has finite extension by definition. In order to get round this inconsistency, the limiting process $\varrho \to 0$ must be interpreted, not in a formal mathematical fashion, but in an informal or pragmatic one, i.e., by saying that, for all practical purposes (FAPP), ϱ may be treated as zero. This is not just a manner of speaking, since the adequacy of the limiting process becomes dependent on experimental, i.e., theory-external conditions, such as the temperature T and the accuracy Δ of the measured data. In this example, it means that a ϱ_0 must be chosen so small that the differences between the results of the two theories are small enough to lie within the error margins of the experimental data. Overall, consistency with the conceptual framework of thermodynamics is attained only for $\varrho = N/V \to \varrho_0(\Delta, T, \ldots) = 0$-FAPP.

Further challenges arise if one asks whether Kepler's theory, conceived as a scientific application of kinematics, can be derived from the gravitational theory which is a specialization of Newtonian mechanics. Actually, this is not directly attainable, because the conceptual systems of the two theories are not consistent with each other. Firstly, Kepler's first law treats the Sun as fixed center, and this is inconsistent with Newton's third law, so a strictly deductive derivation is definitely excluded. Secondly, Kepler's theory is a single-particle theory due to the implicit assumption that the motion of each single planet is determined only by the Sun. In contrast, the application of Newton's gravitational theory to the Solar System results in a complicated many-body problem because interactions between the planets are included explicitly. Thirdly, the term "mass" does not occur in Kepler's theory, but is of crucial importance in the gravitational theory since it determines the strength of the interaction. The strategy for making the two theories comparable with each other consists in the transformation ("reduction") of the gravitational theory into an "effective" single-particle problem. In the first step, the many-body problem is transformed into a two-body problem by neglecting the interactions between the planets. This eliminates the problem that the actual planetary orbits, as derived via the directly measured positions, are not exactly elliptical, as was already recognized by Kepler himself. The second step with the transition to the center of mass system results in the replacement of the planetary mass m by the reduced mass $\mu = m/(1 + m/M_\odot)$, in combination with the assumption $m \ll M_\odot$, where $M_\odot$ is the mass of the Sun. In a strict sense, this corresponds to a shift in meaning of the concept of planetary mass. While this substitution establishes consistency with Newton's third law, it leads to the following modification of Kepler's third law:

$$\left(\frac{a_1}{a_2}\right)^3 = \left(\frac{T_1}{T_2}\right)^2 \frac{1 + m_1/M_\odot}{1 + m_2/M_\odot} .$$

Only after this transformation has been completed, consistent analogs of Kepler's laws are obtained as a function of the ratio $m/M_\odot$. In the last step, Kepler's original laws are eventually recovered through the limiting process $m/M_\odot \to 0$-FAPP, in combination with the range of applicability as a function of the accuracy of the measurements. In summary, the derivation comprises two parts. In the first, the gravitational theory is transformed into a form enabling the derivation of (partially) modified laws that are comparable with Kepler's laws. Although this part represents a largely vertical-deductive procedure, it cannot be rated as strictly deductive due to the additional theory-external and system-specific assumptions which aim to achieve maximal similarity with Kepler's theory. The second part consists in taking the asymptotic limit $m/M_\odot \to 0$-FAPP, thus connecting the dissimilar conceptual levels in the sense that it becomes comprehensible why the term "mass" does not occur in Kepler's theory, for example. This constitutes a correspondence relation that is predominantly of horizontal type and already plays the decisive role in this relatively simple example. This becomes even more apparent for the relation between the thermodynamics of the ideal gas and the kinetic gas theory, conceived as a scientific application of statistical mechanics to the ideal gas, discussed in detail in Sect. 3.3. Assuming the kinetic gas theory as the reducing theory $\mathrm{T}_>$, the reduced theory contains, in the form of the temperature, a term neither occurring in $\mathrm{T}_>$ nor definable in $\mathrm{T}_>$ itself. A partial interpretation only provides the identification $\langle E_{\mathrm{kin}} \rangle = 3k_{\mathrm{B}}T/2$, establishing once again a horizontal correspondence relation.

In tightened form, the challenge of reducibility arises when asking for the global relation between thermodynamics and statistical mechanics. Even restricting to equilibrium thermodynamics, it is not clear which of the different representations of statistical mechanics should be chosen as the reducing theory. Finally, it has already been explained in detail in Sect. 3.3 why there is no deductive relationship between the two theories and neither theory is more fundamental than the other. Furthermore, purely deductive systematizations exhibit a number of inherent shortcomings:

- Each strictly deductive system contains incompletely defined terms whose meaning is determined only by the system as a whole.
- Every strictly deductive system contains assumptions, viz., the premises of the deductions, which are unprovable within the system's framework. For this reason, it inevitably remains hypothetical if it advances any epistemic or ontological claims.

- Strictly deductive systems do not lead to substantially new knowledge: the entire knowledge provided by the system is implicitly contained in the basic assumptions ("axioms").
- The options for selecting both the axioms of the theory kernel and the model assumptions of the scientific applications is only incompletely constrained by a strictly deductive structuring.
- The options for securing the postulated epistemic claims (axioms) and ontological claims (model assumptions) of physical knowledge are severely limited.

For these reasons the securing of knowledge and, in particular, the fulfilment of the claims of validity of the premises is severely limited in a strictly deductive representation. Consequently, the deductive-linear structure reveals itself to be both the strong and the weak feature at the same time: securing by deduction, especially by proof-theoretical methods, is appropriate for theorems, but unsuitable for premises. In order to secure the knowledge postulated through the premises, horizontal structural relations are inevitable, such as interconnections and crosslinks, e.g., via generalized correspondence principles (Fadner 1985) that establish local relations between some terms and laws of the two theories. Therefore, reduction in the spirit of a strictly deductive systematization is not at all a sensible objective when attempting to unify knowledge: unification must be based on both vertically deductive and horizontally corresponding structural relations. The way horizontal structures can be used to scrutinize the consistency of theoretical knowledge is demonstrated by the question of whether there is a consistent classical physical description of reality. The fact that such a description is known to be incomplete does not exclude the possibility that it might be consistent within its claimed range of applicability.

One option for classifying classical physics refers to the four classical "grand theories" that formally may be distinguished by the number of "microscopic" degrees of freedom (DoF) of the respective systems under consideration:

- classical mechanics as the theory of systems with few DoF,
- statistical mechanics as the theory of systems with a very large number of DoF,
- thermodynamics as the theory of systems with a countably infinite number of DoF,
- classical field theory as the theory of systems with an uncountably infinite number of DoF ("continuum theory").

Classical field theory is commonly identified with electrodynamics, but comprises also the continuum theories of matter and generally constitutes the theory of static, stationary, and time-variable fields that are also referred to as radiation fields. The special theory of relativity is an essential constituent of any theory of radiation fields, recognizable by the fact that position and time have equal status and the propagation of effects is described by functions of the argument $x - ct$, where c is the propagation velocity.

There are no elaborate global representations of the relations between these grand theories. One among several reasons is the lack of generally accepted axiom-like reconstructions of these theories. There are actually a number of local correspondence principles that establish local connections between certain laws and terms of classical mechanics, statistical mechanics, and thermodynamics. Due to these local correspondences, it is frequently concluded that there must also be a global, consistent relation between these theories. For instance, classical mechanics (CM) is represented in this vein as the asymptotic limit $\lim_{c \to \infty} \mathrm{SR} = \mathrm{CM}$ of the special theory of relativity SR. This notation not only totally ignores the physical meaning of c as one of the fundamental physical constants, but remains nonsensical even when interpreted as a merely symbolic or metaphorical representation, because a theory is not a function or a single statement, but a cross-linked system of specializations and scientific applications. This excludes any option of defining a meaningful limiting value for entire theories. The paradigm of these superficial notions is the belief that thermodynamics in its entirety should be reducible to statistical mechanics. The starting point is the result that, at the formal mathematical level, some laws of thermodynamics appear as asymptotic limits of the corresponding laws of statistical mechanics when $N \to \infty$ ("thermodynamic limit"), i.e., when the number N of subsystems becomes infinite. In this respect, thermodynamics may be characterized symbolically as the theory of systems with a countably infinite number of degrees of freedom, as given by the classification above. Taken seriously, this produces a number of inconsistencies. Firstly, by definition every thermodynamic system is constrained to a finite space region, so it cannot contain an infinite number of subsystems. Secondly, this comes into conflict with one of the fundamental postulates of statistical mechanics, viz., the equipartition theorem for the energy. This theorem postulates that, in thermal equilibrium, the total energy of a thermodynamic system is equally distributed over all (microscopic) degrees of freedom. If their number becomes infinite, application of the equipartition theorem leads to the following alternative: either every input energy must be equally partitioned into infinitely many parts or a finite amount of energy is attributed to each degree of freedom and the internal energy becomes infinite. As a consequence, either the system has zero temperature or it has a finite temperature and could supply energy indefinitely. In order to avoid these inconsistencies, the thermodynamic limit must be interpreted again as $N \to \infty\text{-FAPP} = N_{\max}$. Under this presupposition, the conceptual systems of classical mechanics, statistical mechanics, and thermodynamics can be made locally compatible by means of the correspondence principles described in Sect. 3.3. Although a proof of global compatibility has never been given, according to common conviction, these three theories provide a consistent description of the relevant part of reality, but without constituting a reduction to mechanics.

Regarding the model systems of a continuum theory, an analogous argumentation is no longer possible. In a continuum, there can occur only internal vibrations that may be decomposed into standing waves, each having an integer number of nodes and corresponding to a single degree of freedom. Since waves with an arbitrarily large number of nodes are possible in a continuum, it always possesses an infinite number of degrees of freedom. The fact that thermodynamics and classical field theory cannot actually be made compatible was shown by Boltzmann for the material continuum and by Planck for the electromagnetic field. In the latter case, the result is called the ultraviolet catastrophe because the shorter the wavelength of the radiation, the greater the deviations from experiment. In fact, a thermodynamic description of the continuum cannot be formulated since it leads to both theoretical inconsistencies and contradictions with experimental results (radiation laws). Although each of the two theories is in itself consistent within its scope, they are not compatible globally. On the one hand, this lack of a consistent, horizontal structural relation provides the proof that a consistent classical-physical description of reality does not exist. On the other hand, this demonstrates how the construction of cross-links via correspondence principles may contribute to securing knowledge. Moreover, such horizontal structural relations are the dominant structural features of grand theories. While descriptive and explanatory theories can to a large extent be reconstructed as deductive systems, a grand theory constitutes a "cluster of deductive systems" (Tisza 1963) whose mutual relations are essentially of the horizontal type. Accordingly, the key issue of systematic representation is the construction of a coherent network of theories based on a unified conceptual system and common ordering principles. For this reason, the grand theories may be referred to as unifying theories in which the unification is oriented toward the conceptual system and methods of a "kernel theory". Regarding the four classical grand theories, these are:

- classical mechanics: dynamics of the mass point,
- statistical mechanics: kinetic gas theory,
- thermodynamics: thermodynamics of the ideal gas,
- classical field theory: electrostatic field of the point charge.

Following Kuhn, such a kernel theory may be thought of as a paradigm in the narrower sense (Kuhn 1974), and the structural analysis of a grand theory primarily consists of the study of intertheoretic relations. Some representative examples will be described in the following.

The possible intertheoretic relations may be classified according to whether they are primarily of vertical type, primarily of horizontal type, or of mixed type. The largely vertical relations are the easiest to reconstruct, although the horizontal ones are more important because they establish the cross-links that are crucial for securing physical knowledge. Since the grand theories are not in a vertical relation to each other, vertical intertheoretic relations occur only within the same grand theory. The prototype of such a relation is the initially described homogeneous reduction. Another primarily vertical relation constitutes the hierarchy of specializations of a theory kernel. Within Newtonian mechanics, for example, such a hierarchy is established by going from the most general force function to increasingly specific ones. The specialization defined by $\mathbf{F}_{rv} = \mathbf{F}(\mathbf{r}, \mathbf{v})$ is then more comprehensive than the one defined by $\mathbf{F}_r = \mathbf{F}(\mathbf{r})$, which is in turn more comprehensive than the specialization defined by $F_n = c \cdot r^n$. This does not, however, permit the conclusion that a more comprehensive specialization T is also richer in empirical content than a more specific one T_0, i.e., that T contains more empirically significant statements than T_0. If one conceives the condition $\mathbf{F}_{rv} = \mathbf{F}(\mathbf{r}, \mathbf{v})$ in such a manner that $\mathbf{F}$ depends on both $\mathbf{v}$ and $\mathbf{r}$, this specialization does not lead to more empirically relevant results than the one defined by $\mathbf{F}_r$. Concerning a force $\mathbf{F}_r$, e.g., the mechanical energy is conserved, but not in $\mathbf{F}_{rv}$, and for a force F_n, the virial theorem applies as well. Actually, in more specific theories, additional laws frequently hold that are not valid in the more general and comprehensive one. Only by subsuming under the comprehensive theory T all of its specializations T_i in combination with their scientific applications does T become trivially richer in empirical content than each of the T_i. Finally, there exists a predominantly vertical intertheoretic relation between a specialization and its scientific applications that may be interpreted as realizations of an abstract theory. Due to the additional model assumptions which establish empirical content and claims of validity, this relation is not strictly vertical either.

Among the scientific applications of a theory frame, a hierarchy is defined through the condition that T is more precise than T_0 if the experimentally comparable data deduced within T are quantitatively more precise. For instance, the thermodynamics of real gases based on the van der Waals equation yields more precise results for their behaviour than the thermodynamics T_0 of the ideal gas. This improved accuracy is achieved through the empirical definition of system-specific parameters for the volume of the gas particles and the internal interactions, both resulting in a differentiation between the various gases. Accordingly, supplementation with these parameters rests on the assumption that a gas consists of tiny particles. Conversely, although less specific, T_0 is more universal in that it is system-independent, since it does not contain any assumption regarding the internal structure of the gas. This universality is lost under the transition to the more precise theory based on the van der Waals equation. More generally, it can be stated that structures resting on greater abstraction are lost under the transition to system-specific theories which in turn produce improved agreement with experimental data. Again, such a relation does not establish a strictly vertical intertheoretic relation.

Another seemingly vertical intertheoretic relation is associated with the concept of a presupposition theory (Ludwig 1978). A theory T_0 is called a presupposition theory of T if terms occur in the axioms of T that carry a physical meaning from T_0. In this vein, physical geometry providing terms like "space" and "distance" is a presupposition theory of kinematics, and kinematics with its terms "time", "velocity", and "acceleration" is a presupposition theory of dynamics. Finally, mechanics ought to be a presupposition theory of electrodynamics because terms like "force", "momentum", and "energy" already carry a physical meaning from mechanics. In analogy to the notion of "protophysics", such a conception is obviously based on the idea that a system of fixed preset terms could be constructed which would enable a hierarchical organization of the conceptual system of physics. The examples show, however, that there is no strictly hierarchical relation in any of these cases. The physical concept of distance, for example, that is appropriate for a theoretically precise foundation of dynamics is by no means already fixed by physical geometry, because such a definition requires concepts like rigid body and inertial system that acquire meaning only within the framework of mechanics. In particular, the concept of space undergoes significant modifications in its physical meaning in going from Newtonian mechanics to the general theory of relativity. The same applies to the general relation between the dynamics of the mass point and the relevant kinematic theory. As shown in Sect. 3.3, this relation is in no way deductive since in particular the fact of equating the dynamical and kinematic velocities constitutes an essentially horizontal structural relation. Overall, a hierarchical relation exists solely in an informal or pragmatic sense: some terms occurring in the axioms of T may be assigned a meaning that comes from another theory, but must be considered only as partial and tentative. In particular, the example of mechanics as a presupposition theory of electrodynamics demonstrates that presupposition theories must be classified as some type of background knowledge and do not establish any intertheoretic relation in the sense of (R1)–(R3). The same applies to the relation between an explanatory theory and the associated descriptive theory. According to Sect. 3.3, the essential difference consists of the ontological claims of validity advanced by the explanatory theory. These, however, become effective only on the level of the scientific applications because, according to Sect. 3.4, ontological claims must not be ascribed to the theory frame.

In summary, the common feature of vertical intertheoretic relations is that they are hierarchical only in pragmatic, functional, or informal respects. The reason for classifying them nevertheless as vertical is that the conceptual system of the superordinate theory T does not change in formal respects under the transition to T_0, since both theories are a part of the same grand theory irrespective of possible modifications of physical meaning. This is the most important difference between the primarily vertical intertheoretic relations and the asymptotic approximations as representative examples of the mixed type. In this case, the transition from T to the asymptotic theory T_{as} is usually accompanied by significant changes in the conceptual system.

A simple, but instructive example is the relationship between the physical geometry on the surface of a sphere and that in the plane (Tisza 1963). Spherical geometry has a natural length unit, viz., the sphere radius R, and for the limit $R \to \infty$, it turns into planar geometry. If a is a typical linear dimension of geometrical objects of the plane, a measure of the quantitative deviation between the results of the two geometries may be defined via the ratio a/R. If $a/R \ll 1$, the planar geometry may be interpreted formally as an asymptotic approximation to the spherical geometry. Initially, a strictly vertical relation seems to exist between the two theories. Actually, however, considerable modifications occur. Not only is the meaning of a number of terms changing, but the planar geometry also contains relations, like scale invariance, which are not valid in the spherical geometry. While all the various spherical geometries are structurally isomorphic in the sense that they differ only by the numerical value of R, the term "radius of curvature" in this role does not occur in planar geometry, so that no natural length unit exists. All the theorems of planar geometry are thus invariant under changes in the length unit, i.e., they are independent of the size of objects. Although in spherical geometry there is an analogous invariance property with regard to the invariance under changes in the curvature radius, the two properties are not hierarchically related to each other. For instance, planar geometry contains theorems that are not valid in spherical geometry, e.g., the theorem that the angular sum in any triangle amounts to $180°$. Therefore, as a physical theory, planar geometry contains additional, empirically testable statements. Conversely, taking the property of having an angular sum of $180°$ as the definition of a triangle, since it is universally valid, the concept of triangle would "break down" with the transition from planar to spherical geometry.

Just as a sphere does not contain the infinitely extended plane, planar geometry is not contained in spherical geometry, so the relation between the two theories is not vertical. Once again, a hierarchy exists at most in some pragmatic respect. If the two theories yield different results, those of the asymptotic approximation are considered to be less reliable. This is formally justified as follows. The theorems of spherical geometry can be brought into a form applicable to the special case where $a/R \ll 1$, but still finite. Such a theory is obviously structurally isomorphic with the general spherical geometry. In the subsequent step, an analogical relation is constructed between this representation of the spherical geometry and the planar one. In doing so, one cannot assume a priori that this relation corresponds to an identity or isomorphism, and according to the preceding analysis this will not be the case either. Rather, this horizontal relation provides a quantitative measure for the assessment, under which conditions the conceptual system of the asymptotic approximation is applicable. Overall, the reconstruction of this intertheoretic relation contains vertical and horizontal parts which are both indispensable in order to encompass the full physical meaning of this relation.

A similarly simple physical example of an asymptotic approximation is provided by the relationship between geometrical optics and wave optics. The asymptotic condition means that the dimension of the obstacles along the light propagation path must be large compared with the wavelength, which is not itself a term of geometrical optics. The horizontal structural element of this relationship consists of the construction of a consistent relationship between the physical concepts of light ray and light wave and the various resulting ways of representing light propagation. Another example is the relationship between the mechanical continuum theory of matter and the corresponding microscopic theory, viz., lattice dynamics. If the wavelength of a lattice vibration is sufficiently long compared to the lattice constants, the atomic structure of the lattice has no influence on the propagation of the vibration. The equation of motion of these acoustic vibrations of the lattice can be transformed into a wave equation of a continuum. It is once again characteristic of the relationship between the two conceptual levels that the lattice constant is not a term of the continuum theory. Further examples of asymptotic approximations are the relationship between Kepler's theory and Newton's theory of gravitation, discussed previously in this section, and the one between thermodynamics and statistical mechanics. Because some laws of thermodynamics can be obtained as asymptotic approximations of the corresponding laws of statistical mechanics, thermodynamics in its entirety is frequently considered to be a part of the latter. In all these cases, it is not only ignored that the transition from T to the asymptotic approximation T_{as} is associated with a change in the conceptual systems, but also that both theories refer to structurally dissimilar conceptual levels. Correspondingly, an asymptotic intertheoretic relationship between two theories T and T_{as} consists of two parts. In the first step, the superordinate theory T is cast into a form As(T) that resembles T_{as} as closely as possible. Subsequently, through an appropriate limiting process, some laws of As(T) are converted or related to those of T_{as}. This step may be associated with changes in the physical meaning of some terms and establishes correspondence relations between concepts of As(T) and T_{as}, e.g., $\langle E_{\mathrm{kin}}\rangle = 3k_{\mathrm{B}}T/2$. For these reasons, there is at best a vertical relation from T to As(T), but not to T_{as}.

Due to the change in the conceptual levels, this type of intertheoretic relationship is of crucial importance for understanding theory development in physics if it can be constructed between different grand theories. Precisely such cases have given rise to the claim that the transition from one grand theory T_1 to a more comprehensive one T_2 should constitute a scientific revolution. Such a view overemphasizes the horizontal components of this intertheoretic relation, and largely ignores the vertical ones. The subsequently diagnosed "breakdown" of the conceptual system in the course of the historical transition from T_1 to T_2 shows only that theory development in physics is not at all a cumulative process and that physical concepts acquire a large part of their meaning through the theory as a whole. The most prominent examples are certainly the relationship of classical mechanics, on the one hand, to the special theory of relativity and to quantum theory, on the other. For the first relationship, a good example is the change of meaning of the concept of mass. In classical mechanics it designates a property of material bodies that is independent, not only of ambient conditions like pressure and temperature, but also of the state of motion. In this meaning, mass is an inherent property and one of the attributes contributing to the recognition of material bodies. In contrast, in the special theory of relativity, the mass depends on the velocity v of the body and becomes a dynamical property. Formally, the expansion of the relativistic mass into powers of v/c (c = speed of light) yields a first, v-independent term called the rest mass m_0, which may afterwards be identified with the mass term of classical mechanics. This second step does not correspond to a vertical relation, but to a horizontal one. In order to recover the classical notion of mass as an inherent property, the rest mass m_0 is introduced as a new concept within the framework of the special theory of relativity. Conversely, this term has no meaning in classical mechanics as far as it is not viewed against the background of this correspondence. The total energy of a free particle undergoes a similar transformation in meaning, which may be represented in the special theory of relativity as the sum of the rest energy m_0c^2, the (classical) kinetic energy $m_0v^2/2$, and terms proportional to powers of $1/c^2$. Considering classical mechanics as the asymptotic limit for $c \to \infty$, as is usually done, the rest energy becomes infinite and must be omitted in order to obtain a meaningful correspondence relation. Finally, the conceptual challenges involved in the relationship between classical mechanics and quantum mechanics are even more obvious. It is not at all predetermined unequivocally and with necessity whether the classical dynamical observables must be correlated with the quantum mechanical operators, their expectation values, or their eigenvalue spectrum. Actually, all three options (quantum mechanical correspondence principle, Ehrenfest's theorem) are used when the relationship between the two theories is established.

Due to the widespread quest to reconstruct theories as deductive systems, the significance of the primarily horizontal intertheoretic relations is frequently underestimated, if not totally ignored, although precisely these contribute decisively to securing physical knowledge. Furthermore, in the context of the notion of grand theories as clusters of deductive systems, their significance becomes evident because the deductive subtheories of a grand theory are not isolated from each other. As already mentioned, the conceptual system of a grand theory is developed with reference to a kernel theory by transferring its concepts and structure to the description of related fields of experience. For this reason, both an intertheoretic relation and the deductive system constructed in this way will be referred to as transference. The formal structure of the dynamics of the mass point, e.g., may be transferred to the dynamical description of rotatory motions represented by the rotation of a rigid body. The equation of motion $\mathbf{M} = \Theta \mathrm{d}\boldsymbol{\omega}/\mathrm{d}t$, although formally analogous to Newton's second law, differs in the meaning of the terms. $\mathbf{M}$ is a torque but not a force, $\boldsymbol{\omega}$ is the angular velocity, and the moment of inertia Θ is a tensor, rather than a scalar like the mass. Irrespective of these dissimilarities, in analytical mechanics a common formal framework is defined for the two theories by introducing generalized forces, generalized coordinates, and generalized momenta. As a consequence, physical properties are subsumed under these generalized concepts, which differ with regard to dimension and symmetry, viz., a generalized coordinate may be either a length or an angle, a generalized momentum may be an angular momentum, and a generalized force may be a torque. With regard to the behaviour under spatial symmetry transformations, they also differ from each other since, e.g., a momentum is a polar vector and thus changes direction under inversion, while an angular momentum is an axial vector which does not. Other examples are the transference of the conceptual system of the electrostatic field to the description of magnetostatic fields within the framework of electrodynamics, and the transference of the conceptual system of a single-particle theory to the description of many-particle systems. Accordingly, in certain respects, a grand theory possesses a similar structure to a theoretical model: the guiding model M_0 corresponds to the kernel theory and the M_i to its transferences.

Further examples of horizontal intertheoretic relations are provided by equivalent, complementary, and competing theories. Qualitatively, two theories may be termed equivalent if they are similar, but not identical. One must then specify in which respect there is similarity. It may be either theoretical or empirical. Two theories T_1 and T_2 may be called theoretically equivalent if each theorem t_1 of T_1 is analogous to a theorem t_2 of T_2, or if for each specialization of the theory frame of T_1 there is an analog in T_2. An example of such an equivalence is the classical mechanical representation of position-dependent forces in the schemes of Newton, Lagrange, and Hamilton. The equivalence relation is constituted by transformation rules, e.g., the Legendre transformation between the Lagrange and Hamilton schemas. Other examples are the position, momentum, and matrix representations of the time-independent quantum mechanics of single-particle systems, or the Schrödinger, Heisenberg, and Dirac pictures of time-dependent quantum mechanics. In particular, the last two examples show that "equivalence" must be conceived in a less formal sense. A simple mathematical analogy may serve to elucidate this. Two mappings f and g are said to be equal if $f(x) = g(x)$ for all x in the domains of both f and g, and if the domains of f and g coincide. Translated to the case of the equivalence of two theories, one would like to postulate that the scopes of the two theories coincide. However, this criterion cannot be applied because the potential range of applicability of a theory is not fixed due to its empirical openness. Consequently, when talking about equivalent theories in physics, only equivalence with regard to their already known scope is concerned. This leaves open the possibility that two equivalent theories could possess scopes of different sizes.

Beside the theoretical equivalence between specializations of the theory frame, empirical equivalences exist on the level of the scientific applications. Qualitatively, two applications ScA_1 and ScA_2 may be described as empirically equivalent if, for each experimentally comparable result in ScA_1, there is a corresponding one in ScA_2, and the two results agree within the error margins of the relevant experimental datum. However, the underlying model assumptions of ScA_1 and ScA_2 must differ from each other, because otherwise two identical applications are concerned. The theory of reflection and refraction of light may be reconstructed as an application of either geometrical or wave optics. Consequently, the equivalence relation is not established by transformation rules, nothwithstanding the fact that some such rules may exist between the two descriptions. Unlike theoretically equivalent specializations, equivalent scientific applications need not be totally transformable into each other. A particularly simple case of equivalence within a scientific application is representation with respect to different coordinate systems.

Two theories are said to be complementary if their scopes are largely disjunct and a superordinate, unifying theory exists for both of them. This may concern both specializations and scientific applications. Two complementary theories then appear as approximations to this superordinate theory, and the connection between them is again established through correspondence principles. Common examples of complementary theories are the valence bond and the molecular orbital theories of molecular physics, and the wave and particle pictures in quantum theory. The correspondence principles in the second case are, e.g., the Einstein and the de Broglie relation. Finally, competing theories are ones with comparable scope as far as they are not equivalent, and which thus rest on axioms or model assumptions that are substantially different or even contradict each other. In physics, competing theories are important only for the study of the process of theory formation, but not for the reconstruction of closed theories, since their persistence is incompatible with the objective of the unification of knowledge. Instead, in the course of theory development, either one of the two theories will be superseded, or they will turn out to be equivalent, or they will arise as asymptotic or complementary approximations to a superordinate theory after reconstruction.

Finally, against the background of this analysis of representative intertheoretic relations, it is interesting to discuss whether all the classical and modern grand theories can be united into a unified conceptual system. The challenges of such a unification become most clearly apparent at the interface between classical physics and quantum theory. Since molecules, gases, and condensed matter consist of atoms whose properties can be described only in the framework of quantum theory, only a quantum theoretical conceptual system could be considered as unifying. According to the results of modern physics, however, a consistent unification of the classically complementary particle and wave pictures requires as a necessary condition that, in contrast to the classical description, one must abandon the assumption that quantum systems possess dynamical properties as pre-existent system attributes: “Quantum mechanical particles do not possess dynamical characteristics” (Landau und Lifschitz 1979). Indeed, this result has been verified by numerous experiments performed since the 1980s. Accordingly, the fundamental challenge of any quantum theoretical description is the question of how to construct a dynamical description of systems which do not possess dynamical properties. The answer can only consist of this alternative:

- The dynamical description of quantum systems and processes is impossible.
- There must be systems, viz., measuring instruments, which may be assumed to be classical in the sense that dynamical properties can be ascribed to them as pre-existent attributes which can serve to define the dynamical properties of quantum systems.

Obviously, only the second option can pave the way to exploring and describing atomic constitution and atomic processes, although it has serious consequences:

1. Since within the framework of a strictly quantum theoretical description dynamical properties cannot be assigned to the systems as pre-existent attributes, there can be no autonomous dynamical conceptual system of quantum theory.
2. Dynamical state variables like position, momentum, or energy are defined, in the first instance, only as properties of the entity "quantum system + measuring instrument + interaction". Such properties, designated in Sect. 2.5 as weakly objectifiable, have similar status to the classical dispositional terms such as solubility or conductivity, i.e., detached from the measuring process the dynamical properties of quantum systems are not defined. Nevertheless, the results of measurements on quantum systems are independent of the observer.
3. The necessity of the instruments for the foundation of quantum theories is in no way due to non-negligible perturbations by the measuring process, but because the dynamical properties of quantum systems are conceptually definable solely via measuring processes with "classical" systems, viz. the measuring instruments.

Due to these consequences, there are two levels in the conceptual description:

- the classical physical conceptual level enabling the definition of strongly objectifiable dynamical properties without running into inconsistencies,
- the quantum theoretical conceptual level where an autonomous dynamical description is impossible in the sense that quantum systems do not possess dynamical properties as pre-existent system attributes.

Accordingly, the quantum theoretical and classical descriptions occur on dissimilar conceptual levels: in this sense classical physics and quantum physics represent different levels of complexity. On the one hand, some relations of classical physics correspond to asymptotic limits of the corresponding quantum theoretical ones. On the other hand, the conceptual system of classical physics is a necessary prerequisite for the formulation of quantum theory. Predominantly due to this conceptual dichotomy, this interpretation has been criticized on the grounds that only a single, uniform reality exists and that this must be described as non-classical. For all physical systems, including instruments and observers, consist of atoms or elementary particles, and the theory for them must provide the unified conceptual framework for the description of the whole of reality. Consequently, not only quantum systems, but also measuring processes, instruments, and observers must be described strictly quantum theoretically, i.e., without referring to the classical conceptual system. These arguments, however, rest on two inadmissible and thus erroneous assumptions:

1. The argument that the unity of reality implies with necessity the unity of its conceptual description corresponds to the identification of reality with its conceptual description, which is the characterizing feature of naive realism and the paradigm of "bad metaphysics".
2. The assumption of a unified theoretical description ("grand unified theory") alleges, at least implicitly, that pre-existent dynamical properties can be ascribed to quantum systems in some way. Otherwise it would be impossible to have a self-contained dynamical description by a unified theory that is grounded on a strictly quantum theoretical conceptual system.

Due to this erroneous view, a central issue in attempts to interpret quantum mechanics, until about the late 1970s, was considered to be the "measurement problem", comprising items like the state of the observer, the collapse of the wave function, and the "section problem", i.e., at which point in the chain "quantum system – measuring instrument – observer" the transition occurs from quantum theory to the classical description, as the factual description with ascertainable measuring results. The extreme variant results in the subjectivist interpretation of quantum theory assuming that this transition takes place only in the consciousness of the observer. These discussions were largely constrained to the conceptual level and have led to a number of pseudo-problems, e.g., the collapse of the wave function rests implicitly on an inadmissible "realistic" interpretation of the wave function (see Sect. 2.6). Due to substantial advances on the experimental side during the past 40 years, the discussion has shifted to a more object-oriented level with the result that the transition from quantum to classical behaviour is decisively associated with the disappearance or destruction of coherence ("decoherence"), especially in entangled systems. As a consequence, the current state of discussion focuses on topics such as:

- When is an interaction between quantum system and measuring instrument a measurement: the key role is played here by the irreversible destruction of coherence which may occur either within the measuring instrument or due to coupling with the environment, but definitely not in the consciousness of the observer.
- The transition from the quantum regime to the classical one: this is not only a question of the size of the system. The key role is played by the preservation of coherence, as exemplified among other things by "macroscopic" quantum phenomena such as superconductivity or suprafluidity.
- The interpretation of "demolition-free" and delayed-choice experiments which demonstrate convincingly that quantum systems do not possess dy-

namical properties as pre-existing system attributes and elucidate the relation between wave–particle duality and the uncertainty relations.
- The interpretation of experiments on entangled systems demonstrating among other things that there are non-dynamical (non-local or "statistical" as defined in Sect. 2.6) correlations in quantum many-particle systems. Once again, coherence plays the decisive role.

The corresponding theory of decoherence (Giulini et al. 2013) shows among other things that the "section problem" is a pseudo-problem since the transition from quantum to classical behaviour does not occur suddenly or abruptly and is only marginally related to the measurement problem. Overall, this approach must be judged as the currently most promising and reasonable attempt to comprehend the oddities of quantum systems.

Beside the impossibility of defining dynamical properties as pre-existing system attributes, another serious argument exists against the possibility of a strictly quantum theoretical description of reality. It is similarly impossible to define the concepts of molecular and crystal structure within the framework of a strictly quantum theoretical conceptual system. This definition necessitates the Born–Oppenheimer approximation, treating the atomic nuclei as classical systems, in that well-defined positions are assigned to them in combination with the assumption of vanishing momenta. This is at variance with Heisenberg's uncertainty relation as one of the fundamental quantum theoretical principles. Taken together, the two basic concepts of any physical description of matter, viz., structure and dynamics, cannot be defined within the framework of a strictly quantum theoretical description. Notwithstanding that many people may dream of a "theory of everything" where all existing accepted theories appear as specializations or scientific applications of one and the same theory kernel, unification according to such a naive notion must be considered as an illusion.

In summary, in the course of the historical development of physics, phases of unification repeatedly occurred when theories with an originally autonomous status appeared as specializations of the same superordinate theory. Representative examples are Newtonian mechanics with the unification of terrestrial and cosmic types of motion, and electrodynamics with the unification of theories about electric, magnetic, and optical phenomena. Actually, the historical development of physics is dominated by the formulation of new theories aiming to explore new universes of discourse. Due to this increase in the number and variety of theories, physics in its entirety must be characterized as a pluralistic or poly-paradigmatic science, consisting of a network of grand theories which are reducible neither to one another nor to a single theory. This is already demonstrated by the classical physical description of reality, which is also indispensable for formulating the quantum-theoretical conceptual system. Among the grand theories, there exist relations that are established through generalized correspondence principles between certain concepts and laws and show the degree to which local consistencies exist between different grand theories. Even if the unification of the fundamental interactions turns out to be successful, this result demonstrates that such a "theory of everything" will by no means be a theory that unifies all the grand theories of physics under a single unifying conceptual system. As a realizable objective, it remains to establish a unified physical world view framed by some fundamental, supra-theoretical principles such as symmetries and conservation laws.

References

Born M (1921) Kritische Betrachtungen zur traditionellen Darstellung der Thermodynamik. Phys Zeitschr **22**:218, 249, 282

Caratheodory C (1909) Untersuchungen über die Grundlagen der Thermodynamik. Math Ann **67**:355

Fadner WL (1985) Theoretical support for the generalized correspondence principle. Amer J Phys **53**:829

Falk G and Ruppel H (1972) Mechanik, Relativität, Gravitation. Springer, Heidelberg

Giulini D, Joos E, Kiefer C, Kupsch J, Stamatescu IO, Zeh HD (2013) Decoherence and the Appearance of a Classical World in Quantum Theory. Springer, Heidelberg

Hamel G (1949) Theoretische Mechanik. Springer, Heidelberg

Heisenberg W (1948) Der Begriff "abgeschlossene Theorie" in der modernen Naturwissenschaft. Dialectica **2**:331

Heisenberg W (1972) Die Richtigkeitskriterien der abgeschlossenen Theorien in der Physik. In: Scheibe E, Süßmann G (1972)

Hempel CG (1952) Fundamentals of Concept Formation in Empirical Science. Chicago UP, Chicago, Ill.

Hermes H (1964) Zur Axiomatisierung der Mechanik. In: Henkin L, Suppes P, Tarski A (eds) The Axiomatic Method. Elsevier, Amsterdam

Hertz H (1965) The Principles of Mechanics. Dover, New York

Hilbert D (1899) Die Grundlagen der Geometrie. 9th edn supplemented by P Bernays. Stuttgart 1962

Kuhn TS (1970) The Structure of Scientific Revolutions, The University of Chicago Press, Chicago, Ill.

Kuhn TS (1974) Second Thoughts on Paradigms. In: Suppe F (1977)

Lakatos I, Musgrave A (eds)(1974) Criticism and the Growth of Knowledge. Cambridge UP, Cambridge, Mass.

Landau LD, Lifschitz EM (1979) Lehrbuch der theoretischen Physik. Vol III: Quantenmechanik. Akademie Vlg, Berlin

Lebernegg S, Amthauer G, Grodzicki M (2008) Single-centre MO theory of transition metal complexes. J of Phys B **41**, 035102:1

Ludwig G (1978) Die Grundstrukturen einer physikalischen Theorie. Springer, Heidelberg

Nagel E (1982) The Structure of Science. Routledge & Kegan Paul, London

Ritter J, Gründer K, Gabriel G (eds)(1971-2007) Historisches Wörterbuch der Philoso-

phie. 13 vols. Schwabe AG, Basel-Stuttgart
Mach E (1919) The Science of Mechanics. 4th edn, Open Court, Chicago, Ill.
Planck M (1915) Das Prinzip der kleinsten Wirkung, and: Verhältnis der Theorien zueinander. In: Planck M (1983) Vorträge und Erinnerungen. Wiss Buchges, Darmstadt
Scheibe E, Süßmann G (eds) (1972) Einheit und Vielheit. Vandenhoeck & Ruprecht, Göttingen
Sneed J (1971) The Logical Structure of Mathematical Physics. D Reidel, Dordrecht
Stegmüller W (1973) Probleme und Resultate der Wissenschaftstheorie und Analytischen Philosophie, vols I–IV. Springer, Heidelberg
Suppe F (ed)(1977) The Structure of Scientific Theories. U of Illinois Press, Urbana, Ill.
Tisza L (1963) The Conceptual Structure of Physics. Rev Mod Phys **35**:151
Toulmin S (1974) Does the Distinction between Normal and Revolutionary Science Hold Water? In: Lakatos, Musgrave (1974)

Chapter 4
Theory and Experience

In summary of the two preceding chapters, the general structure of physical knowledge is presented. The analysis of representative examples of physical laws demonstrates that they can be neither discovered nor falsified by experiment alone. The subsequent definition of the concept of physical law enables its demarcation from empirical statements and hypotheses. Since the relationship between theory and experience is not unequivocal, there are various options available to react to discrepancies between theory and experiment. The scientific applications are the largest units subject to refutation in an actual comparison between theory and experiment, and they constitute the contribution of the theory to the empirical content. The methods of consistent adjustment are elucidated with several examples, emphasizing their variety in establishing the relations between theory and experience.

4.1 The Structure of Physical Knowledge

The relations between theory and empirical experience are of crucial importance for the comprehension of physical knowledge. Firstly, the corresponding methods of consistent adjustment contribute decisively to the securing of knowledge. Secondly, these methods describe how concepts, laws, and theories acquire meaning and empirical content. The original objective of logical empiricism, e.g., consisted in the reconstruction of these relations via the reduction of theoretical knowledge to direct sensory perceptions. More generally, the empiricist approaches to reconstructing the relationship between theory and experience rest on a number of assumptions:

(E1) The whole of empirical-scientific knowledge originates from experience: a theory can be considered as genuine knowledge only insofar as its origin from experience is unambigously traceable.

M. Grodzicki, *Physical Reality – Construction or Discovery?*,
https://doi.org/10.1007/978-3-030-74579-0_4

(E2) The entire meaning of empirical-scientific terms and statements originates from experience: a scientific term is meaningful only if it corresponds to a possible experiential act (empiricist criterion of meaning).
(E3) An empirical science must confine itself to those problems having a solution that is either logically traceable or empirically testable.
(E4) Irrevocable theoretical knowledge does not exist: no theoretical statement is secured against revision due to novel experiences (Quine), or an empirical-scientific theory may fail on experience (Popper).
(E5) The sole truth criterion for theoretical assertions and knowledge is empirical experience: experiment is the sole judge of theory.

From (E1), it follows immediately that autonomous, empirically and epistemically significant theoretical knowledge does not exist. (E2) constitutes the basis of the empiricist theory of meaning, which relies on the unambigous correlation between an object and its conceptual description. Accordingly, the meaning of empirical terms rests on their immediate reference to directly perceptible entities, as established by ostensive definitions (Przelecki 1969). Moreover, all theoretical terms should be reducible to empirical terms, and are thus dispensable, in principle. This constitutes the basis of the empiricist program for eliminating theoretical terms. Finally, (E4) and (E5) are meaningful only under the presupposition that a theory consists of a collection of statements each of which can itself be confronted with experience. This assumption is known as the "statement view" of empirical-scientific theories.

The development of the analytical philosophy of science may be interpreted as a chronology of the vain endeavour to reconstruct the relation between theory and experience on the basis of these assumptions (Bayertz 1980). In a first attempt, directly oriented to (E1) and (E2), each theoretical term should be reconstructed as an explicit definition from elementary terms corresponding to the direct perceptions of the individual subject ("sense data") (Carnap 1928). This would enable the elimination of the theoretical terms without any difficulty, because explicitly defined terms can be replaced by their defining equivalent, and are thus dispensable, in principle. It was soon recognized, however, that the complete reduction of scientific terms to sense data as the constituents of direct experience through exclusively explicit definitions could never be achieved. As a consequence, Carnap modified his approach by assuming that the experiential basis of science does not consist of the subjective perceptions themselves, but of their conceptual descriptions, viz., the "protocol sentences" (Carnap 1931). The reconstruction of the relation between theory and experience must then furnish the proof that the theoretical terms are definable by the observational ones contained in the protocol sentences, and that the theoretical sentences are subsequently deducible from the protocol sentences. This approach does not lead to the theoretical terms either, as Carnap himself later admitted (Carnap 1966):

> We observe stones, trees, and flowers, perceive various regularities, and describe them by empirical laws. But regardless how long and how carefully we observe these things, we never arrive at the point where we observe molecules. The term

> "molecule" never occurs as the result of observations. Therefore, the generalization of observations will never yield a theory of molecular processes. Such a theory must be generated in a different way.

Due to these insights the next attempt consisted in reconstructing theories in such a way that theoretical terms and sentences do not appear as generalizations of empirical experiences, but as constituents of a hypothetical, abstract calculus. The connection with experience is then established via the deduced, empirically checkable results. Associated with this hypothetical-deductive empiricism is inevitably the abandonment of one of the basic assumptions of logical empiricism, viz., the conception of a unified scientific language, in favour of a dualistic one (Leinfellner 1967, Stegmüller 1973). Referred to as the double-language model, this partition into an observational language O and a theoretical language T was understood semantically, entirely in the sense of (E2): the terms of the O-language receive their meaning through reference to the directly given objects and facts. The meanings are thus theory-free, completely interpreted, and directly comprehensible, so that the O-language, although not necessarily identical with the vernacular, fulfills the same function, because its sentences ought also to be understandable to non-scientists. In contrast, the terms of the T-language do not possess an autonomous meaning. They receive meaning solely via the connections with the O-language, which are mediated by correspondence rules containing both theoretical and observational terms. In this way, a partial interpretation is assigned to those theoretical terms occurring in the correspondence rules, while all the others receive meaning only indirectly by occurring, together with the partially interpreted terms, in theoretical sentences. The intention of this conception is thus to demarcate that part of science available in the professional language and comprehensible only to experts from that part which is theory-free and ought also to be understandable to the layperson, and to establish this part as the conceptual basis of science.

It is obvious that such a conception is at best applicable to the reconstruction of scientific knowledge, since in the development of science, there is no stage in which an observational language is exclusively used. The failed attempts to formulate precise semantic demarcation criteria between the two languages showed, however, that such a theory of meaning is unfeasible, even to reconstruct the scientific language. These attempts passed through several stages. In the first, the empirical–theoretical distinction was identified with the differentiation into the observable and the non-observable. According to the results of Sect. 2.3, this must necessarily lead to difficulties since observability depends on the current state of instrumental and theoretical development. The objections raised against such an identification rest precisely on this circumstance: an unequivocal demarcation is only possible if one prescinds from the use of instruments of any type. This would be at variance with the pre-scientific understanding of observable because, e.g., objects and properties which can be recognized, not with the naked eye, but at least with a magnifying glass ought to be theoretical. The subsequent idea of admitting "simple instruments" somewhat blurs the demarcation. Whether an instrument is considered to be simple depends again on the technical development and the skills of the observer. Consequently, a semantic distinction between the two languages cannot be established on the dichotomy observable/non-observable.

For this reason, in a second stage, the double-language model was modified by searching pragmatic demarcation criteria for the distinction between empirical and theoretical terms, but still preserving the existence of an essential semantic difference between the two languages. The condition for terms of the observational language was now that different observers under suitable conditions should be able, by means of direct observation, to strike a high measure of agreement if the term applied to the given situation (Hempel 1965). A further criterion for an observational term was that it should be a descriptive (non-logical) term corresponding to a quickly arbitrable sentence (Maxwell 1962). However, the difficulties remained much the same because the demarcation between the two languagues became dependent on the differences in the individual cognitive abilities of the observers. Consequently, a "standard observer" had to be introduced, thought of as an unbiased layperson, with the sole intention of establishing a separation between an understandable and a professional language. Such a procedure transferred not only the subjectivist concept of experience of empiricism to the conceptual domain, but also produced an additional inconsistency by introducing a pragmatic criterion for a semantic differentiation because it was still maintained that a principal semantic difference should exist within the scientific language. Meanwhile, it is generally agreed (Suppe 1977) that the double-language model was unsuitable for providing an adequate description of the relationship between theory and experience because all attempts to construct a reasonable demarcation between the two languages were confronted with the general problem that it depends on the context what can be made directly observable in the field of action and what is comprehensible at the conceptual level.

The impossibility of establishing precise semantic demarcation criteria between empirical and theoretical terms led, in a third stage of the double-language model, to a differentiation into non-theoretical and T-theoretical terms (Sneed 1971), which resembles the concept of a presupposition theory (Ludwig 1978) discussed in Sect. 3.5. Apart from the circumstance that this approach refers only to the theoretical-conceptual level, so that the original objective of the double-language model is abandoned to reconstruct the connection between theory and experience, it is subject to the same objections enumerated in Sect. 3.5. This differentiation returns to the idea that position and time ought to be non-theoretical relative to mechanics, in the sense that they can be defined without reference to mechanics. That this is impossible has been argued in Sect. 3.5: a theoretically precise and operationally realizable definition of length or distance of physical bodies necessitates concepts of mechanics like rigid body and inertial system. Indeed, the actual development of physics has disproved the assumption of such a distinguished status of the length as non-theoretical relative to mechanics. The operational definition of length, as valid since 1983, is based on the speed of light and time. The term "length" thus became a derived quantity and the velocity has taken over the status of a basic one. Hence, the selection of basic quantities is largely a matter of convention, so each quantity of mechanics may appear in principle as a non-theoretical basic quantity or as a derived, T-theoretical one. This is just a change in the conceptual basis, similar to the transition between equivalent axiom systems of a theory.

All things considered, the differentiation into non-theoretical and T-theoretical follows the spirit of the double-language model. It is just that the hierarchically conceived relationship between theory and experience is shifted to the theoretical level and replaced with a hierarchical relation within the system of the physical theories. Consequently, this distinction also has a semantic background: the cognitive status of the non-theoretical terms is conceived in such a way that their physical meaning can be determined without referring to T. As exemplified by the status of the term "length" in mechanics, this distinction corresponds solely to the functional differentiation into old and new terms, or into familiar and less familiar ones. A separation, by semantic arguments, of the physical language into non-theoretical and T-theoretical is, like the separation into observable and non-observable, largely a matter of convention, individual skills, or taste.

The unsuccessful attempts to define semantic demarcation criteria between theoretical terms and empirical, non-theoretical, or observational ones suggest that neither the empiricist theory of meaning nor the empiricist assumptions provide an appropriate basis to analyse the relationship between theory and experience. In addition, the postulates (E1)–(E5) already contain an internal methodological inconsistency. On the one hand, it is denied according to (E1) that theoretical methods lead to substantially new knowledge. On the other hand, according to (E5), experiment ought to be the judge of theory. If, however, the whole of knowledge should actually originate from experience and experiments, i.e., theoretical considerations yield nothing new anyway, it is unintelligible why this must be checked experimentally: experimental examination of theoretical results would rather constitute a redundant activity. In other words, postulates like (E4) and (E5) are only reasonable if it is *not* the case that the whole of knowledge comes from empirical experience. Actually, the relations between theory and experience are far more complex and multifaceted than imagined in empiricist philosophies of science. The relations between them are instead established through an extensive network of theory–experience relations which will be described in this chapter on the basis of a structural model of physical knowledge that follows from the results of the two preceding chapters and is depicted schematically in Fig. 4.1. The basic feature is the distinction between an empirical and a theoretical level of physical knowledge that is, however, in no way related to the double-language model, but rests on fundamental dissimilarities in the tacit assumptions, the methods, and the notion of reality. It involves neither an ontological differentiation into existent and non-existent or into objective and conceptual, nor a semantic one in the spirit of the double-language model.

Starting with the empirical domain, it must be remembered that singular perceptions are not a part of physics. Although they may be interpreted or explained with the aid of physical knowledge, the empirical basis of a theory contains only reproducible events and observations. Yet the basic assumption of the double-language model reveals itself as incorrect because the empirical basis of physics is not constituted by single perceptions and it does not express the conceptual basis in a theory-free observational language. The conception of experience in physics rests rather upon instrumental, representative, reproducible and, in this sense, "law-like" experience, and an essential part of experimental methodology thus consists in generating reproducible data. A necessary prerequisite is the creation of systems providing reifications of the model systems as ideally as possible, and characterizing them physically. Due to the resulting non-eliminable theory determinateness of the empirical basis, the conceptual description of its elements is not an epistemologically unproblematic verbalization of direct experience. Consequently, the language of the empirical domain is not an observational language in the spirit of the double-language model. Such an O-language does not appear as a part of the physical language because the transition from a theory-free O-language to the professional language, like the transition from single perceptions to reproducible observations, does not occur within physics, but precedes it. Consequently, the double-language model constructs a demarcation, at best, where the transition to physics as science occurs, but not within physics.

This is also demonstrated by the question of the observability of properties. The necessary condition for them to belong to the conceptual system of physics is their measurability. The double-language model takes this into consideration in the sense that "simple" to measure properties like volume and mass are deemed to be directly observable. However, physical quantities as quantified properties are never directly observable, but are initially defined operationally by measuring procedures. Since operational definitions always contain conventions and theoretical aspects, they are never theory-free. Just like the conceptual description of the empirical basis, all physical quantities, according to the double-language model, must be attributed without exception to the T-language. Furthermore, in order to structure observations and experimental data sets, empirical hypotheses are framed by means of which effects are interpreted, data sets are reduced to empirical correlations, and groups of objects are structured through the construction of empirical classifications. This requires some theoretical ideas, but not a theory as defined in Chap. 3. Overall, the empirical level is characterized and dominated by a combination of experimental and largely inductive methods. The notion of reality corresponds to an ordered or structured reality that may be claimed to be subject-independent, if not in its entirety, at least in its essential traits.

In contrast, the theoretical level is characterized by primarily deductive methods and a notion of reality which is dominated by constructive aspects

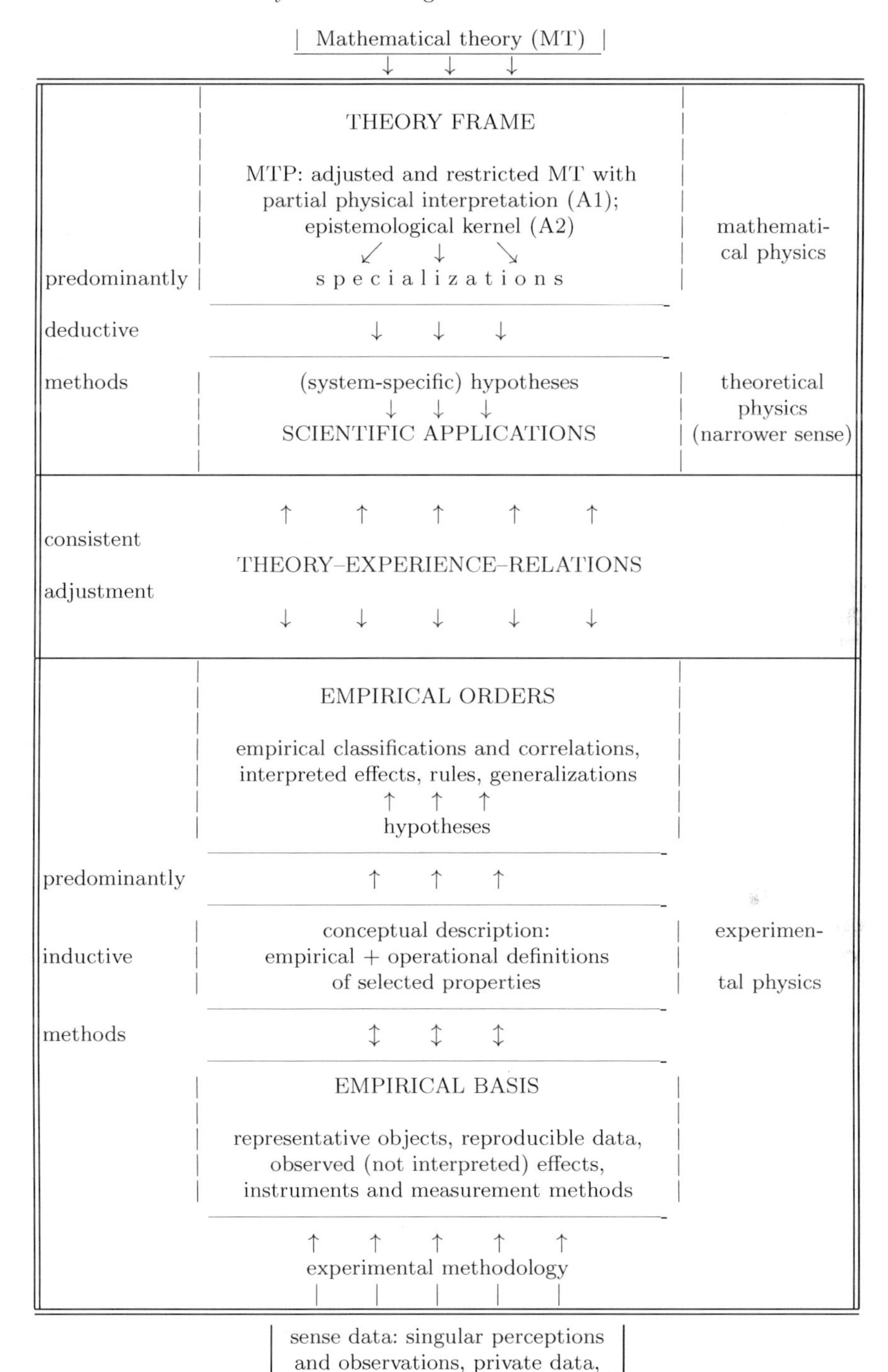

Fig. 4.1 Structural model of physical knowledge

because the underlying structures are not unequivocally correlated with experience, although not freely invented. In the case of a closed theory, the theoretical knowledge is given in the form of a system-independent theory frame and system-specific scientific applications. The theory frame is non-empirical, in that it is not subject to refutation in the comparison between theory and experiment, but is nevertheless not a formal calculus without physical meaning. It comprises a partly physically interpreted mathematical theory which itself is presupposed as given and whose study is not the subject matter of physical theory formation. The epistemological kernel of the theory contains a theoretical structural principle that is represented as a mathematical equation (equation of motion, field or state equation) and defines a theoretical order. This order and the structural principle are not arbitrary, conventionalist or instrumentalist constructs, but constitute theoretically autonomous and epistemically significant knowledge. Reducing the number of independent variables in the mathematical equation of the structural principle provides the specializations whose properties are studied purely mathematically. They are likewise system-independent and do not contain any experimentally comparable, quantitative results. These are obtained by supplementing a specialization with system-specific model assumptions ⟨M⟩ defining a scientific application which may be considered as a concrete example of the general theory conception or as a $\langle C, C\rangle$-model. However, it is not a model in the spirit of formal model theory because it is established by additional theory-external assumptions resting on empirical knowledge. By specifying the contingent conditions (system parameter, initial or boundary conditions) and with further assumptions if necessary, statements on observable facts can eventually be deduced which are comparable with quantitative, experimental data. The terms of the theoretical level are theory-oriented conceptual constructions, but not generalized or specified empirical terms. They receive their meaning, firstly, via the connections to the empirical level through operational definitions and observable effects and, secondly, via their position within the theory, viz., through the formal mathematical definition, through the physical laws and the scientific applications in which they occur, and through the deduced, experimentally comparable results.

In short, against the background of this structural model of physical knowledge, the challenge which the double-language model aimed to solve does not exist. Neither is there an observational language as a part of the physical language, nor exists within this language a basic semantic difference between empirical and theoretical terms. The empirical terms always possess a theoretical component; through the reference to observations alone, they are not exhaustively interpreted, and receive in this way only a part of their meaning. Another part is provided by the relationship with the theoretical level. The theoretical terms receive their meaning both via the relation with the empirical level and through their position within the theory. For these reasons, the empirical terms do not possess a monopoly regarding interpretation or meaning of the kind ascribed to the observational terms of the double-language model. The autonomy of theoretical knowledge becomes manifest in the circumstance that the theoretical terms acquire an important part of their meaning irrespective of the reference to reality, so they are not eliminable. For instance, the discussion of the velocity terms in Sect. 3.3 exemplifies the fact that a strict separation of the physical language into observational and theoretical terms is not meaningful and contradicts the intention to construct a conceptual system unifying the theoretical and operational aspects in a single concept. As a consequence, not only is the search for semantic demarcation criteria between empirical and theoretical, or observable and non-observable, or between the understandable and gobbledegook obsolete, but so also is the entire program of elimination of theoretical terms. Instead, the empiricist demand for elimination is to be replaced with the consistent adjustment between the empirical and theoretical conceptual levels. Accordingly, the differentiations important for the structure of physical knowledge are entirely different:

1. An ontological differentiation into objective and conceptual which is based on fundamental dissimilarities between objects, observable effects, and perceivable regularities on the one hand, and their conceptual descriptions on the other. Regarding the ontological status of objects, effects and regularities, it does not matter whether they were initially characterized experimentally or theoretically, nor whether the description took place on the empirical or theoretical level, because experimental, empirical, and theoretical methods are all concerned with the same reality.
2. An epistemic differentiation at the conceptual level which is not conditioned by semantic dissimilarities, but by essential methodological differences between empirical and theoretical formation of concepts, models, and hypotheses.

The preceding chapters described in detail the different status of empirical and theoretical concept constructions, empirical and theoretical hypotheses, and also empirical and theoretical orders. A further aspect of this epistemic differentiation is the difference between the physical laws, as the scientific statements at the theoretical level, and the various types of statements at the empirical level. In addition, the physical laws provide an important contribution to the relations between theory and experience because they appear as premises in the theoretically deduced propositions and data. In this respect, the relationship between laws and experience is representative for the general relationship between theory and experience. Accordingly, one must first define what is meant by physical laws in combination with a thorough analysis of their status and function.

4.2 Physical Laws

In Chap. 2, it was argued in detail why physics is not concerned with the description of direct perceptions and the investigation of unspecified objects of an "inanimate" nature. A central topic of fundamental research in physics is rather the structures of reality that are described conceptually by physical laws. Therefore, the physical laws are not only indisputable and fundamental constituents of physical knowledge that form the basis for understanding natural phenomena, but also essential parts of the relationship between theory and experience, since they appear as premises in the experimentally comparable results which are deduced within a theory. In spite of this paramount importance, there is no consensus about the meaning, function, and epistemological status of physical laws, even among physicists. This is reflected in the broad spectrum of opinions:

- Naive realism: physical laws are constituent parts of nature ("natural laws") which are discovered by careful observation and experiment, similarly to things and phenomena.
- Inductivism: physical laws are inductive generalizations from direct perceptions and observations.
- Critical rationalism: laws are empirically highly confirmed hypotheses, i.e., propositions that are valid with very high probability, but remain tentative and never represent irrevocable knowledge since they may reveal themselves to be false by virtue of new experiences or experiments.
- Conventionalism: laws are solely economic representations of experimental data without providing any new additional knowledge.
- Instrumentalism: laws serve as a mere auxiliary means to deduce and predict experimentally comparable data ("facts"), without any ontological or epistemic claims.
- Radical constructivism: physical laws are largely free inventions of physicists.

These broadly differing opinions can be put down to a number of things. Firstly, physical laws are not distinguished from empirical propositions, viz., rules, algorithms, causal hypotheses, empirical generalizations, and correlations, as discussed in Sect. 2.4. Secondly, there is no clear distinction between laws and hypotheses. Thirdly, no distinction is made between an observable regularity as a part of external reality and its conceptual representation by a physical law. An essential presupposition for understanding the relationship between theory and experience is the precise specification of the concept of law in physics by determining its constitutive properties. These comprise its formal structure, its status within physical knowledge, and its methodological function, especially concerning the comparison between theory and experiment. Only afterwards can the demarcations between hypotheses and the various types of empirical statements be reasonably discussed.

While there are rarely any serious doubts that concepts and theories are conceptual constructs, a widespread belief prevails that physical laws are discovered (Peierls 1955, Feynman 1967, Rothman 1972, Holton und Brush 1985), although they themselves rest crucially on conceptual constructs. Actually, this belief rests on the naive-realistic identification of reality with its conceptual description. Although the observation of repetitions, regularities, invariances, and relatively constant structures may serve as indications for describable law-like connections, these qualitative structures of reality are not identical to the law itself. If this dissimilar ontological status between an observable regularity and its conceptual description is ignored, it may seem that physical laws can be discovered. Actually, even the reduced view is untenable, that laws could be inductively inferred solely from observations. Two representative examples may serve to disprove both views.

The first example is the law of inertia or first law of Newton as one of the fundamental principles of classical mechanics: "Every body remains in a state of resting or moving uniformly in a straight line unless forces on it compel it to change its state." If this law could actually be discovered or directly induced from observations, the question immediately arises whether Aristotle did not find this law simply due to insufficient observations. That this is not at all the case is demonstrated by the presuppositions required for its formulation. In the first place, it is the theoretical construction of the process of a uniform rectilinear motion; theoretical because it is not a possible empirical experience, but is in fact rather at variance with direct experience. For instance, on Earth, every "freely moving" body, i.e., undergoing no forces according to common understanding, comes eventually to rest. Moreover, due to the rotation of the Earth, no reference frame rigidly connected to the Earth corresponds strictly to an inertial system. Consequently, a strictly uniform rectilinear motion cannot be observed on Earth, so it can be inferred solely by extrapolation in thought, i.e., by theoretical construction. The extensive explanations in textbooks which aim to make this principle comprehensible confirm this fact. If this law were obvious through direct experience and without any theoretical considerations, such efforts would be unnecessary. They are necessary precisely because understanding the law of inertia necessitates a basic rethink, very different to notions of motion that rest on direct experience. On this basis, a second step is necessary in this thought process, because it must be shown how the seemingly contradictory everyday experience that every body eventually comes to rest is to be interpreted within this conceptual framework. To this end, the coming to rest is reinterpreted within the framework of Newtonian mechanics as the result of an external influence, viz., friction, rather than as an internal property of the moving body (impetus theory) or as a natural property of terrestrial motions which does not need an explanation. This interpretation can be supported by experimental demonstration by gradually reducing the influence of friction. It is obvious, however, that such experiments merely demonstrate that the principle of inertia is consistent with direct empirical experience, and that this type of interpretation of motion is also feasible. As a result, these experiments can neither be taken as proof of the law of inertia nor as its inductive inference from observations, because the experimental demonstration presupposes knowledge of this law as its theoretical basis. In summary, it has been shown that the law of inertia can neither be discovered, because it is grounded on theoretical conceptual constructs in the sense that it postulates a non-existent model process, nor can it be inductively inferred from direct experience, since it contradicts it.

A number of further reasons are provided by the derivation of Newton's gravitational law as an example of a quantitative mathematical relation. The phenomenon of mass attraction may be considered as a corresponding observable regularity. Actually, there are just certain events that may be interpreted in this way, such as the falling of apples from a tree or the motion of the planets around the Sun. However, from such events, there is no unequivocal route to the mechanism of mass attraction, let alone to the gravitational law. Even before Newton, it was known that apples fall from trees, but nobody viewed this as an effect of a gravitational force or as an analogy to the Moon falling towards the Earth. Rather this necessitates additional constructive thought in the form of theoretical ideas and model assumptions. Within the framework of the theoretical order established by Newton, deviations from uniform rectilinear motion must be ascribed to the action of external forces. In the case of planetary motion, it follows that a force must be constructed perpendicular to the direction of motion of the planet, i.e., a centripetal force, corresponding to the decomposition of the motion into a tangential and a radial component. This decomposition cannot be justified solely on the basis of empirical arguments, because any other decomposition is also conceivable, e.g., it could be said that zero is composed of 5 and -5, but the same applies to a and $-a$ for any real number a. Any such decomposition is thus convention. In this respect, the centripetal force is not the cause of the central motion, but the result of its description within the framework of Newton's schema (Hertz 1965).

Having constructed the centripetal force as the mechanism of attraction between the Sun and the planets, a mathematical expression for this mechanism must be formulated, because quantitative and experimentally comparable results are not deducible otherwise. In the first step, Newton showed that, for circular motion with orbital radius r and orbital velocity v, the centripetal force is proportional to v^2/r. Secondly, the modulus of v is constant for a circular motion, so v is proportional to r/T with the orbital period T. In combination with Kepler's third law, $T^2 \propto r^3$, one obtains v^2 proportional to $1/r$, so finally the centripetal force must be proportional to $1/r^2$. This ingenious consideration is in no way comparable with the setting up of empirical correlations described in Sect. 2.4. There are a number of reasons why this mathematical expression for the gravitational force does not follow from observation:

- The existence of a central force arises only within the theoretical framework established by the principle of inertia and the Newtonian schema defining change of motion and attributing it to the action of external forces.
- Moreover, the mathematical expression for the centripetal force cannot be found by empirical methods because the decomposition of the planetary motion into a tangential and radial component cannot be inferred inductively from mere observations.
- The simple derivation, as described above, applies only to circular orbits. The derivation of the analogous relation for elliptical orbits requires considerably more effort, as can be found, for example, in Newton's *Principia*.
- As already mentioned in Sect. 3.5, the assumption that the Sun is the fixed center of force is inconsistent with Newton's third law. In fact, the planets also exert a force on the Sun and accelerate it. This leads to the following modification of Kepler's third law:

$$\left(\frac{a_1}{a_2}\right)^3 = \left(\frac{T_1}{T_2}\right)^2 \frac{M_\odot + m_1}{M_\odot + m_2},$$

  containing, in addition, the masses m_i of the planets and the mass $M_\odot$ of the Sun.
- As already recognized by Kepler himself, the actual planetary orbits deviate to varying degrees from the ideal elliptical shape, so they do not agree exactly with the calculations of the gravitational theory. From these deviations, one may estimate the limits within which the power of $1/r$ in the gravitational force can vary around the value of 2 if Kepler's third law alone is taken as basis. For example, Eddington showed that the perihelion rotation of Mercury's orbit can be reproduced when $1/r$ has an exponent of 2.000 000 16. Actually, elaborate calculations have shown that these deviations are predominantly caused by the interactions between the planets.
- The derivation rests on the assumption that the masses of the Sun and the planets are concentrated in single points. Actually, the deviations from the ideal spherical shape and the inhomogeneity of the mass distribution associated with finite extensions requires modifications of the gravitational force that are non-negligible when calculating, e.g., the orbit of an artificial satellite around the Earth.

Due to these numerous constructive elements a quantitative physical law must not be identified with a qualitative, observable regularity. This also stresses the way experimentally comparable data are deduced from the gravitational law, which reveals its methodological and epistemic significance. To this end, the force in the second law is replaced by the expression for the gravitational force. The subsequent integration yields conic sections as solutions that correspond to the possible trajectories. Although the general structure of the *possible* trajectories is now known, the one realized in a particular case is not. Hence, quantitative results comparable with measurement data are not yet available, and all the more so in that the directly observed planetary trajectories are not ellipses. This requires information about system parameters like the masses of the planets, orbital semi-axes, and orbital inclinations, and the specification of initial conditions like position and velocity at a certain point of time. Finally, one must take into account that the actually observed planetary trajectories are obtained as the result of the combination of the motions of the planet and the Earth. All this must be included as *supplementary* information that is not contained in the law itself. Consequently, the crucial point of the law-like description consists in the decomposition of an observed phenomenon into a general part and a particular part: the general one is represented by the law, assumed strictly valid, while the particular one comprises contingent attributes such as initial conditions and system-specific parameters. Such a decomposition cannot be justified on the basis of purely empirical arguments because it depends on the theoretical context. For example, considering the harmonic oscillator, friction is a contingent attribute to be minimized in the experimental design. Concerning the damped oscillator, friction is part of the equation of motion, i.e., the law. It may also happen that such a separation is not feasible. In such a case, a law-like description is either impossible or only possible to a limited extent. As a consequence of this separation, the law may contain more general structures and may provide additional connections and insights that cannot be obtained by the mere summary of data in the form of empirical correlations. Regarding the gravitational law, this greater universality is manifest in the higher symmetry that it possesses compared with particular solutions. The equation of motion of a central force is invariant under spatial rotations around the center of force. In contrast, any possible trajectory given as its solution lies in a spatially fixed plane and does not exhibit this symmetry. The angular independence of the central force and the associated symmetry of the physical law are not reflected in the particular solutions, i.e., the trajectories. Therefore, they cannot be derived from observations, but constitute an additional theoretical structural element on a higher level of abstraction. On the other hand, for this reason the law yields a spectrum of possible solutions, but does not provide any statement regarding the solution realized in a special case. All things considered, the abstraction carried out by separating the law from contingent conditions is methodologically entirely dissimilar from the type of abstraction underlying the classificatory concept formation of empirical methodology. Actually, this difference, reflected in the tacit assumptions (3A)–(3C), is not only the reason for the effectivity of a law-like description for gaining, structuring, and securing physical knowledge, but it is also a fact that, without this type of abstraction, physics in its contemporary form would not exist.

A physical law is thus neither the mere summary of empirical data nor the conceptually unproblematic description of observable regularities, because the decomposition of an observable phenomenon into a general law and contingent attributes creates additional structures not amenable to direct observation. Firstly, these structures represent theoretically autonomous knowledge in that they constitute at the outset a constructed reality. Nevertheless, they are not at all considered to be fictitious, but are associated with the ontological claim of validity to correspond to a structure of reality. It is only the inadmissible identification of this structure with the conceptual representation by the law that creates the impression that physical laws are parts of nature and can be discovered by observation. Secondly, the structuring of knowledge in the framework of a law-like description leads to a unification which exceeds by far any economical summary of empirical data and which cannot be justfied solely by empirical methods, but takes place on the basis of theoretical concept constructions, model assumptions, and theoretical orders. Without this theoretical background, a law such as the law of inertia cannot be inferred from even the most careful observations, just as the gravitational law cannot be discovered by simply watching apples falling.

Due to this background a physical law cannot be discussed in isolation from the theory in which it is embedded. For this reason, the second and eventually constitutive property is the embedding or incorporation in a closed theory, either in the theory frame or in one of the secure scientific applications ScA. This largely determines the further properties of laws. Firstly, physical theories are formulated in mathematical language, so that formally a physical law represents a quantitative functional relation given as a mathematical equation (Feynman 1967):

> Every one of our laws is a purely mathematical statement.

Secondly, since theories are concerned with the investigation of physical properties, a law is a relation between physical quantities which is entirely independent of particular systems or processes and thereby acquires its universality (Planck 1983):

> A physical law is any proposition which expresses a definite, steadfastly valid interrelation between measurable physical quantities.

Thirdly, as a proposition within a theory, it is a conditional statement whose range of validity is determined by the theoretical context as defined by the axioms of the theory or the model assumptions in an ScA: the law is strictly valid only relative to these model assumptions, but not in an absolute sense. Finally, theoretical embedding in a closed theory has important methodological consequences because the law, as a part of secured knowledge, is non-hypothetical in the sense that it is not subject to refutation in the comparison between theory and experiment. As a consequence, the law assumed as strictly valid, defines a theoretical expectancy attitude and acts as a theoretically well-grounded norm, relative to which experiments are interpreted, explained, and assessed. This normative function, however, must not be understood as saying that nature obeys the laws or that nature is governed by them. Physicists act in their experimental and theoretical practice in accordance with the laws, but not nature. In this respect, physical laws serve as guidance for the design of experiments and as guidelines for both their interpretation and the gaining of new knowledge. According to these properties, a physical law may be defined as follows:

Definition 4.1 A physical law is a scientific statement, conceptually embedded in a closed theory or in one of its secure applications, either as an axiom or as a deducible theorem. It constitutes a functional relation between physical quantities, independent of contingent conditions and parameters, and represents theoretically autonomous, epistemically significant knowledge. As a part of secured knowledge it is not subject to refutation in the comparison between theory and experiment, but serves as a theoretically well-grounded guideline for the interpretation, explanation, and assessment of experiments. It is associated with the demand of being strictly valid regarding the model assumptions of the theory, and to correspond to a structure of reality.

This definition implies a number of consequences both for the status and the function of laws and for the distinction from empirical statements. Firstly, laws are concerned with structures, but not with facts, and as verbal formulations, they are statements *about* nature, but not a part of it. Even if the structure of reality postulated by the law is assumed to exist objectively, it is not identical to the conceptual representation by the mathematically formulated law. Such an identification conforms again to naive realism, which does not distinguish between reality and its conceptual description. Secondly, physical laws do not represent causal relations. As mathematical equations they are equivalence relations regarding their formal structure. Due to the symmetry between the two sides, the law itself does not permit any conclusion which side should be assigned to the cause and which one to the effect, nor whether the associated structure of reality is causal. Thirdly, although physical laws embody highly intersubjective knowledge, they do not represent objective, theory-independent knowledge and nor are they unequivocal and necessary descriptions, let alone mere reflections of structures of reality. Actually, inter-subjectivity is achieved by virtue of methodological standards, rather than through a property enforced by nature. In particular, the mathematical presentation as a relation between physical quantities rests on conventions which are non-eliminable since they are already contained in the definition of these quantities. Fourthly, physical laws do not satisfy a principle of economy which would postulate their simplicity, in neither a conceptional nor a formal respect. Regarding the mathematical presentation, it does not apply, because a law as a quantitative statement can be improved, in principle, in order to obtain better agreement with experimental data. The functional relation then becomes more complicated as the demands of quantitative accuracy grow greater. In the conceptional respect, simplicity is related not to the mathematical form, but to the unification of knowledge in the framework of a law-like description, in the sense that, for example, a greater number of observable facts can be deduced from a single functional relation. Moreover, the associated structures of reality need not actually be simple: the successes of a law-like description do not admit the conclusion to the simplicity of nature, but at most indicate limitations of physical methodology or human cognitive capabilities. Finally, since physical laws, due to the abstraction from direct experience, rest crucially on thought constructions, they constitute theoretically autonomous and epistemically significant knowledge, in that they provide important contributions to structuring and securing knowledge. For all of these reasons, physical laws are not largely arbitrary conventionalist constructs and their function cannot be reduced to the purely instrumentalist aspects of explanation and prediction of empirical facts. The formulation of laws rather constitutes, together with concept formation and the construction of theoretical orders, the central part of theoretical methodology.

All three aspects of the concept of law, viz., formal structure, status, and methodological function must be taken into account for the demarcation from the empirical propositions. These dissimilarities provide another concrete example of the structural differences of empirical and theoretical methodology and knowledge. First of all, the separation of the law-like relation from the contingent conditions that is constitutive for the concept of law cannot be justified by purely empirical arguments. Accordingly, it does not occur among most of the empirical statements. In particular, the rules demonstrate that, without this separation, the statements obtained are only valid with a certain probability. Empirical generalizations differ from physical laws in assigning a property to an object. In contrast, laws are relations between properties, so that between law and empirical generalization there is a structurally similar difference to the one between a theory and an empirical classification. In particular, the identification of empirical generalizations with laws, as usually occurs in empiricist philosophies of science, leads to entirely misleading accounts of the relations between theory and experience. Collectively, causal hypotheses, rules, and empirical generalizations differ from laws in that they are not given in mathematical form throughout. Finally, the normative function of physical laws is absent from all empirical statements. Since they do not represent secured knowledge, they possess a more heuristic function.

As already mentioned in Sect. 2.4, the securing of the knowledge and the claims of validity associated with the empirical statements is not accomplished by empirical methods, but mostly via physical laws. This type of securing thus constitutes one of the connections between theory and experience. Regarding rules, securing takes place by isolating the particular conditions that lead to the exceptions, i.e., by identification and subsequent separation of contingent conditions. In the form of the problem of induction, the transformation of empirical generalizations into secured knowledge played a central role in the empiricist philosophies of science. In physics this is not at all the case, because the assertion of universality of empirical generalizations is not proven by inductive methods. They are transformed either into system definitions, i.e., into model assumptions of a secure application, or into deducible theorems. For instance, the statement "all metals conduct electric current" arises as a deducible theorem in the framework of the electronic theory of solids, according to which a metal is a solid whose occupied and unoccupied electronic states are not separated by an energy gap. Based on this definition, this empirical generalization becomes a theoretically deducible statement. For this reason, empirical generalizations are only a transitional state on the way to secured knowledge. Causal hypotheses occurring frequently in the phenomenological interpretation of physical effects are similarly transformed into deductive arguments, but without establishing a causal connection, for the following reason. With numerous effects, there is also an "inverse" effect, so that cause and effect are arbitrarily interchangeable. This can be exemplified by the piezoelectric effect, according to which electric fields are generated at surfaces or interfaces in certain crystals by a mechanical deformation. Conversely, applying an electric field to these crystals induces changes in the geometrical structure corresponding to a mechanical deformation. Consequently, the structure of reality consists in the mutual influence of mechanical deformation and electric field. In contrast, the distinction between cause and effect is defined only through the experimental context, i.e., by the decision of an experimenter, so that it does not correspond to a structure of the subject-independent reality. Finally, regarding algorithms, securing is unnecessary. Although they exhibit a similar formal structure to laws, they define a purely computational procedure without any demand of epistemic or ontological significance, i.e., the functional relation does not need to correspond to a structure of reality and nor is there any need for theoretical embedding. Therefore, in contrast to empirical correlations, they are rarely theoretically deducible and are not designed with regard to this consideration, since they pursue just the representation or computation of data as precisely as possible.

The greatest similarity, and thus the closest relation, between physical laws and the empirical level comes about through the empirical correlations that may be considered as the theory-independent representation of measured data, in the sense that their construction takes place according to empirical methodological rules, but not on the basis of theory-specific assumptions. In contrast to algorithms, the arbitrariness of the functional relation is constrained by the claim of being as simple as possible, because its choice is guided against the background of future deducibility from a theory. When such a deduction is possible, the empirical correlation corresponds, as a hypothesis at the empirical level, to a law as a statement within a secure application of a theory. Due to this theoretical embedding, the empirical correlation is transformed into secured knowledge. In particular, it is proven that the inductively obtained correlation is not accidental, but does indeed represent an epistemically significant relation, thus being a constituent of the physical knowledge of nature. Ohm's law, for example, can be deduced in the framework of the electronic theory of solids. This establishes not only the precise functional relation in combination with the range of validity, but also the epistemic significance and universality. Another historically important example is provided by the development of Planck's radiation law. In the first step, Planck attempted to construct an interpolation formula between the two known limiting cases of the radiation laws of Wien and Rayleigh–Jeans. The result he obtained in the autumn of 1900 already had the functional form of his radiation law, but the status was merely that of an empirical correlation. Due to the missing theoretical embedding, it had thus to be considered only as preliminary, but not as secured. This required a second step whose importance was absolutely clear to Planck (Planck 1983):

> The question of the law of the spectral energy distribution in the radiation of a black body could thus be considered as finally accomplished, but there now remained the theoretically most important problem: to provide an appropriate justification of this law, and this was definitely a more difficult task.

For this theoretical deduction he needed Boltzmann's statistical interpretation of the entropy (his "act of desperation"), as well as the additional assumption that energy is exchanged only in quantities of finite size during the interaction between radiation and matter. The fact that the latter assumption is commonly considered to mark the point of inception of quantum theory demonstrates the crucial importance of the theoretical embedding in order to arrive at secured knowledge which may itself lead to novel insights. Therefore, as a general rule, an empirical correlation is subject to refutation as long as it has not been deduced from a theory. In turn, the theoretically deduced law is itself directly connected to a statement at the empirical level. Accordingly, the empirical checking of laws by significant empirical correlations would be ideally suited for checking theoretical knowledge. Apart from the circumstance that not all laws can be related to empirical correlations, this procedure is in general so involved and all in all such an inefficient method of gaining and securing knowledge that it does not usually precede the theoretical deduction of laws. However, empirical correlations that are deducible within a theory represent a special kind of physical law, and between the two of them is no longer distinguished. For instance, one speaks of Ohm's law regardless of whether the empirical correlation between applied voltage and measured current is concerned, or a law of electrodynamics, or a law of solid state theory. Therefore, the distinction between law and empirical correlation is meaningful only as long as there is no embedding in a theory. In this case, in particular, the normative function is absent because such a correlation only has a tentative and hypothetical status as part of physical knowledge. A representative example is the Titius–Bode series.

Finally, the theoretical embedding of the laws, in combination with the normative consequences, also provides the crucial difference with hypotheses. If a single statement, e.g., in an ScA, is subject to refutation, so that it is a matter of hypothesis rather than a law, this rests according to Sect. 3.2 on its incomplete embedding in the theory, e.g., since the hypothesis is insecure with regard to the ScA. In turn, the argument for assessing a scientific statement as a law rather than a hypothesis rests neither on a high degree of empirical confirmation nor on its resistance to attempts at falsification, but on theoretical embedding. Accordingly, a theoretically entirely isolated hypothesis at the empirical level does not become a physical law due to a high degree of empirical confirmation or because it has not yet been falsified. As a consequence, there is no continuous transition from hypothesis to law that might be reconstructed as increasing empirical confirmation. Actually, the notion that laws are more or less well confirmed hypotheses can be traced back to the improper identification of physical laws with empirical propositions. The crucial step from hypothesis to law is rather the embedding in a secure ScA of a closed theory. For these reasons, laws possess a significantly different function and status in the research process than hypotheses since, as secured knowledge, they are not subject to refutation in comparing experiment and theory.

This result emphasizes that the problem of induction does not play any role, either in establishing the relations between theory and experience, or in securing knowledge. Consequently, it is not a problem of physical methodology. The claims to validity of both empirical statements and physical laws are not met by repeatedly citing confirming instances. In Sect. 3.2, it was mentioned that the methods for securing knowledge are largely complementary to those used to acquire it. The securing of empirically obtained knowledge occurs primarily by theoretical methods, viz., by embedding in a theory, because any attempt to secure empirically obtained knowledge by empirical methods leads to a methodological circle. Conversely, the securing of theoretical knowledge, represented by the physical laws, occurs partially with empirical methods, but in a manner that is likewise unrelated to the problem of induction. Most importantly, this concerns the issue of securing claims to strict validity and claims of ontological significance. Regarding strict validity, this is not the case because this claim is not absolute, but tied to the condition that the model assumptions of the theory are precisely fulfilled. A law is strictly valid only for model systems that are non-existent in the sense that they are not an element of reality. Consequently, this claim to validity cannot be justified or refuted experimentally. The claim of ontological significance, viz., that a law corresponds to a structure of reality, is considerably more difficult to justify, for two reasons. Firstly, although a physical law is a functional relation between measurable properties, the carriers of these properties may be non-existent model systems. Secondly, a term contained in the law may not have a counterpart in reality, as exemplified by inertial forces. It must then be asked in what sense a law containing an inertial force corresponds to a structure of the subject-independent reality. The securing of both claims necessitates some effort, but provides important insights into the relationship between laws and experience and is the subject matter of the two following sections.

4.3 Theory–Experiment Comparisons

Since the claim that physical laws are strictly valid and epistemically or even ontologically significant, as well as the securing of the knowledge represented by them, cannot be provided by the proof that laws can be discovered by observations or inductively inferred from them, it must necessarily be carried out after their formulation. This corresponds to the approach of hypothetico-deductive empiricism, so one of its key issues is the justification of the claims of validity of general theoretical statements. According to (E5), experiment is ascribed the key role in this process, i.e., securing of theoretical knowledge ought to be carried out exclusively through empirical testing, and agreement with experiment is the sole criterion of truth. In order to put this into practice, theoretical propositions like physical laws must be formulated in such a way that they are directly testable by comparison with experimental data. This testing of theoretical knowledge should even be the main task of experimental methodology, because the success and general acceptance of physical knowledge should rest precisely on the circumstance that theoretical speculations are permanently scrutinized by experiments. Although the comparison between theory and experiment constitutes an important part of the connections between empirical and theoretical knowledge, it is far from being the only one. Actually, many views about the structure of physical knowledge are misleading because the connection between theory and experience is reduced to this aspect. In contrast, these connections involve a number of quite different aspects. Among the most important ones are the assessment of a discrepancy between theory and experiment in combination with the strategies to remove it. Both are crucially dependent on the status of the relevant theory, viz., whether a secure application of a closed theory is concerned or not. In order to make the subsequent analysis more precise, two assumptions are made at the outset:

- The theory is closed: this means, that the theory frame in the form of the axioms (A1)–(A3) is not subject to refutation.
- The scientific application within which the statement to be tested is deduced is secure: this means that in addition the model assumptions $\langle \mathrm{M} \rangle$, i.e., the hypotheses of type (H1)–(H3), are not subject to refutation, and discrepancies between experiment and theory can be localized relatively easily.

The second assumption will later be abandoned. An example of a secured application of classical mechanics is the motion of the non-existent model system known as the "mathematical pendulum", defined by the following model assumptions:

(H1) The system is a mass point of mass m hanging on a massless string of constant length ℓ.

(H2) The sole force acting on the mass point is the gravity, assumed constant.

Massless strings do not exist, of course, so (H1) means that a "thread pendulum", i.e., the experimental realization of the mathematical pendulum, must be designed in such a manner that the mass of the string is negligible compared with the mass of the oscillating body. (H2) implies that friction forces, e.g., air friction or at the mounting of the string are negligible. Since the motion is planar and the trajectory is a circular arc under these premises, the introduction of polar coordinates corresponding to an assumption of type (H3) enables one to express the position vector in terms of the measurable coordinates (ℓ, φ). If in a first step the inertial mass m_t and gravitational mass m_s are distinguished, and g denotes the gravity acceleration, the equation of motion is obtained as

$$m_\mathrm{t}\ell\frac{\mathrm{d}^2\varphi}{\mathrm{d}t^2} = -m_\mathrm{s}g\sin\varphi \Longrightarrow \frac{\mathrm{d}^2\varphi}{\mathrm{d}t^2} = -\frac{m_\mathrm{s}}{m_\mathrm{t}}\frac{g}{\ell}\sin\varphi\ . \tag{4.1}$$

The general solution of this differential equation is rather complicated and leads to elliptic integrals of the first kind. Therefore, in elementary treatments, a simplifying mathematical assumption of type (H7) is made, namely that the amplitude of the oscillation is small. In this case, $\sin\varphi$ can be approximated by φ, and one obtains

$$\frac{\mathrm{d}^2\varphi}{\mathrm{d}t^2} = -\frac{m_\mathrm{s}}{m_\mathrm{t}}\frac{g}{\ell}\varphi\ . \tag{4.2}$$

This equation has the general solution

$$\varphi(t) = A\sin(\omega t) + B\cos(\omega t)\ , \tag{4.3}$$

with $\omega^2 = (m_\mathrm{s}/m_\mathrm{t})(g/\ell)$. The constants A and B can be determined through assumptions of type (H5) regarding initial conditions, chosen in accordance with the experimental design. Then, for given time points, the corresponding points of the trajectory can be calculated and compared with the measured ones. According to an assumption of type (H6), the oscillation period T can be introduced as a directly measurable quantity which is related to the angular frequency ω of (4.3) by

$$T = \frac{2\pi}{\omega} = 2\pi\sqrt{\frac{m_\mathrm{t}}{m_\mathrm{s}}\frac{\ell}{g}}\ . \tag{4.4}$$

This dependency of the oscillation period on the masses may be discussed from two different viewpoints. Assuming the validity of the general theory of relativity, the oscillation period is independent of the mass due to the equivalence of gravitational and inertial mass, i.e., $m_\mathrm{t}/m_\mathrm{s} = 1$. Conversely, one could adopt the position that, due to the experimentally observed independence of the oscillation period from the mass, the relation $m_\mathrm{t}/m_\mathrm{s} = 1$ can be inferred from experiment inductively. A simple consideration demonstrates that such an argument is not conclusive. Formally, the dependence of the oscillation period on m_t/m_s, as follows from (4.4), is a deductive hypothesis of type (H9) that is directly comparable with experiment. If the experiments show that T varies somewhat with the mass, which can be expected, at least due to the approximations contained in (H2) and (H7), one must conclude that $m_t/m_s = 1$ is not exactly valid because none of the other assumptions concerns the mass. Such differences, however, will be justified in a different way, e.g., by the insufficient accuracy of the experimental data for T or the inadequate experimental realization of the non-existent model system, the mathematical pendulum, by the thread pendulum. This demonstrates once again that, solely on the basis of experiments, precise statements are impossible. Indeed, the hypothesis "T is independent of m" is not subject to refutation in the case of a possible discrepancy with experiment, since the conviction that $m_t/m_s = 1$ has a higher rank than the belief in the obtained experimental data. The fact that this theoretical background knowledge actually dominates is confirmed by the treatment of the thread pendulum in most textbooks, where the factor m_t/m_s in (4.4) does not appear, i.e., the equivalence of the masses is assumed anyway. The same applies when the functional dependence of T on the length ℓ of the pendulum should be tested experimentally. This is necessary in principle, because (4.4) is a formally deduced hypothesis. Actually, discrepancies between measured and calculated results are again ascribed to the non-ideal experimental conditions. This pattern of argumentation can be generalized to the statement:

> In a secure scientific application, the secureness of the theoretically deduced results is more highly ranked than the reliability of the experimental data.

This normative attitude, which may be considered to be an additional defining property of a secure application, means that discrepancies between experiment and theory are ascribed to the experimental side. Since in such a case theory–experiment comparison obviously does not consist in the empirical test of a theoretical result, the question arises as to which function it may have in a secure application.

Most of the scientific applications of closed theories deal with non-existent model systems, such as the mass point, the mathematical pendulum, the rigid body, or the ideal gas, since only theoretically constructed systems are completely characterized in the sense that their properties are defined by the theory and can be calculated, in principle, to any desired degree of accuracy. The theory provides an exhaustive description of the model system, and the assertion of strict validity refers only to this system. The experiment seems to be superfluous, if not meaningless, because the theoretically defined model system does not exist in reality and can be realized only approximately. For this reason, the function of experiment cannot consist in verifying or falsifying a theoretically deduced result, but rather in exploring the scope of the theory when applied to real systems, since this scope cannot be defined within the theory itself. Actually, the model assumptions of the application define a theoretical standard, and the comparison with experiment informs as to how close and under which conditions a particular real system meets this standard. For instance, the knowledge, available only experimentally, as to how accurate and within which range the behaviour of various real gases agrees with the theoretical standard, defined by the model of the ideal gas, not only contributes to physical knowledge about reality, but also supplies data to formulate a state equation of real gases. All this is entirely unrelated to the verification or falsification of a law or a theory by an experiment.

The same appplies to the experimental demonstration of physical laws that are strictly valid only for the model systems. The experiment provides the necessary connection with the empirical level by demonstrating that the general statements of the theory regarding non-existent model systems are also applicable to real systems. Examples are the demonstration of the mass-independence of the oscillation period of the thread pendulum, Galileo's experiment with the inclined plane, Atwood's demonstration of the law of falling bodies, or the experiment by Cavendish demonstrating mass attraction directly. The demonstration experiments used for teaching purposes have this function: knowing the law, it is demonstrated experimentally in order to convince students of the correctness of the theoretical description, but not to prove the law. Disregarding the fact that the law is not at all subject to refutation, such experiments are unsuitable as empirical proof, because they are distinctly theory-oriented. The law to be "proven" must be presumed as known, and the experimental design rests crucially on the model assumptions and results of the theory in which the law is embedded. Consequently, the interpretation of such experiments as seriously providing verification or an empirical proof of the law not only leads to a methodological circle, but also gives the impression that only the results of those experiments that are carried out just in order to obtain these results could be interpreted by means of the given physical theory.

In all of these cases, quantitative discrepancies occurring in the comparison between theory and experiment are not ascribed the function of falsifying theoretically deduced results, but rather of imparting the feeling of what must be judged as "reasonable agreement" with respect to the model assumptions of the theory and the measuring methods (Kuhn 1961). According to the discussion in the context of the thread pendulum, this is necessary because the comparison of experimental and theoretical results never consists in the simple comparison of two numbers, but always involves an assessment against the theoretical and experimental background. This ability of assessment must be learnt in order to be able, for example, to judge the degree to which the model system is realized in a particular experiment.

In contrast, the comparison between experiment and theory is entirely different when investigating particular real systems that are usually so complex that they can be theoretically characterized only incompletely because the theoretical treatment requires too many simplifying assumptions. On the other hand, the experimental data of complex systems can only be incompletely interpreted, if at all, without the help of theoretical results. The experiment–theory comparison then pursues the goal of combining in a consistent fashion the theoretically not very precise results and the incompletely interpreted experimental data (Grodzicki et al. 1988). This is exemplified by the various methods used to study complex molecules and solids. By combining several experimental and theoretical methods, one attempts to arrive at knowledge about the particular system that is as comprehensive as possible. Accordingly, the function of the theory–experiment comparison consists in obtaining a consistent picture of the real system or process under study via mutual supplementation of experimental and theoretical results. Such an activity is not fundamental research in the common sense, but the application of both empirical and theoretical physical knowledge to collect factual knowledge about particular real systems and processes that is as comprehensive as possible. This is the essential attribute of normal science (Kuhn 1970): the investigation of particular problems ("puzzle solving") within the framework of a secured application represents the typical activity of normal science. Nevertheless, this may lead to substantial novel developments and results as exemplified by semiconductors and high-T_c superconductors.

In summary, none of these functions corresponds to the notion of the theory–experiment comparison that considers experiment as judging the theory or as a falsifying instance in the case of discrepancies with a theory. Both require that some statements on the theoretical side are subject to refutation. In general, this will be the model assumptions $\langle M \rangle$ of a scientific application not yet accepted as secured. In this case, it must be clarified how discrepancies between theory and experiment are assessed, and which strategies are applied to remove them. Afterwards one must investigate which conclusions can be drawn from the theory–experiment comparison with regard to the premises entering the theoretical derivation, and what has actually been falsified in the case of a discrepancy. To begin with, these challenges will be illustrated by two simple examples:

- The first is the law of falling bodies: the experimental test may proceed using a number of qualitatively dissimilar objects such as a sphere of iron, a piece of wood, a sheet of paper, or a feather. Observations demonstrate that this law applies within the measurement accuracy to some of the objects, but not to others. Due to these results the question arises whether this law, as postulated by Galileo, is actually universally and strictly valid.
- The second example concerns the discovery of the planet Neptune: in 1781 Herschel discovered the planet Uranus by accident. Over the following years distinct deviations were observed from the trajectory calculated by the gravitational law. If the reliability of the measured data is not in doubt, the question again arises whether this law is strictly valid.

Various options are conceivable in reaction to these discrepancies between theory and experiment. Each implies a different strategy to remove the deviations, and thus a different research methodology:

1. "Empiricist reaction": a theoretically entirely unbiased assessment of the experimental results for the law of falling bodies yields that it is valid only with a certain probability. Since experimental tests of other physical laws, like the gravitational law in the case of the anomalies in the Uranus orbit, yield analogous results, it follows immediately that all physical laws are ultimately of a statistical nature and do not differ structurally from rules. Accordingly, the law is accepted in its given form as a rule, considering the deviations as inherent, non-isolatable parts. The deviations are only stated: they are not considered to be in need of explanation, and they are not removed, because all physical laws are valid only with a certain probability anyway. Such a conception of law is advanced by some doctrines in the analytical philosophy of science where, in a next step, one attempts to define a degree of confirmation for laws or hypotheses (Leinfellner 1967), given that the prediction of events is possible only with a certain probability. Without any question, this is a conceivable option for describing natural phenomena, but it is not the one adopted by physics. The resulting research heuristic cannot be said to be constructive because it leads neither to an improved understanding nor to new knowledge. At least, the restriction to such an empiricist heuristic, rejecting as theoretically biased the decomposition into general and contingent parts, demonstrates in an immediately comprehensible way that the existence of strictly valid laws cannot be justified solely by observations and experiments. Instead, such a heuristic leads merely to rules or statistical expectations.
2. "Reaction of critical rationalism": the discrepancies are not just stated, but are considered to show that the law must be false. This implies the general conclusion that all physical laws and then all physical theories are false. Such an attitude may be justifiable in the framework of a naive logicism, but is hardly convincing. It does not seem plausible that the numerous practical successes of physics should rest on basically false knowledge. Moreover, such an attitude does not comply with the methodological practice of physics. Indeed, a discrepancy between experiment and theory virtually never leads to the rejection of the law under study or of the scientific application which the law is embedded in, i.e., it is *not* rated as falsification. For this reason, put forward are usually variants of a "sophisticated falsificationism", developed in particular by Lakatos (1974). These do not provide an adequate representation of either the relationship between theory and experience or scientific progress, mainly due to the fact that none of the varieties of falsificationism differentiates between laws and empirical statements, nor between "limited applicability" and "false".
3. "Instrumentalist reaction": deviations from measured data are reduced where possible by modifying the law, e.g., by introducing correction terms. In the case of the orbit of Uranus, this could be the assumption of a force law that gets modified for large distances from the Sun and tends to the known gravitational law for the previous range of validity. Such a modification is, at this stage, purely empirical, since it is based on the experimental

data without theoretical justification. It may be argued, for example, that the law under discussion is only qualitatively valid because it has been derived on the basis of highly simplifying model assumptions, or because unknown perturbing influences may play some role. Consequently, the deviations are classified as insignificant for research progress; they are not in effect considered to be in need of explanation, and the procedures for removing them are quite arbitrary. This reflects the instrumentalist attitude that physical laws are just an epistemically insignificant auxiliary means for the deduction of factual knowledge, whereupon they possess a similar status to algorithms. Such a heuristic regarding empirical corrections is not unscientific a priori and is indeed applied in order to improve agreement with experimental data. A representative example is the extension of the ideal gas law to the van der Waals equation by empirical corrections for the particle volume and the interactions between the gas particles. Whether any physical relevance can be ascribed to such correction terms remains undetermined in general. Such a strategy definitely becomes unscientific if for every new experimental datum a new correction factor is introduced that aims exclusively at reproducing that datum. The transition from this type of correction to a reasonable one is blurred, however, and certain techniques are still considered by some scientists to be permissible, while others rate them as unscientific.

4. "Metaphysical reaction": starting from the belief in the strict validity of the law, reasons are sought for these deviations. Actually, this attitude provides the most efficient heuristic, although it may look the most naive or even dogmatic from an epistemological point of view due to its belief in the existence of strictly valid laws. For such an attitude means that the assessment of the measured data occurs under the "prejudice" that the law applies strictly, and not just with a certain probability. In the next section, it will be shown that there are indeed good reasons for adopting this attitude, and that it is not at all dogmatic. As soon as discrepancies arise, one does not enquire *whether* the law applies, but one searches for the reasons why it does not appear to be valid. It is only through this belief, acting as a methodological norm, that deviations are rated as physically significant and in need of explanation, and it is only the goal of understanding and explaining them that can eventually lead to new knowledge. The discovery of Neptune elucidates the constructive aspect of this heuristic. In 1840, believing in the strict validity of the gravitational law, Bessel made the ad-hoc assumption, i.e., a permitted hypothesis of type (H8), that there is another planet beyond Uranus that should be the cause of these anomalies. Until 1846, Leverrier and Adams independently calculated the orbit of this hypothetical planet and could thus specify its position in the sky at certain points of time. In this way, in combination with an ad-hoc assumption, a theoretically existent hypothesis which was directly comparable with an observation was thus deduced from the gravitational theory. On the basis of these calculations, about 52 arc minutes

from the calculated position, Galle found a faint object that quickly revealed itself to be the postulated planet. Neither the mere statement of the deviations of the orbit of Uranus from the calculated one nor empirical modifications to the gravitational law, let alone its rejection, would have led to this discovery and thus to new knowledge.

Another impressive success of this heuristic is provided by the discovery that the speed of light is finite. When observing the motion of the satellites of Jupiter, Ole Rømer recognized that the precisely determinable moment of disappearance behind the planet did not agree with the calculations according to Kepler's laws, but occurred sometimes earlier and sometimes later than calculated. Due to his belief in the strict validity of these laws, Rømer considered this discrepancy to be in need of explanation and thus began to look for reasons. First, he noted that this moment regularly occurred too early when Jupiter was close to the Earth, and too late when it was more distant. He thus introduced the ad-hoc hypothesis that light needs a finite time to make the journey from Jupiter to Earth. Consequently, not only must the moment of disappearance as observed on Earth be corrected appropriately, but further knowledge is obtained because a quantitative value of the speed of light can be derived from this time difference. The remarkable thing about this is that it seems to exhibit the logical structure of a circular argument. Through the assumption of the existence of a finite speed of light, an ad-hoc hypothesis is introduced in order to explain an observed effect. Afterwards this hypothesis is presupposed as given for the calculation without having furnished an independent proof of its legitimacy. Actually, this constitutes a consistency argument: observed effect, hypothesis, and calculated result are adjusted in such a way that they are consistent with each other. Further research then aims to provide additional evidence for both the finiteness and the value of the speed of light, which is independent of the original evidence. Only the combination of arguments, independent of each other and consistent with one another, finally gets a proof-like structure.

In contrast, the difficulties involved in assessing discrepancies between theory and experiment when the validity of the underlying model assumptions and laws is not accepted as secure are illustrated by the nineteenth century discussion of the specific heat capacity of gases within the framework of the kinetic gas theory. In this theory the ratio c_P/c_V of the specific molar heat capacities at constant pressure P and constant volume V, respectively, is obtained as $(c_V + R)/c_V$, where R is the molar gas constant. According to the equipartition theorem, each degree of freedom f contributes an amount $R/2$ to c_V. For a gas model consisting of mass points ($f = 3$), this ratio should be $5/3$, and $4/3$ if the gas particles are assumed to be rigid bodies, since the three rotational degrees of freedom must be taken into account. The values of the rare gases measured at ambient temperature are in exact agreement with the result deduced from the mass point model, while those for the polyatomic gases are at least consistent with the assumption of rigid bodies (see Table 4.1).

Table 4.1 Ratio c_P/c_V of the specific molar heats of gases

Rare gases	H_2	O_2	N_2	CO_2	NH_3	CH_4
1.666	1.405	1.396	1.399	1.288	1.306	1.303

If the kinetic gas theory, interpreted according to the historical situation as a scientific application of mechanics, is rated as secure, one can immediately conclude from the values for the diatomic gases that, at ambient temperature, only five degrees of freedom contribute to the molar heat capacity corresponding to the ratio $c_P/c_V = 1.4$. Long before quantum theory was known, this result enabled conclusions to be drawn about the internal structure of the gas particles. However, at the time of Maxwell, who first discussed the specific heat of gases, the situation was fundamentally different. Firstly, the theory had been developed just recently and was by no means considered to be secure owing to its numerous unprovable assumptions. Secondly, at this time (1860–80), the rare gases were not yet known, so experimental data were absent precisely for those systems that best realized the model of the ideal gas. In contrast, the diatomic gases were the best studied experimentally, so the empirical standard for the ratio of the specific heats was taken to be somewhere around 1.4. Accordingly, this was the value that was supposed to be deduced from the theory. The assumption of five degrees of freedom that is actually correct and removes this discrepancy had to be rejected at that time as an unprovable ad-hoc assumption, according to common methodological standards, because it was introduced just to explain this single experimental datum. Actually, this discrepancy between theory and experiment persisted right up until the early twentieth century, not only discrediting the atomistic world view, but contributing also to the attitude that physics should be constrained to the description of directly observable facts. This example shows the importance of the consensus about whether a theory is secured or not, but also how misleading incomplete or incompletely understood experimental data can be for the assessment of theoretical results.

In summary, the standard case of a real theory–experiment comparison exhibits the following general structure. Within the framework of a scientific application, a statement (H9) is deduced from the model assumptions $\langle M \rangle$ in combination with hypotheses of type (H4)–(H8) establishing the consistency with the particular experimental conditions. Only this deduced statement (H9) is directly comparable with measured data. Model assumptions and physical laws appear only as premises that are not directly confronted with experiments. A discrepancy between theory and experiment shows that there is something amiss, but the mere comparison of a deduced hypothesis of type (H9) with experience does not permit any conclusion about which one of the premises contained in the derivation was incorrect. Before analyzing, in the next section, the general structure of this central issue of any theory–

experiment comparison, some properties will be summarized and exemplified here:

1. In the case of the anomalies of the Uranus orbit, only the *qualitative* existence hypothesis for the new planet was confirmed empirically, in a strict sense, while according to purely logical criteria the deduced *quantitative* hypothesis should have been considered as falsified due to the discrepancy of 52 arc minutes between the calculated and observed positions. This would imply the conclusion that one of the hypotheses entering the deduction was incorrect. The absurdity of such an argument emphasizes the fact that the agreement between theoretical and experimental data is not the same thing as the equality of two numbers, and that the specification of error margins is of basic importance in order to judge whether a reasonable agreement obtains. Accordingly, comparison between theory and experiment always contains an estimate based on experience and assessment by the judging subject regarding measurement methods, model assumptions, mathematical approximations, etc., which are designated across the board as background knowledge. Therefore, the determination of the agreement between experimental and theoretical data is not a value-free and objective decision resting exclusively on purely logical reasoning.
2. For none of the laws under discussion was the claim proven to be valid without exception within the scope determined by the model assumptions of the theory. The determining factor for assessing secureness, and thus the decision about whether to believe in the strict validity of a law, is the degree of theoretical embedding. In contrast, criteria of empirical confirmation play a subordinate role in most cases. The example of the general theory of relativity clearly illustrates this. In particular, this claim of validity is not justified by continuously citing confirming instances, because validity is considered to be the normal case and does not require additional support. Confirming instances for a law whose validity is not doubted are scientifically dull. In the case of discrepancies between theory and experiment which are in need of explanation, of course, the justification of the claim of validity consists rather in offering evidence that the necessary contingent conditions were not met or that the model assumptions of the theory were insufficiently realized. This is well exemplified by the second law of thermodynamics. One of its formulations reads that, in an isolated physical system, the entropy cannot decrease, i.e., the system cannot go spontaneously into a state of higher order. However, ideally isolated systems do not exist in nature. If a transition into a more highly ordered state is observed or documented, such as the existence of life on Earth, it is not concluded that the law of entropy is violated and must be considered falsified. Instead, it is argued that the system in question was not isolated, so that one of the necessary presuppositions for the applicability of the second law was not met. By proving that this is indeed the case, a discrepancy is transformed from a seemingly falsifying instance into a confirming one.

3. Especially instructive are those cases where this strategy has not been successful. Such an example is provided by another anomaly in our planetary system, viz., the perihelion advance of Mercury's orbit. Due to the success in assuming the existence of a perturbing planet in the case of Uranus, a planet orbiting somewhere between Mercury and the Sun was postulated as the cause of this anomaly. This emphasizes the preference for methods that have already been successful (see Sect. 1.5). Remarkably, towards the end of the nineteenth century, this inner planet, named "Volcanus", was "observed" twice, although it is now considered certain that it does not exist. In fact, this anomaly was eventually removed by a "modification" of Newtonian mechanics. According to the general theory of relativity, space is curved by the presence of masses, and the perihelion rotation of Mercury's orbit is interpreted as arising from this curvature of space, rather than due to an inner planet. Occasionally, this has led to the conclusion that Newtonian mechanics has been falsified. However, such a conclusion contradicts the common view in physics. According to this view, the scope of Newtonian mechanics has been restricted: it applies only with respect to the model assumption of a strictly Euclidean geometry, not affected by the mass distribution.
4. Another interesting question is what would have happened if Neptune had not been found close to the calculated position. In the first instance, neither the law of gravitation nor, a fortiori, the theory of gravitation would have been falsified, since the hypothesis of the existence of an additional planet was not theoretically existent, so the opposite was also permitted. In order to explain the discrepancy, several options would have been conceivable: the planet being sought might exist, but just be too faint to be detected by the then available instruments; furthermore, the calculation could have been flawed or the planet could have been at an entirely different position, i.e., the deduced hypothesis might not have been determinate. In fact, there were a number of options that could have explained the discrepancy without there being general methodological rules about how to remove it in any particular case.
5. As an isolated statement, a hypothesis would be verifiable or falsifiable solely by experiment. In physics, however, this is irrelevant because a physical theory is not a collection of isolated statements. In this case, any conclusions regarding the experimentally not directly comparable model assumptions of the theory are obviously impossible. Actually, in comparing experiment and theory, it is not the hypotheses of type (H9) that are concerned, but the assumptions and laws entering the deduction as premises. It is precisely because these experimentally directly comparable hypotheses are not isolated from the other parts of the theory due to their theoretical deduction that it becomes possible to test the model assumptions and laws. On the other hand, the question arises not only as to what may be incorrect in the case of a discrepancy between theory and experi-

ment, but also to what extent the securing of the knowledge contained in the laws is possible by experimental methods.

4.4 The Duhemian Theses

The discussion of representative examples in the preceding section has revealed the difficulties involved in justifying the claims of laws as strictly valid and ontologically significant by a mere comparison with experiments, and in providing the secureness of the autonomous knowledge they establish. It will be demonstrated below on a more general level that such attempts must remain unsuccessful. Although it is widely acknowledged that physical laws as generalizing propositions cannot be definitively verified by experiments, a widespread opinion still prevails that they should be falsifiable by experiments. On this belief rests the methodology of falsificationism. Instead of verifying the law by quoting a large number of confirming instances, one attempts to falsify it by counterexamples, and the more tests the law passes, the better it is confirmed. The pattern of argument here is in principle based on a syllogism of type *modus tollens*:

$$(A \implies E_{\mathrm{th}}) \wedge \neg E_{\mathrm{exp}} \implies \neg A \,.$$

Accordingly, if a theoretical claim E_{th} about an observable effect E_{exp} is deducible from the assumption A and a measurement yields the result that it does not occur, it necessarily follows that A was false. This argument rests on a number of implicit assumptions which a careful analysis proves to be untenable.

1. All falsification methodologies rest on the implicit assumption that the experimental result E_{exp} is definitive in that it is not subject to revision. This assumption is untenable for several reasons. Firstly, quantitative, experimentally obtained data are always subject to improvement or revision. Therefore, it is not clear according to which criteria E_{exp} should be assessed as finally confirmed and definitive. Secondly, this requires exact knowledge of the operating mode of the measuring devices, i.e., a theory of the relevant instruments that should enable, in particular, the identification of systematic errors. Thirdly, as a targeted action, any experiment rests on theoretical presuppositions and general background knowledge. It may happen, for example, that what is measured in an experiment is not what was expected according to the preliminary theoretical ideas. Finally, if the incidence of a certain effect is postulated, e.g., the splitting of spectral lines in an external field, a negative experimental result may just mean that the splitting is too small or the applied field is too weak to detect the effect. A representative example is the Stark effect that was initially not detectable because the applied electric fields were too weak. In short, even

the reliable assessment of experimental results is not a purely logical problem, but contains background knowledge and necessitates detailed expert knowledge about both the performance of the measuring instruments and the evaluation procedures.

2. Examples given in the last section showed that the comparison between quantitative theoretical and experimental results is not just the unproblematic comparison of two numerical values since the two values will never agree with each other precisely, but only within certain error margins determined by both experiment and theory. Whether a discrepancy is to be assessed as falsification or as qualitative agreement depends on the particular context, so the comparison of E_{th} with E_{exp} rests on conclusions and judgements that are not at all unequivocal and beyond any doubt. Finally, corresponding terms of the empirical and theoretical level need not necessarily have the same meaning. Consequently, comparing experiment and theory requires compatibility of the conceptual systems, and this cannot be taken as given a priori, as exemplified in Sect. 4.5.
3. The argument rests on the view that physical laws possess a structure analogous to empirical generalizations. If it is postulated, for example, that all swans are white, this claim is falsified by the discovery of a single non-white swan. Physical laws, however, possess a fundamentally different structure. Firstly, they establish relations between measurable properties, not between objects and properties. Secondly, they are independent of contingent conditions such as system-specific parameters and initial or boundary conditions. Therefore, a law is only experimentally testable in combination with additional contingent assumptions, so that within a physical theory an experimentally testable result E_{th}, i.e., a hypothesis of type (H9), is never deduced from a single assumption A, but rather from the conjunction of law L and contingent conditions C_i:

$$\Big[A :\Longleftrightarrow (L \wedge C_1 \wedge \ldots \wedge C_n) \Longrightarrow E_{\mathrm{th}}\Big] \wedge \neg E_{\mathrm{exp}} \Longrightarrow \neg(L \wedge C_1 \wedge \ldots \wedge C_n) \, .$$

As a consequence, A is false if just a single C_i is incorrect. Accordingly, even in formal respects, the experimental testing of physical laws is already entirely dissimilar from that of empirical generalizations which do not contain the decomposition into L and C_i, and may thus be falsifiable by counterexamples. Without localizing the discrepancy with respect to E_{exp} within A, the law L itself could not be falsified anyway.
4. A closely related fallacy of all falsification methodologies consists in misinterpreting the claim of strict validity, according to which physical laws, by analogy with mathematical theorems, exhibit the structure of "for all" sentences, i.e., "for all x, one has ...". The claim of strict validity, however, rests crucially on the condition that all model assumptions of the theory which the law is conceptually embedded in have actually been met. Since none of the known physical theories possesses an unrestricted scope, the

corresponding laws are also only conditionally valid, and never have the form “for all x, one has ...”, i.e., they are *not* “for all” sentences.

5. At least the theories of classical physics refer virtually exclusively to non-existent model systems like mass points or the ideal gas. The associated laws are strictly valid only for these model systems. In contrast, every experiment is performed on real objects representing just approximate realizations of the model systems. Accordingly, the claim of strict validity is not at all refutable empirically. This is the decisive reason why physical laws as statements within a scientific appplication are not experimentally falsifiable. At best, what is experimentally falsifiable are empirical statements that must be strictly distinguished from the laws.
6. Even more laws of the theory frame are not falsifiable, such as Newton’s equation of motion or Maxwell’s equations. For example, in order to solve the equation of motion, not only must a specific force law be provided, but so also must initial conditions and system-specific parameters such as the value of the acceleration due to gravity in the case of free fall, or mass and force constants for the harmonic oscillator. If theory and experiment do not agree with each other, it won’t be the equation of motion that is considered to be falsified. Instead, another force law will be constructed if an error cannot be found in the contingent parameters. Basically, laws of the theory frame are not falsifiable because they are, like the proposition “X is a student”, not at all “truth-capable” and without direct relation to experiment.
7. Analogous conclusions apply to the universal, in the sense of supra-theoretical, principles, like the law of energy conservation. On the face of things, most direct experiences would appear to contradict this law. A freely rolling sphere, e.g., being force-free according to everyday understanding, comes eventually at rest, i.e., it loses kinetic energy, as does a freely oscillating pendulum, and colloquially, energy is consumed. These direct observations are not interpreted as falsification, but are reinterpreted in such a way that they become consistent with the law of energy conservation, e.g., by introducing friction. Consequently, this law is preciseley the opposite of an experiential fact, but has a *normative* function:

   > Observed phenomena and processes must be interpreted in such a way that the law of energy conservation is not violated. If it seems to be violated, either the system was not isolated or a new type of energy must be introduced.

   An analogous function is ascribed to the above-mentioned entropy law. Such examples demonstrate that these universal principles are not falsifiable, because they represent the fundamental postulates of the physical description of reality, which are not subject to refutation. According to Popper, statements of this kind are identical with dogmas and ought to be rated as unscientific.

All in all, at the outset, discrepancies between theory and experiment provide only an indication that something is wrong, because it might be that:

- one of the contingent conditions with respect to which the law applies is not met,
- the real object is too different from the theoretical model system,
- the experimental design does not realize the model assumptions of the theory with sufficient accuracy,
- influencing factors due to the environment, such as the T-dependence of material properties or the P-dependence of boiling points, are not recognized or accounted for,
- calculated and measured physical quantity do not refer to the same quantity,
- the experiment contains systematic or accidental measurement errors,
- the instrument does not work as would be expected according to the theoretical background knowledge,
- the error margins of experimental and/or theoretical data are incorrectly estimated,
- the effect corresponding to the deduced hypothesis cannot be detected with currently available devices,
- the experiment is incorrectly evaluated or interpreted,
- individual errors of the experimenter are the reason ("N-rays").

In the case of a discrepancy between theory and experiment, what remains undetermined in the first instance is whether on the theoretical side assumptions must be changed, or whether on the experimental side the interpretation, significance, quality, or completeness of the data may be questioned, or again whether there are inconsistencies in the relationship between theory and experience. None of the itemized reasons is refutable by purely logical or purely empirical methods, and they come to the fore in particular when there is a discrepancy between experiment and the result of a theory accepted as secure.

Against this background, the general question arises regarding the extent to which experimental methods may contribute, if at all, to securing law-like theoretical findings. To this end, one must examine the functions of theory–experiment comparison for the identification and removal of discrepancies:

- An essential function is the experimental control of contingent conditions. Firstly, this enables one to decide whether the assumptions contained in the theoretical deduction have been realized by the experimental design. Secondly, through systematic variation of these conditions, the reason for the discrepancy can frequently be identified. This constitutes the essential difference with the purely observational sciences, where such a control of contingent conditions is rarely feasible. Accordingly, the function of experiment is not the verification or falsification of the law, but rather the checking of the contingent conditions, in order to show indirectly that in the case of an observed discrepancy the law is consistent with the seemingly contradictory observations.

- Strictly, the scope of a theory and its associated laws extends only to the model systems as defined within the framework of the theory. In contrast, the range of applicability to real systems cannot be defined by theory-internal methods, but only via the comparison with experiment. Correspondingly, theory–experiment comparison does not aim to verify or falsify the law, but has the function of determining the scope of the theory when applied to real systems. Furthermore, possible discrepancies may serve as the starting point to develop a modified law that is better adapted to real systems. A representative example provides the transition of the state equation of the ideal gas to the van der Waals equation, as described in Sect. 3.5.
- In contrast, theory–experiment comparison is faced with a number of serious challenges if it is considered as a method for securing theoretical knowledge. Firstly, there is an essential structural dissimilarity between an experimental datum and a physical law. While an experiment corresponds to a local event, since it is performed on concrete systems under specified conditions, the law is according to its structure of global type, because it does not refer to a particular system and possesses numerous connections to other parts of the theory in which it is embedded. This structural dissimilarity becomes particularly apparent with respect to the empiricist notion of experience based on sense data. As spontaneous, one-off, and irreducible elements of experience, individual perceptions and singular facts are unsuitable to test law-like statements. This requires at least reproducible experience. Although theory–experiment comparison is meaningful, in the first place, due to the circumstance that the empirical basis of a theory does not consist of singular, but of intersubjective, reproducible, and in this sense law-like experiences, this does not change anything regarding this structural disparity. Experiments are always being performed on concrete systems and are in this respect of local type. On the other hand, due to the theory-determinatenesses of the empirical basis, it is not clear to what degree its elements are still suitable to test and confirm theoretical knowledge. Indeed, many experimental data which a law is confronted with cannot be obtained independently of the theory in which the law is embedded (Duhem 1908):

> This fundamental impossibility of separating a physical theory from experimental procedures which ought to serve to check this same theory makes such checking particularly complicated.

Consequently, the interpretation of the confrontation of laws with experimental data as verification or falsification is subject to the principal objection that it rests on circular reasoning. Finally, it is not the law itself that is compared with experiment, but only a result deduced from the combination of the law and contingent conditions which must be consistently

> adjusted to the experimental design. For all of these reasons, it is impossible to definitely falsify the whole network in which the law is embedded by local confrontation with an experiment.

According to Duhem, it must be concluded from these interdependences within the physical knowledge that any empirical testing of a law or a proposition always consists in principle in testing the entire network the law is embedded in. In chapter ten of his book *Aim and Structure of the Physical Theories*, Duhem (1908) summarized this insight as headings of sections two and three which are designated here as the first and second Duhemian theses:

(DT1) A physical experiment can never lead to the rejection of an isolated hypothesis, but only to the rejection of a whole theoretical group.

(DT2) An *experimentum crucis* is impossible in physics.

The second thesis differs structurally from the first in that it is not a theory that is confronted with an experiment, but the experiment has the function of deciding between two competing theories. The first thesis is once again set out towards the conclusion of the second section:

> All this summarized yields that the physicist can never subject an isolated hypothesis to checks by experiment, but always only a whole group of hypotheses. If the experiment is at variance with his predictions, it teaches him that at least one of the hypotheses forming this group is inadmissible and must be modified.

The best evidence for the first thesis is the confrontation of a theoretically existent, determinate hypothesis of type (H9) with an experiment. Discrepancies do not affect the hypothesis itself, but go back directly to the premises. In turn, it is precisely these interdependences within the theoretical level that enable an indirect experimental check of the premises.

While Duhem leaves undetermined the size of the group of hypotheses affected by the comparison between theory and experiment, Quine drew from these interdependences the radical conclusion that in any theory–experiment comparison the whole of scientific knowledge is in principle affected (Quine 1951):

> Any statement can be held true come what may, if we make drastic enough adjustments elsewhere in the system [...] Conversely, by the same token, no statement is immune to revision.

Since the whole of scientific knowledge is indeed somehow interlinked, this radical "Duhem–Quine thesis" (Lakatos 1974) cannot be disproved logically. Structurally, however, it conforms to hypothetical realism due to its irrelevance concerning the methodology of testing, justifying, and securing physical knowledge. Actually, not even the whole of physical knowledge is systematically subject to refutation in any real theory–experiment comparison with the corresponding methodological consequences, let alone the whole of scientific knowledge. That "drastic adjustments" may be possible in special cases is illustrated by the development of a quantum logic for the interpretation of quantum mechanics. It must be emphasized, however, that firstly this development does not go back to discrepancies between theory and experiment, and that secondly its practical irrelevance is an argument more against than in favour of the methodological relevance of Quine's claim.

Actually, in research practice neither isolated hypotheses nor the whole of knowledge are subject to experimental testing, and this raises the question of what are the largest units that are subject to refutation in a real theory–experiment comparison. In a closed theory, the hypothetical elements come into play at the transition from a specialization to a scientific application ScA, and only statements in ScA can be directly compared with experiment. Consequently, the theory is experimentally testable only through its ScA. Therefore, these are the potentially largest units that are affected in a real comparison between theory and experiment. In the first instance, not even the entire ScA is subject to refutation, but only the contingent conditions recorded in hypotheses of the type (H4)–(H8). In the case of discrepancies with experiment, these are scrutinized first. Regarding (H4) and (H5), this occurs by experimental methods via the improved experimental realization of the contingent conditions, in the case of (H7) by mathematical methods via the stepwise improvement of the approximations, and the low priority of ad-hoc hypotheses (H8) is obvious. In contrast, in order to assess the empirically not directly comparable model assumptions (H1)–(H3) by means of experimental data, the conditions contained in (H4)–(H8) must be controllable and known, so that they can be assumed as non-hypothetical. Under this premise, the model assumptions can also be checked experimentally to a certain degree. The setting up of such a priority list thus rests on a "principle of minimal perturbation" of the theoretical system: even though changes at other places within the theory would be possible, their arbitrariness is rather constrained in practice due to the interconnections within the theoretical system.

For these reasons, the strategy of searching for the causes of possible discrepancies in the contingent conditions, rather than in the law, is reasonable not only in a pragmatic sense, since corrections to these conditions are simpler than the modification of the law embedded in the theory, but is even compulsory, because the strict validity of the law depends on the presupposition that the model assumptions of the theory are realized in the experimental design. This provides the rational justification for the "metaphysical reaction" to discrepancies described in the last section, when comparing theory with experiment. In contrast, the characterization of this strategy as "immunization" against experimental falsification (Lakatos 1974) reveals little appreciation of physical methodology, since it suggests that physical laws ought to be experimentally falsifiable, and this strategy aims precisely at preventing this. Altogether, this analysis exemplifies the fact that falsifiability is not an appropriate demarcation criterion between scientific and non-scientific statements.

In contrast to the scientific applications, the theory frame is not subject to refutation in any real theory–experiment comparison because it does not contain any experimentally testable statements, so it defies direct experimental checking. If a confirmed experimental result can be interpreted consistently neither by modifying the model assumptions of an existing ScA nor by supplementing a specialization with hypotheses to establish a new ScA, it is concluded that this result lies outside the scope of the theory. Therefore, a failed ScA does not falsify the theory frame, but is excluded from the circle of possible applications. This highlights the methodological irrelevance of Quine's claim. Experiments that seem to scrutinize directly laws of the theory frame actually always refer to a test within an ScA. For example, the interpretation of Cavendish's experiment to determine the gravitational constant as a proof of Newton's law of gravitation is untenable. Firstly, this is a methodological circle because this law is presupposed in the experimental design, but not subject to refutation. Secondly, this experiment refers only to a particular system, i.e., corresponds to an ScA of the theory of gravitation, and represents at most the experimental demonstration of a realization of the law of gravitation where all C_i are experimentally under control. Changes to the theory frame of a closed theory are not excluded, of course, but signify the transition to a new theory, and result most notably from theoretical considerations, because the theory frame itself will not be changed solely due to local discrepancies with experimental results. This is exemplified by the change to Newton's first law made by the general theory of relativity, according to which force-free motion occurs along a geodesic, which becomes a straight line only in the case of Euclidean geometry. Hence, changes to the theory frame are not located in the realm of the relationship between theory and experience.

The methodological significance of the first Duhemian thesis rests, among other things, on two insights. Firstly, it makes explicit the reason for the ambiguity in the relationship between theory and experience. Secondly, it reveals the limitations of experimental methods in securing theoretical knowledge. In particular, the relation between law and experiment is not univocal, but constrained only by consistency conditions. Nevertheless, the law-like description is not at all arbitrary as is alleged, at least implicitly, in conventionalism or instrumentalism. Due to these ambiguities the strategies to remove discrepancies are exceedingly multifaceted, with very different results, as demonstrated by the following representative selection:

1. The usual and mostly successful strategy for removing a discrepancy is its localization in the contingent conditions. This unspectacular option is generally ignored only because it does not require modifications of the theory. As a result, a discrepancy is transformed from a seemingly falsifying one into an instance confirming the law or the theory.
2. The problem of the ratio c_P/c_V of the specific heats of gases was solved by advances on both the theoretical and experimental side. On the theoretical side, one of the model assumptions of the kinetic gas theory, conceived as

an ScA of mechanics, was modified by ascribing the gas particles additional degrees of freedom. On the experimental side, the empirical basis was extended by the discovery of the rare gases.

3. Discrepancies between the thermodynamics of the ideal gas and empirical experience, such as the fact that real gases liquefy below a certain temperature, were removed by empirical corrections to the ideal gas law which led with the van der Waals equation to an improved theoretical approach. The result was the construction of a hierarchy of theories, but it did not lead to the rejection of the previous theory, which was thus not considered to be falsified.
4. The problem of the anomalies in the orbit of Uranus was solved by the introduction of an ad-hoc assumption, viz., the existence of a hypothetical planet. A similar example is the hypothesis about the existence of the neutrino in order to explain the continuous spectrum of the electrons in β-decay.
5. In the case of a theoretically existent but not determinate hypothesis, the ad-hoc exclusion of a possible solution may be justified if the corresponding effect is not observed. In this way, one excludes, e.g., mathematically possible solutions that do not meet certain regularity demands or plausibility arguments, as in the case of elastic collisions. In elementary particle physics, a common strategy for excluding theoretically possible processes that are not observed consists in introducing new quantum numbers and conservation laws by means of which these unobserved pocesses become "forbidden".
6. The discrepancy between the calculated and measured dissociation energy of the hydrogen molecule, described in detail in the next section, was removed by revising the experimental datum. Here, knowledge of the theoretical value played the decisive role.
7. Finally, the anomaly of the perihelion rotation of Mercury's orbit was removed by the general theory of relativity, i.e., by a new grand theory.

This selection of options for removing discrepancies which have all occurred in the historical development of physics not only proves the gist of the Duhemian theses, namely that the relationship between theory and experience is not unambiguous, but also demonstrates that Kuhn's anomalies as the starting point of scientific revolutions are fiction. Before the anomaly has been removed, one cannot assess how far-reaching the necessary modifications will be on the theoretical side because there is no guarantee that all options to remove it have been exhausted.

Against the background of these ambiguities, one must also understand Duhem's second thesis, that an *experimentum crucis* is impossible. This does not mean, of course, that experiments that decide between competing theories are impossible, but only that the assessment of an experiment as an *experimentum crucis* is untenable in the following sense. According to Bacon, there should be experiments that enable a final, definitive decision between competing theories and exclude alternative options. Based on the example of

Foucault's experiment, which furnished the proof that the speed of light is smaller in water than in air, Duhem criticizes the assertion that this should be considered as an *experimentum crucis* because it falsifies Newton's conception of light as a particle flow, and also furnishes a definitive proof of the wave nature of light. According to Duhem, this experiment falsifies only a group of hypotheses, because the statement of Newton's emission theory, that light would propagate faster in water than in air, is deduced from a group of assumptions. The hypothesis that light is a particle flow is only one of several premises, viz., concerning the interaction of the light particles with the medium they are moving through. Accordingly, those who subscribe to the emission theory could react to Foucault's result by changing other assumptions of this theory. Secondly, even if every assumption of the emission theory except the particle hypothesis has been proven to be true by other means, Foucault's experiment does not definitively prove the wave nature of light. Indeed, it is impossible to prove that the alternative that light corresponds to either a particle flow or a wave must be exclusive, and that a third possibility does not exist. It is precisely in this respect that the development of modern physics has corroborated Duhem's argument, impressively confirming his critique. According to Duhem, an experiment can be assessed as an *experimentum crucis* only when it definitively eliminates all except one group of conceivable hypotheses. The fact that there are no such experiments that accomplish this task constitutes the content of Duhem's second thesis. Careful assessment of other examples cited for its refutation (Franklin 1987) do indeed show that no experiment can accomplish this, so Duhem's conclusion is confirmed as entirely correct.

Based on this analysis of the relationship between experiment, law, and theory, one can answer the initial question of how the theoretically autonomous knowledge, represented by the physical laws, gets secured, and how the claims of strict validity and ontological significance are justified. Firstly, the justification for accepting the knowledge contained in the law as secured rests on a combination of various procedures. These include the deduction of hypotheses that are directly comparable experimentally, the transformation of seemingly falsifying instances into confirming ones, the connection with statements at the empirical level, and above all, the theoretical embedding in an ScA in combination with the interconnections between the various ScA. Secondly, it has been demonstrated that the claim of strict validity with respect to the model assumptions of the ScA is justified neither by an inductive derivation, nor by a high degree of empirical confirmation, nor by continuous falsification attempts, because this claim of validity initially refers only to non-existent model systems. Consequently, discrepancies obtained when comparing experiment and theory may at best disclose the limits of applicability to real systems and processes. For this reason, the essential requirement is the theoretical embedding: as a statement in a closed theory or in one of its secure ScA, the law itself represents secured knowledge, and is thus not subject to refutation in the comparison between theory and experiment. Therefore, challenging the validity of a law is not the same thing as attempting falsification. Due to the theoretical embedding, the given ScA is also concerned since the scope of the law and the ScA mutually determine one another. Discrepancies with experiments do not lead to rejection of the theory, but either to a restriction of its scope and the construction of a hierarchy of theories, or to the exclusion of the given ScA from the scope of the theory. Finally, the claim of ontological significance must not be identified with the naive-realistic notion that laws are a part of nature. Instead, it means that the law corresponds to a structure of reality, notwithstanding the fact that the law is strictly valid only for non-existent model systems. Observable regularities may serve as justification in particular cases, but in general form, this ontological claim is no more proven than the existence of a subject-independent reality. In short, neither the notion that reality is structured nor the existence of strictly valid laws are results of the law-like description of reality, but they are the starting point and must be classified as tacit assumptions which determine both the relation between theory and experience and the methods of comparison between theory and experiment.

Just as for the particular case of the relationship between law and experimental data, it is even more true that the general relationship between theory and experience is also not unambiguous, because theories are empirically underdetermined in the sense that they offer only a spectrum of possibilities. It is precisely the empirical openness due to this underdetermination that is exploited when an ScA is established by making additional, system-specific assumptions, and it is only afterwards that experimentally comparable results are deducible from the theory. In combination with the central statement of the Duhemian theses, that the relationship between theory and experience is not unambiguous, this precludes the verification or falsification of theories exclusively by experimental methods. This brings out their limitations regarding the securing of theoretical knowledge, and also disproves the empiricist assumptions (E4) and (E5): a theory is rejected not just because it contradicts experience when faced with an observation or experiment. Consequently, the claim that an empirical-scientific theory ought to fail on experience (Popper 1962) does not establish a criterion to be fulfilled by a physical theory, and nor can experiment stand as judge over theory. Using a similar metaphor, experiment may serve as a witness or may provide evidence for the shortcomings of a theory, but the judges are always physicists or a scientific community. Accordingly, the objective in establishing relations between theory and experience is not the verification or falsification of theoretical knowledge by experiments, but the determination of the scope of the theory when applied to real systems, and thus the establishment of its empirical content, as well as the creation of secure knowledge via the consistent adjustment of theory and experience.

4.5 Consistent Adjustment: Examples

Starting from the Duhemian theses that the relationship between theory and experience is not unambiguous, the function of comparison between theory and experiment does not consist in verifying or falsifying theoretical results, but in testing consistency. Such testing includes not only empirical testing of theoretical results, but also theoretical checks on experimental results. The interrelation between these two aspects will now be elucidated on a number of representative examples of the interaction between theory and experience. The first of these examples describes the determination of the dissociation energy of the molecule H_2.

The covalent chemical bond, i.e., the fact that two or more neutral atoms form stable molecules, was not understood theoretically until Heitler and London (1927) applied quantum mechanics to the problem. Along with a number of additional assumptions, they were able to deduce theoretically the existence of a stable state of two bound hydrogen atoms. This result was taken to be an important success of this novel theory, irrespective of the fact that the bond length they obtained was about 8% too large and the bond energy almost 30% too small. This demonstrates that even a discrepancy of 30% between theory and experiment may not be rated as a falsification, but as qualitative agreement, and it emphasizes the fact that exact agreement with experimental data is by no means the sole criterion for the quality of a theoretical result. Whether a discrepancy is rated as falsification or qualitative agreement depends essentially on the context. In this case, the assessment rested on the circumstance that quantum mechanics had been developed just a year before, and that the calculation contained numerous simplifying assumptions. Consequently, quantitative agreement could not be expected and there was the justifiable hope that systematic improvement of the calculation would yield results within the error margins of the experimental values. In other words, the applicability of quantum mechanics to the description of the covalent bond could not be questioned on the basis of this first result, and the removal of these discrepancies between theory and experiment was considered merely as a technical problem, not a fundamental one.

Subsequent historical development, however, shows how long the way can be from qualitative to quantitative agreement, because not only instrumental developments, improved measuring methods, new computational methods, and computers turned out to be necessary, but so also did certain conceptual adjustments. Initially, the expectation seemed to be confirmed. Indeed, by improving the atomic basis functions (inclusion of contraction and polarisation of the atomic orbitals in the molecule compared with those of the free atoms), the numerical discrepancy could be decreased to 8% for the bond energy and to 0.4% for the bond length. Prior to any further numerical improvement, however, a *conceptual* adjustment between empirical and theoretical terms turned out to be necessary, involving a typical quantum mechanical effect, which made it unforeseeable. Contrary to the original expectation, the calculated and measured bond energies are conceptually not identical, i.e., what is calculated is not what is measured. The theoretical bond energy is obtained by subtracting the total energies of the atoms in their ground state from the total energy of the molecule at the minimum of the potential curve computed as a function of the distance between the nuclei. The empirical "bond energy" is called the dissociation energy and is the measured energy difference between the molecule and its fragments. If the fragments are in their ground state, the two definitions should yield the same result. Actually, they differ by the zero point energy. According to Heisenberg's uncertainty relation, even at the absolute zero point, the molecule executes vibrations around a (fictitious) equilibrium position, accordingly referred to as zero point vibrations. Therefore, the ground state of the molecule is determined not by the minimum of the potential curve, but by the lowest vibrational level, i.e., the vibrational ground state. The measurement yields the difference between the vibrational ground state and the fragments, so that the calculated bond energies must be reduced by the zero point energy when comparing them with experiment. In the hydrogen molecule, this energy amounts to about 6% of the bond energy, so it must necessarily be accounted for in the subsequent numerical improvements.

The first stage of this development came to an end in 1935 with a calculation by James and Coolidge (1933, 1935) and an improved measurement by Beutler (1935). This decreased the difference between the measured and calculated bond energies to 0.5%, an excellent agreement at that time, although in absolute numbers this corresponds to an energy difference of 192 cm^{-1}, which leaves the theoretical result well outside the experimental error margin of 6 cm^{-1}. It was not until 1960 after the appearance of electronic computers in science, that the calculated bond energy could be improved by an order of magnitude (Kolos and Roothan 1960). In the same year this was achieved on the experimental side (Herzberg and Monfils 1960), decreasing the deviation between experiment and theory to just 5.4 cm^{-1}. This was still well outside the experimental error margin of 0.5 cm^{-1}. Herzberg and Monfils ascribed this to uncertainties in the determination of the zero point energy. On the theoretical side, however, it was clear that, due to several approximations, the calculations should produce corrections of the same order of magnitude as this deviation. During the subsequent years, the calculations were further improved by accounting for adiabatic and relativistic effects, and eventually also radiation corrections. The result obtained in 1968 exhibited not only a discrepancy with experiment of 4.3 cm^{-1}, but was now larger than the experimental value, i.e., the absolute value of the calculated bond energy, reduced by the zero point energy, was larger than the measured dissociation energy. This contradicted one of the methodological principles of quantum mechanics according to which the absolute value of a bond energy derived via a variational calculation cannot be greater than the experimental value, so the discrepancy had a more fundamental significance.

Since further theoretical attempts at improvement did not change the situation, Herzberg (1970) undertook an examination of the experimental value with a series of new measurements. There he recognized that, in the immediate neighbourhood of the edge of the continuum, corresponding to the dissociation energy, there is a sharp line that had been erroneously measured in 1960. Without the background of the theoretical result in combination with the conviction that it was correct, this would not have been noticed. As a consequence, the previous experimental value had to be shifted up by 4.7 cm^{-1} and once again became larger by 0.9 cm^{-1} than the then best theoretical value. Subsequent improvements on the theoretical side up until 1983 resulted in another reduction of the difference to 0.3 cm^{-1}, now within the experimental error margin of 0.5 cm^{-1}. In absolute numbers, this means $D_e(\mathrm{exp}) = 36118.3 \pm 0.5\ \mathrm{cm}^{-1}$ compared with $D_e(\mathrm{th}) = 36118.01\ \mathrm{cm}^{-1}$. Up until the mid-1990s, the difference between the two values was further diminished. The relevant values are $D_e(\mathrm{exp}) = 36118.06 \pm 0.04\ \mathrm{cm}^{-1}$ (Eyler et al. 1993) and $D_e(\mathrm{exp}) = 36118.11 \pm 0.08\ \mathrm{cm}^{-1}$ (Balakrishnan et al. 1994), as well as $D_e(\mathrm{th}) = 36118.069\ \mathrm{cm}^{-1}$ (Wolniewicz 1993) with an estimated error margin below 0.01 cm^{-1}, i.e., the theoretical value is more accurate than the experimental ones. More generally, it may be noted that the calculations of spectroscopic data of molecules with few atoms usually produce results that are considered to be better confirmed than the corresponding experimental values. Indeed, there are many examples where measured data have been corrected by computed ones. The scientific applications of quantum chemistry, understood as a specialization of quantum mechanics for the structure of molecules, have thus attained the state of secure ScAs. If the theoretical accuracy is not reduced by simplifying additional assumptions, the theoretical results have a higher rank regarding secureness than the experimental data, which can frequently be obtained only under difficult conditions. Although this example should not suggest that the development of physical knowledge always occurs in this way, it contains a number of representative aspects:

1. It demonstrates an almost 60 year long process of increasing precision and improvement of both the experimental and the theoretical value. The increase in precision consists in the systematic reduction of the error margins, and the improvement in the decrease in the deviation. In fact, the experimental datum is by no means definitively given. Rather its quality left much to be desired initially, so it underwent changes and it was as much subject to revision as the theoretical datum. This shows that experiment did not at all play the role of judge over the theoretical result.
2. Progress occurred by the mutual adjustment of empirical and theoretical knowledge, not only from a numerical point of view, but also from a conceptual one. Initially, it was not quantitative agreement that was decisive, but the qualitative deduction of the existence of a bound state. Experiment served as reference point and motivation to improve the theoretical result. On the other hand, the theory helped to recognize and interpret the facts correctly. Consequently, the empiricist assumption of an unrestricted

priority of experiments in the process of securing knowledge is untenable and is proven to be the result of a naive belief in facts that characterizes the patterns of thought and action in physics and must be rated on the same level as the tacit assumptions.

3. The impetus to check the experimental result from 1960 with the subsequent correction came after:

   - the theoretical options of improvement were exhausted in relation to the size of the existing discrepancy,
   - a qualitative inconsistency remained when the theoretical efforts were concluded,
   - the status of molecular physics as a secure application of quantum mechanics was not subject to refutation due to the numerous successful calculations made during the 1960s, but primarily for theoretical reasons.

   In this example the spectrum itself was remeasured, i.e., an old raw datum of the empirical basis was replaced by a new one, but it would not have made any difference in this context if only its interpretation had been revised. The uninterpreted spectrum is without methodological and epistemic significance: only through its interpretation does it become a datum relevant for the theory. The revised interpretation of an otherwise unchanged raw datum then provides an example of what it means to have recourse to the empirical basis of uninterpreted facts in case of doubt.

4. At no time was the discrepancy interpreted as falsification in any form, so that not even molecular physics, let alone the whole of quantum mechanics, was ever considered subject to refutation. This is not surprising in view of the fact that formally every quantitative comparison between theory and experiment leads to a discrepancy since no theoretical result is ever mathematically identical with the corresponding experimental value. In addition, any theory–experiment comparison constitutes only a local test, so a discrepancy reveals at most a local inconsistency, usually without consequence in practice. A comprehensive theory like quantum mechanics will not be rejected just due to some local inconsistencies or "anomalies", because one cannot say about a discrepancy at the moment of its occurrence whether it is just a temporary disagreement or a fundamental failure of the theory, i.e., an anomaly that, according to Kuhn, ought to lead to a scientific revolution. Anomalies as decisive moments of scientific development are thus a fiction: when theories are abandoned, this occurs on the basis of theoretical arguments. Indeed, at any given time, every theory contains local anomalies that may even be long-standing just because their elimination is not considered as particularly urgent. For instance, regarding the problem of the dissociation energy of the hydrogen molecule, in a review article from 1967 the authors mention in passing (Hirschfelder und Meath 1967): "The reason for this small discrepancy is not understood at the present time." Therefore, even the lifetime is not a criterion for a

"revolution-generating" anomaly. In the case of the hydrogen molecule, this becomes particularly apparent because in the end there was even a non-local anomaly that violated one of the methodological principles of quantum mechanics.

The second example concerns the historical development of the measurement and concept of length, and describes the process of the mutual adjustment of the theoretical, empirical, and operational definition of one of the most basic physical quantities (Bayer-Helms 1983, Giacomo 1984). The first empirical operational definition of the metre as length unit goes back to the French Academy of Sciences, fixing it as one ten-millionth part of the quarter of the meridian running through Paris from the North to the South Pole. This length had been determined by Delambre und Méchain around 1791 by measuring the meridian arc between Dunkirk and Barcelona. This definition was generally binding, as decreed in 1795 by the French National Assembly, and realized materially in 1799 by the fabrication of two platinum prototypes by the Parisian engineer and physicist Fortin that may be considered as the precursor of the later metre prototypes. After the Restoration from 1815 on, this decision was revised, and this length definition was forbidden as an undesirable byproduct of the revolution. It was not until 1840 that the metric system was reintroduced in France, and the idea of an internationally binding length unit was advanced due to increasing cross-border trade, i.e., for economic rather than scientific reasons. This ended in the interstate treaty of 1875 (the "metre convention") which aimed to introduce such a length unit. In the meantime, due to refined measurements by Bessel, it turned out that, not only had the previous determination of the Earth's quadrant been 2.3 km too short, but also and more importantly the definition of the metre via the Earth's circumference was not operationally reproducible with sufficient accuracy. As a consequence, the definition via the Earth's circumference was abolished, and in 1889 a new metre prototype was manufactured in the form of an X-shaped platinum–iridium bar, in 30 copies, as materialization of the metre. The choice of shape and material was already guided by the theoretical insight that the length of solid bodies depends on external conditions, such as temperature and mechanical stress. The differences in length between the individual prototypes amounted to something like a thousandth of a millimetre, which was sufficient precision for the needs at that time. Also in 1889, Michelson and Morley already suggested using light for length measurements. This was realized in 1960 at the 11th General Conference of Weights and Measures by the definition of the metre as 1 650 763.73 times the wavelength of the radiation in vacuum between the energy states $5d_5$ and $2p_{10}$ of ^{86}Kr. The theoretical basis for this operational definition is even more apparent. The relative accuracy was about 10^{-9}, corresponding to an error of 4 cm in the measurement of the Earth's circumference. This accuracy soon proved to be insufficient since, e.g., for intercontinental interferometry in radioastronomy, the distance between two telescopes must be accurate up to a fraction of the relevant detection wavelength.

Almost simultaneously with this redefinition of the metre in 1960, the laser was invented. Through the stabilization of the frequency and reduction of Doppler broadening over the subsequent years, the relative accuracy of the measurement of frequencies, and thus of time, could be improved up to 10^{-13}, so the second was operationally realizable four orders of magnitude more precisely than the metre. On 20 October 1983, this led to the following definition by the 17th General Conference of Weights and Measures:

> The metre is the length of the path traveled by light in vacuum during a time interval of 1/299 792 458 of a second.

Through the speed of light, the measurement of length thus goes along with the measurement of time. Since then, the speed of light, as one of the fundamental physical constants, has by definition the fixed value of 299 792 458 ms^{-1}. Regarding its operational definition, length has now changed its status from one of the basic quantities of physics to a derived one, and is replaced by velocity.

This example demonstrates that the definition of a physical quantity occurs neither by measurements nor by theory alone, but is developed in a process of consistent adjustment. Concerning the operational definition this is obvious: instrumental and theoretical development in physics necessitates at least from time to time improvement in the methods and precision of measurements for the quantitative determination of physical quantities, because they must be determined much more precisely than is possible by direct sense perception. The determination of differences not perceivable by the senses, however, is unattainable without theory. In the course of this development, measurement procedures become increasingly elaborate in both technical and theoretical respects, so that, from a certain stage on, operational definitions and measurement processes can be constituted only by presupposing physical theories. Conversely, these theories rest on the terms that are made metrologically more precise on the basis of theories. Hence, the exact operational and theoretical definition of a fixed set of basic quantities which may serve as an invariable basis for physics is actually an illusion. There is neither a secured empirical basis nor an exact conceptual one, both definable a priori, which might enable the reconstruction of the whole of physical knowledge. This is the crucial reason for the necessity of consistency-generating methods, both to secure physical knowledge and to reconstruct research progress in physics.

The indispensability of consistency-generating methods has consequences for the epistemic status of physical knowledge. The original operational length definition, based on the objectively existent Earth circumference, conformed to the idea that physics as an exact science ought to be the objective and true description of nature. Accordingly, it was seen as a natural task to assign to physics through this object-oriented definition a basis esteemed as objectively given. On the other hand, every operational definition of a physical quantity contains necessarily non-eliminable conventions, because at least fixing the unit is pure convention. This debunks as illusion the belief in the possibility of convention-free and, in this respect, objective operational definitions. Indeed, the revised definition of 1889 shows unambiguously what matters regarding operational definitions: they are not true and objective settings, but preferably simply reproducible, consensus-suited, generally binding determinations. The resulting change in the epistemic status of physical knowledge from objective to intersubjective and from true to consensus-suited demonstrates that modern physics as a collective enterprise possesses a non-negligible sociological component and that the status of physical knowledge cannot be assessed in a way that is completely detached from the scientific community. Moreover, when even operational definitions cannot be established as objective determinations, this applies even more to the formation of theoretical concepts. Consequently, the reduction of deviations between theory and experiment or the convergence of measured data towards a certain value with increasing accuracy cannot be interpreted as evidence for the stepwise approach towards objective or true knowledge. Consistency-generating methods are in no way related to an approach towards truth or to the production of objective knowledge, but refer exclusively to the decrease in discrepancies between measured and calculated results and to the adjustment of two types of knowledge obtained by dissimilar, viz., empirical and theoretical methods. In this way, in particular, the number of conventional elements can be reduced by the simultaneous anchoring of the conceptual definitions in the operational, the empirical–conceptual, and the theoretical levels with mutual consistent adjustment.

The character of the consistency-generating methods is most clearly apparent in the theory-oriented development of measurement procedures and instruments. This is exemplified by the development of spectroscopic methods to determine the binding energies of electrons in atoms and molecules. The first of these methods was optical spectroscopy, developed around the middle of the nineteenth century by Kirchhoff and Bunsen. The central and for a long time unrecognized problem regarding the theoretical interpretation of the measured spectra was the fact that the spectral lines, contrary to the theoretical understanding at that time, do not correspond directly to the energy states of the system, but to energy differences between two states. This was first recognized by Bohr and later confirmed by quantum mechanics, with the result that the electrons in atoms and molecules occupy discrete stationary states. Radiation is only emitted during the transition of an electron from an occupied to an empty ("excited") state, while electrons in stationary states do not emit radiation. Hence, there is no direct connection between theory and experiment: the energies of the stationary states are calculated directly, but measurements yield energy differences between occupied and excited states. Since the calculation of excited states is considerably more involved and less precise, the question arises whether the energies of the occupied states could be made measurable. This is not directly possible because electrons in stationary states do not emit radiation, so no detectable information is obtained. On the other hand, such an adjustment of experimental techniques to the needs of the theory would lead to a substantially improved comparison between theory and experiment because the influence of the excited states is eliminated.

The basis for the development of an experimental method able to accomplish this task was provided once again by theory. When an electron is entirely detached from the atom or molecule (ionization), this ionized electron may be considered as free, and its final state is determined solely by its kinetic and binding energy. Consequently, if one succeeds in developing an instrument to measure the kinetic energy of the ionized electrons, a direct experimental access to the theoretically more easily calculable energies of the occupied states becomes possible through such a measurement. The development of such an experimental method was accomplished in the 1960s using X-ray electron spectroscopy, supplemented a few years later with photoelectron spectroscopy exhibiting substantially better resolution, and thus greater precision. These methods, developed on the basis of the theoretical results of quantum mechanics, enable virtually direct experimental access to the occupied states of the electrons in atoms and molecules and thus the option of direct comparison with quantities computed within the framework of the same quantum mechanics, viz., the corresponding binding energies. Over the course of time, subsequent instrumental development of increasingly higher resolution techniques enabled the accumulation of comprehensive empirical knowledge regarding the interpretation of these spectra, which could in turn be correlated with the various theoretical approaches, with relevance also for fundamental research. This knowledge was gained, for example, from the fine structure of the bands, the line widths, and the intensities, so that it could be concluded without any computation whether a certain orbital in a molecule was localized or delocalized, whether it contributed significantly to the chemical bond or not, and so on. This is also representative of research progress in the experimental sciences. The development and refinement of a certain measurement technique is accompanied by the establishment of a collection of empirical rules enabling, in the sense of a fingerprint technique, the comprehension and interpretation of experimental data without applying quantitative theory. In this example, for small molecules, it enabled the virtually complete and secure interpretation of the spectra, and led temporarily to the belief that theory was dispensable for that purpose. This seemed to be confirmed by the fact that the experimental results were used on the theoretical side to calibrate or test quantum chemical approximation methods used to calculate these energies. Somewhat later it turned out, however, that the spectra of certain molecules were theoretically not entirely reproducible. Subsequent studies, now on the theoretical side once again, led to the surprising result that this was not due to shortcomings of the approximate computational methods, but had to be attributed to the failure of two basic theoretical approximations, viz., the single-particle and the Born–Oppenheimer approximations. As a consequence of this interacting development of experimental and theoretical procedures, a virtually complete comprehension of the electronic states in atoms and molecules was eventually attained, wherein theory-oriented instrumental developments were of central importance.

In summary, these representative examples elucidate how empirical and theoretical knowledge develop in parallel in a process of continuous, mutual adjustment. Consequently, the empiricist models of the justification of theoretical knowledge prove to be obsolete. On the one hand, in contrast to the assumption of logical empiricism, there is no fixed empirical basis which is not subject to revision and should serve for the inductive development of theories through an increasingly detailed definition of empirical terms and the extension of empirical classifications. On the other hand, in contrast to the model of hypothetico-deductive empiricism, experiment does not act as judge over a freely invented theoretical calculus. Physical theories cannot fail on empirical experience, and the complete deduction of empirical terms and orders from the theory is not possible without additional system-specific assumptions. This type of development of knowledge, as illustrated by these examples, is rather so formative for the knowledge, methods, and objectives of physics that, beside the inductive and deductive methods, the consistency-generating methods must be considered to be a third autonomous class of methods. This concerns, in particular, the domain of the interaction between theory and experience which involves predominantly the securing of empirical and theoretical knowledge.

4.6 Consistent Adjustment: Methods

The second important function of consistency-generating methods, after their contribution to the securing of knowledge, consists in determining the empirical content of a theory. How it should be defined is controversial. According to a definition in the analytic philosophy of science, the empirical content of a theory T should encompass the entirety of basis propositions, both the true and the false ones, that are deducible in T (Detel 1986). This is dictated by the attempt to define the empirical content independently of the connection with experience. In this case, although it becomes a "sub-structure" of the theory (van Fraassen 1980), it contradicts any intuitive understanding of the empirical content to designate a theory as empirically rich in content when it contains exclusively empirically wrong statements. Instead, the empirical content of a theory T ought to be defined as the entirety of the secured findings which T provides about reality. Based on the theory conception as elaborated in Chap. 3, this suggests identifying the empirical content of a theory T with the entirety of its secure, successful scientific applications ScA. The success of an ScA is assessed essentially by its range of validity and applicability to reality, and this cannot be determined by theory-internal considerations, but only via the comparison with experience. The most notable reason is that, on the one hand, the scientific applications refer throughout to non-existent model systems that are not a part of reality. On the other hand, the empirical content results from the claim of also adequately describing truly existing systems, which should reflect the limits of the theory and its applicability to reality. Consequently, the empirical content, e.g., of the theory of the ideal gas as an ScA of thermodynamics does not consist of the fact that it provides a complete and precise description of the non-existent model system "ideal gas", but rather concerns the secure and empirically relevant findings that may be gained on real gases from that theory, such as the following:

- Real gases in thermodynamic equilibrium can also be completely described by the state variables pressure P, volume V, and temperature T: comparison with experience shows that further state variables are not required.
- Real gases in thermodynamic equilibrium behave as approximately ideal within certain ranges of (P, V, T), depending on the particular gas: how precise and in which range can be determined only by comparison with experiments.
- Universality of the behaviour of highly dilute matter: comparison with experience shows that all real gases behave asymptotically in the same way for $P/T \to 0$.
- Existence of a lower bound for the temperature and the resulting definition of the absolute temperature scale by extrapolation of the law of Gay-Lussac. The permissibility of this extrapolation can again be proven by comparison with experience.

The decisive point is, firstly, that the scope and empirical relevance of these findings cannot be determined by theory-internal methods, but only by comparison with experience, i.e., through relations between theory and experience which are not a part of the theory. Secondly, since a theory makes statements *about* reality, but is not a part of it, the empirical content is neither a section of reality nor a sub-structure of the theory. Its determination is possible only via the interplay of the various ScA, consistency-generating methods, and the relations between theory and experience which are considerably more multifaceted than assumed in the empiricist philosophies of science. Actually, they are not restricted to the mere comparison of experimental and theoretical data, nor do they consist of the construction of theories inductively from experience, and nor do they involve the verification or falsification of theories by experiment. The following detailed description and exemplification of the multifaceted relations between theory and experience will show, as a summary of the results of this chapter, in what sense and to what extent these relations and the corresponding consistency-generating methods determine the empirical content of a theory.

The first type of relations between theory and experience comprises the methods of direct comparison between experimental and theoretical data or facts as already partly described in Sect. 4.3. Since experimentally comparable data can be obtained only within an ScA, such a comparison always refers to an ScA:

(1a) In a secure ScA, the first function of the quantitative comparison between experiment and theory consists in determining reasonable error margins. While all discrepancies vanish with respect to margins estimated as too large, irrelevant or misleading ones will be created if the margins are estimated as too small. Only with respect to reasonable error margins can one assess the degree to which measured and calculated results agree with one another. Therefore, the purpose of a quantitative experiment–theory comparison is primarily to make the error margins as small as possible in order to enable the identification of significant discrepancies. On the experimental side, this requires the specification of the statistical and systematic measurement errors limiting the informative value of the experimental datum. To this end, theoretical knowledge is needed about the measuring procedure and the mode of operation of the measuring instruments. On the theoretical side, this requires assessment of the calculated result on the basis of the employed model assumptions and mathematical approximations. Every step towards improved accuracy is associated with increasing theoretical effort. The reduction of the error margins on both sides may readily be understood as mutual adjustment of experimental and theoretical data, because the development of improved instruments and measuring procedures is frequently based on theoretical advancements. Since the error margins depend crucially on the current state of instrumental and theoretical development, the result of a theory–experiment comparison can only be assessed with respect to the historical situation. Against an improved experimental or theoretical background, an originally satisfactory agreement may turn out to be a significant discrepancy. For instance, against the background of Brahe's precise measurements, the deviation of the Mars orbit by eight arc minutes from an assumed circular orbit represented to Kepler a significant discrepancy that had not previously been recognized or had been considered irrelevant. In the case of the dissociation energy of the hydrogen molecule, a deviation of 192 cm^{-1} was rated as an excellent agreement in 1935, while in 1968 a difference of 4 cm^{-1} represented a significant discrepancy. These examples clearly demonstrate how the reduction of error margins increases the informative value of the theory, and is associated with increased empirical content.

(1b) Another part of the comparison between experiment and theory in a secure ScA is the experimental realization of the theoretical statements deduced within that ScA, in addition to the goal-oriented experimental exploration of the scope of the ScA based on previous theoretical development. According to Kuhn (1970), this is referred to as consolidation or the empirical filling out of a theoretical scheme. In practice, this consolidation involves exhausting the conjectured scope of ScA and creating a network as dense as possible between theory and experience. This enlarges the empirical content in a directly comprehensible manner.

(1c) While in a secure ScA, a quantitative theory–experiment comparison does not contribute to securing knowledge, the experimental testing of hypotheses of type (H9) within an ScA that is not yet accepted as secure represents the classical case where a theoretical proposition may be confirmed or disproved by experiments. Even a significant discrepancy will not lead automatically to the refutation of the ScA.

(1d) A quantitative theory–experiment comparison is frequently preceded by a qualitative test to see whether a theoretically deduced effect is detectable, or whether a postulated object exists. A negative result of the given experiment does not indicate a falsification of the theoretical claim, but solely an inconsistency whose reasons must subsequently be sought, and this not only on the theoretical side. Note also that the improvement of instrumental techniques may lead to the detection of a theoretically deduced effect, as in the case of the Stark effect.

(1e) Finally, distinct attributes of consistency-generating methods are exibited by the application of an ScA to the study of complex real systems. The objective here is the consistent combination of approximate theoretical and incompletely interpretable experimental data in order to get a comprehensive picture of the phenomena and systems under study (Grodzicki et al. 1988). Although the data themselves are not matched with the theory, their interpretation and representation is. This application-oriented feature becomes manifest in the largely cumulative extension of factual knowledge by gaining system-related data without the primary goal of formulating law-like relations. Although this activity does not represent proper fundamental research, it is of crucial importance regarding the determination of the empirical content of the theory, because it furnishes the proof that the theory is not just applicable to the model systems designed according to its requirements, but also to truly existing parts of reality. Here it becomes most apparent how the creation of connections between theory and experience by consistency-generating methods leads to the enlargement of the empirical content.

The second type of relations between theory and experience comprises the methods for securing the knowledge postulated by statements and assumptions associated with a general claim of validity. This concerns, on the one hand, the function of experiment to provide the secureness of both the model assumptions of an ScA and the physical laws, and on the other hand, the function of theory to secure the knowledge postulated via the various empirical orders, especially the claims to validity of the order principles. The common feature of these propositions is that they are not concerned with facts, but with structures. Hence, the connection with experiment is less direct:

(2a) The first connection of the model assumptions and physical laws of an ScA with the empirical level is through the deduced hypotheses (H9). As Bohr's theory demonstrates, agreement with experiment does not furnish a proof of the correctness of these assumptions. Conversely, discrepancies with experiments merely prove the existence of inconsistencies which, especially in a secure ScA, are not interpreted as falsification, and hence do not lead, for example, to the rejection of a law. In contrast, strategies for removing the discrepancies rest on the belief in the strict validity of the law. Even in the case of an ScA considered not to be secure, a discrepancy does not lead automatically to its rejection, as exemplified by the ratio of the specific heats of gases, where the discrepancy vanished by completion of the experimental data. Finally, experiment has a selective function for securing non-determinate hypotheses or comparing competing theoretical assertions without ascribing to such experiments the predicate *experimentum crucis* in the sense of Bacon.

(2b) One function of experiment in the confrontation of a law with experience is the transformation of hypotheses concerning the contingent conditions into statements in accordance with experimental design. This creates the premise for the experimental demonstration of the law and is characterized by a tight theory-orientation in designing the experiment. Accordingly, the aim is not verification, but demonstrating the consistency of empirical experience with the theory assumed already known. As a consequence, an additional empirical facet of meaning is assigned to the law, which was initially a primarily theoretical construct. The experimental demonstration creates the impression that the law exists as a part of nature and might be discovered like an effect, at least in principle, whence it attains an autonomous meaning detached from the theoretical context. The aim of establishing as tight a connection between theory and experience as possible may then lead to violation of the norm of the strict distinction between reality and its conceptual description, as postulated by the constructive realism in Sect. 1.2. One ought to be aware of this fact in order to avoid naive-realistic fallacies.

(2c) Another part of the comparison between theory and experiment is the determination of the scope of an ScA due to simplifying assumptions of the type (H6), which facilitate or enable a mathematical treatment. In principle, their implications may be determined theory-internally, but because the quantitative consequences of such assumptions are not easy to assess, the more common method for assessing an approximation consists in a pragmatic comparison with experimental data that are rated as reliable.

(2d) Another function of experiment that may be misunderstood as experimental testing of laws consists in the controlled variation of the conditions under which the law applies, and the identification of perturbing factors in order to control or eliminate them. Actually, these methods aim to study the scope of the law when applied to real systems. For such goal-oriented activities are impossible without knowledge of the relevant law, which is thus being presumed rather than tested. Although this may lead to the restriction of its scope, the empirical content is enlarged in the sense that the scope of the theory is made more precise.

(2e) As elaborated in detail in Sect. 4.2, the function of the theory in the comparison between theory and experience consists in securing the knowledge postulated by the propositions of the empirical domain, e.g., rules and empirical generalizations. In particular, this concerns the empirical correlations whose range of applicability, significance, and mathematical form are justified theoretically by the connection with the relevant physical law, so that they are transformed into secured knowledge. Conversely, the connection of a law to an empirical correlation derived independently of the theory contributes to securing the law.

Both of these types of relations between theory and experience presume the comparability and consistency of the two conceptual systems, which is not necessarily given. In this case, the relevant presuppositions must be established:

(3a) Of basic importance is the question of how the physical quantities forming the conceptual basis of theories and laws receive their physical meaning. This is accomplished through the combination of experimental, empirical, and theoretical methods, namely:

- through operational definitions by the specification of measurement procedures, especially by making measurable those properties to which a meaning is assigned initially only with respect to a certain theory,
- through observable effects and conceptual descriptions at the empirical level,
- through formal definitions by mathematical equations establishing the precision required for theoretical arguments without representing mere nominal definitions,
- through the physical laws frequently providing further measuring procedures,
- through their status within the theory, i.e., in axioms, model assumptions, and scientific applications,
- through the deduced experimentally comparable data,
- through the meanings they possess in different theories.

Firstly, due to this simultaneous embedding of the conceptual definitions in the operational-experimental, empirical, and theoretical conceptual domains, the number of conventional elements can be reduced to a minimum, with the final goal of arriving at a “natural” system of units based on fundamental constants, such as the speed of light, elementary charge, and quantum of action. The necessary successive improvements in the numerical values of these constants is attained by applying various experimental procedures, themselves based on theories containing these constants. Secondly, these different definitions must be consistent with each other. That this cannot be assumed a priori has been demonstrated by the comparison of the theoretical binding energy and experimental dissociation energy of molecules. Another example is the difference between calculated vertical and measured adiabatic ionization potentials of molecules. Thirdly, the operational realization frequently utilizes results of other theories when, e.g., measurements of temperatures are reduced to length measurements or measurements of lengths are carried out by optical methods. More generally, the procedures for the successive improvement of operational definitions rest on theories based themselves on the relevant quantities. On the one hand, inconsistencies must not be introduced via these interconnections. On the other, these crosslinks may serve as a consistency check because the same quantity is measured by qualitatively dissimilar methods.

(3b) Another type of necessary adjustment consists in establishing the relations between the space of the theoretical description and the measuring space where the experiments are performed, described, and evaluated, and which is represented mathematically by a three-dimensional Euclidean vector space. In general, this requires additional postulates of the type (A3), which must be introduced axiomatically. Their necessity is particularly obvious when the theoretical space is not identical with the measuring space, like the phase space in statistical mechanics or the Hilbert space in quantum mechanics. Actually, this problem already occurs in mechanics in the form that the dynamical variables calculated in the theory are transformed into geometrical or kinematic data. On a more technical level, this comprises the transformation of theoretical results into a form making them directly comparable with experimental data. Previously mentioned examples are, in Kepler's theory, the conversion of the calculated positions of the planets to the actually observed ones, or in scattering processes, the conversion from the center of mass system to the laboratory system. Although such conversions may sometimes be laborious, they are not a part of the theory, but are assigned to the relations between theory and experience.

(3c) Especially in modern physics, a characteristic problem follows from the fact that, in many experiments, the system under study is perturbed in a non-negligible way due to energy supply, since information is obtained only through the radiation emitted or absorbed during the transition between different energy states. On the theoretical side, it is then not sufficient to treat only the unperturbed system, but one must also describe how the system reacts to a perturbation induced by the experiment. Regarding the given experiment, additional theoretical concepts and model assumptions are required which are particularly difficult to understand because their physical meaning becomes intelligible only in the combination of theory and experiment. A representative example is the concept of the reciprocal lattice in solid state physics: this would be unnecessary for the theoretical description of a crystal that is not subject to any experimental studies, but transforms the theoretical results into a form where they become directly comparable with the outcome of diffraction experiments.

(3d) Another adjustment problem between the empirical and theoretical levels rests on the circumstance that the development of qualitative, empirically oriented terms and notions frequently precedes theoretical elaboration. Common experience shows that the empirical conceptual system, even after the completion of theory development, is not replaced by the theoretical one, because it is not entirely deducible from the latter. The objective is then the consistent adjustment of the two conceptual systems, but not to eliminate the empirical terms or to disqualify them as meaningless because they do not match with the conceptual system of the theory. This is another indication of the methodological autonomy of empirical and theoretical research, because this problem would obviously not occur if theories could be developed inductively from empirical experience. Examples are the concept of the chemical element discussed in Sect. 2.5, or the explanation of dispositional terms by their deduction from a theory. Numerous instructive examples are provided by chemistry, with its sophisticated empirical system of concepts. Examples are the theoretical foundations of electronegativity (Parr et al. 1978) and isomerism (Slanina 1981), or the discussion about the consistency of the concept of molecular structure with the principles of quantum mechanics (Woolley 1978). Such a harmonization of empirical and theoretical conceptual systems is basically different from the adjustment between empirical orders and experimental data where, in a limited form, consistency-generating methods occur, but not the problem of a possible incompatibility between the conceptual systems.

In addition to these three classes of relations between theory and experiment, which may be described as direct in the sense that they refer to the connections between the scientific applications and the empirical level, there is also a number of indirect relations that are similarly typical for the consistency-generating methods. On the one hand, these comprise the adjustment of the empirical level to theoretical needs which are primarily reflected in the theory determinacies of the empirical basis. On the other, they concern the structuring of the theoretical level with regard to empirical relevance

(4a) The first of these theory determinacies concerns the object domain of the empirical basis due to the norm of representativeness of the objects under consideration. Since a system is defined conceptually through properties within a theory, all theoretically deduced results implicitly rest on the assumption that the system exclusively possesses these properties. As a consequence, the quantitative comparison between theory and experiment necessitates the fabrication of systems that come as close as possible to this ideal, i.e., which represent those properties postulated by the theory as precisely as possible, while others, also present of course, should not play a perturbing role. Correspondingly, one of the functions of experiment consists in this theory-oriented realization of the theoretically defined model systems, e.g., the ideal gas or monochromatic light, in order to perform experiments enabling a more precise theory–experiment comparison. Concerning quantitative measurements, this results in the reduction of the error margins, while the detection of effects will be facilitated. Although this constructive experimental practice rests on theoretical knowledge, the empirical counterparts of these model systems are added, as elements of reality, to the empirical basis, so that with advancing theory development, the original, only moderately representative objects are replaced by the empirical counterparts of the model systems. In addition, extension of the object domain is possible in the course of theoretical progress, when a theoretically postulated object is discovered by experiment. In short, there is a theory-oriented adjustment of the object domain which not only decisively improves the consistency between theory and experience, but also enlarges the empirical content.

(4b) By analogy with the object domain, the data basis, according to the norm of reproducibility, is adjusted to the current state of theory development by improving the accuracy of already existing data, by replacing older data with more precise new data, and by adding more representative or better reproducible data. All this rests on improved experimental techniques in combination with the valuating selection of data which removes results that are only poorly reproducible or not at all. Although this process of valuating selection of data may be controversial due to the possibility of manipulation, it is nevertheless necessary and legitimate because the experimental realization of the contingent conditions and model assumptions set by the theory is rarely entirely successful. In particular, actual fundamental research is usually situated on the frontier of what is technically feasible. In order to restrict the possibility of manipulations, the selection should be made according to generally accepted methodological criteria. Broad agreement exists, for example, that the direct manipulation of raw data is inadmissible, and that the best way is the improvement of the experimental design to obtain more precise data sets. More controversial, however, is the question of how far it is methodologically legitimate to "rationalize away" experimental results, and all the more so as theoretical objectives and background notions may influence this selection process. This is exemplified by the Millikan–Ehrenhaft debate concerning the experimental proof of the existence of an elementary charge unit (Holton 1973). Depending on different assessments and selection of data, it can be concluded that such a charge unit does exist (Millikan) or that it does not (Ehrenhaft). In any case, the belief in its existence was a strong motivation for Millikan to gradually improve the experimental conditions. Irrespective of these discussions, it may be concluded that the valuating selection of data is necessary and enlarges the empirical content, while the arguments for its justification must be checked in each particular case.

(4c) The last among the theory determinacies of the empirical basis concerns the construction of instruments based on theoretical knowledge. These instruments are subsequently used for the study and measurement of properties and the experimental testing of hypotheses that have been deduced within the same theory. That such a procedure is not at all circular is illustrated by the development and application of spectroscopic methods in modern physics.

(4d) Finally, a number of theory determinacies occur in the empirical-conceptual domain. The construction of empirical orders, for example, does not necessarily occur purely inductively by exclusive use of empirical knowledge, and for the formulation of hypotheses at the empirical level, all available experience, including theoretical knowledge, is put to use. An example is the hypothetical extrapolation of the electrical conductivity down to low temperatures at the beginning of the twentieth century, whose experimental examination led to the discovery of superconductivity. This proves that such a procedure is by no means necessarily circular, but may lead to novel factual knowledge, i.e., an increase in the empirical content.

The other type of indirect relations between theory and experience, which is in certain respects complementary to the theory determinacies of the empirical level, consists of the implicit empirical connections of the theoretical level due to which the theoretical knowledge cannot be reconstructed purely deductively. These references furnish the proof that even the experimentally not directly amenable theory frame is neither free invention nor a conventionalist construct.

(5a) Since there are no a priori criteria to distinguish between significant and insignificant properties, the selection of those properties establishing the conceptual basis of the theory is goal-oriented and goes back to empirical and experimental knowledge and experience.

(5b) The model assumptions (H1)–(H3) establishing an ScA rest on externally introduced empirical experience, frequently originating from pictorial analogies.

(5c) The direct experimental demonstration of physical laws of the theory frame is feasible if it is exclusively the experimentally controllable contingent conditions that enter this demonstration, and no system-specific model assumptions.

(5d) Those specializations of the theory frame that are elaborated theoretically are usually selected, although not always, on the basis of empirical relevance. Every successful ScA enforces the belief in the capability and adequacy of the theory frame and enlarges the empirical content.

In summary, the relationships between theory and experience in combination with the corresponding consistency-generating methods for securing knowledge fall into five classes:

1. Direct comparison between experimental and theoretical results.
2. Securing the claims of validity of the model assumptions and laws.
3. Harmonizing the empirical and theoretical conceptual systems.
4. Theory orientation of the empirical basis and knowledge.
5. Structuring the theory frame with regard to empirical knowledge.

Beside their diversity, the enumeration and elaboration of these relations has shown that they constitute an autonomous part of physical knowledge because they can be reconstructed neither inductively, starting from the empirical level, nor deductively from the theoretical one. Since likewise the procedures for establishing them are not reducible to inductive or deductive methods, the consistency-generating methods constitute the third, although little recognized, autonomous class of physical methods. The need for these rests on the fact that neither the safe empirical basis nor the exact conceptual basis can be defined a priori in order to reconstruct from them the whole of physical knowledge. The function of the consistency-generating methods consists primarily in their contribution to securing knowledge by establishing a comprehensive network of interconnections between the empirical and theoretical conceptual levels. This concerns both the mutual support and the mutual control of the two types of knowledge. On the one hand, a number of limitations of empirical knowledge with regard to precision, significance, and conventional elements can be overcome by theory. On the other hand, the relationship with empirical methods and results prove that theories are neither free inventions, nor instrumentalist or conventionalist constructs. Accordingly, mutual control must be viewed as a methodological norm linking theoretical knowledge with the empirical level as comprehensively as possible, and conversely adapting the empirical level to the theory if this leads to closer connections. The more relations exist for the mutual control of theory and experience, the more highly esteemed will be the reliability and secureness of knowledge, and the same goes for its empirical content. This is precisely the basic idea and rationale of the methodology of consistent adjustment: the secureness of knowledge is increased to a greater extent if empirical and theoretical knowledge are gained largely independently of one another and subsequently made consistent, than if one attempts to obtain one type of knowledge from the other either inductively or deductively.

As a further result, the empirical content of a physical theory may be described as follows:

- Empirical content and empirical basis must be carefully distinguished: the empirical basis represents a part of the subject-independent reality, while the empirical content encompasses the secured knowledge *about* the relevant part of reality.
- The empirical content is not a substructure of the theory: since a theory, even at the level of its scientific applications, offers only possibilities, their entirety may be designated as potential empirical content, but it is only by comparison with experience that the "factual" scope is obtained in the form of secured knowledge about reality.
- Due to its theoretical surplus, every theory contains parts that do not correspond to elements of reality. These must not be interpreted "realistically", and do not belong to the empirical content.
- Even the empirical content of closed theories is never definitive due to empirical openness: new discoveries of objects and effects, and novel, especially unexpected applications enlarge the empirical content. The empirical content, e.g., of Dirac's theory of the electron was enlarged by the discovery of the positron.

It is now apparent how the potential empirical content contained in the scientific applications is realized and concretized through the relations between theory and experience. In this context, the consistency-generating methods play the key role in all cases, in the sense that the improvement of the consistency between theory and experience enlarges the empirical content.

Finally, the consistency-generating methods decisively define the objectives and capabilities of physical methodology, and in connection with the establishment of the empirical content, the conception of physical reality. Firstly, the objective of a consistent adjustment of empirical and theoretical knowledge is meaningful only in terms of the conception of a unified reality which is not decomposed into a reality of phenomena and another consisting of ideas or principles hidden behind them. The methods of consistent adjustment are in this respect representative of the unity of reality. The autonomy of empirical and theoretical kinds of knowledge and methods must be understood in such a manner that the same reality is explored and described from two different vantage points that supplement each other. This reality is not fixed, but is created and designed by physics, to a large extent not only conceptually, but also materially. Nevertheless, the resulting theory determinacies of the empirical basis do not justify the conclusion that physical theories are related to reality just because this reality is constructed according to their guidelines. Indeed, the consistency-generating methods furnish the proof that the gaining and securing of physical knowledge is not a necessarily circular process, although possible restrictions remain, in principle, due to the limitations of human cognitive faculties. Since, in particular, there is no absolute measure for assessing the secureness of physical knowledge, the possibility of collective errors can never be excluded. Accordingly, the methods of consistency generation must not be misconceived as an approximation towards truth: gaining true knowledge cannot be a cognitive objective of physics because this is unattainable. Rather the objective consists in building a unified, consistent physical world view with a minimum of conventional determinations. The establishment of connections between theory and experience required for that purpose to be as tight as possible will result in a far-reaching leveling of empirical and theoretical knowledge. Although intended to a certain degree in the course of establishing a unified physical reality, there is a risk that this reality may eventually be interpreted in a one-sided fashion either as naive-realistic or as radical-constructivist. All in all, a methodology resting on consistency generation decisively determines the capabilities and constraints of knowledge, and is thus of fundamental methodological, epistemological, and didactic importance because it paves the way for a proper understanding of physical methodology, the structure and secureness of physical knowledge, scientific progress, and the aims of physics.

References

Balakrishnan A, Smith V, Stoicheff BP (1994) Dissociation energy of the hydrogen molecule. Phys Rev A **49**:2460

Bayer-Helms F (1983) Neudefinition der Basiseinheit Meter im Jahre 1983. Phys Blätter **39**:307

Bayertz K (1980) Wissenschaft als historischer Prozeß. W Fink, München

Beutler H (1935) Die Dissoziationswärme des Wasserstoffmoleküls H_2, aus der Rotationsstruktur an der langwelligen Grenze des Absorptionskontinuums bei 850 Å bestimmt. Z Physikal Chem B **29**:315

Carnap R (1928) Der logische Aufbau der Welt. Weltkreis, Berlin

Carnap R (1931) Die physikalische Sprache als Universalsprache der Wissenschaft. In: Erkenntnis, vol 2

Carnap R (1966) Philosophical Foundations of Physics. Basic Books, New York

Detel W (1986) Wissenschaft. In: Martens, Schnädelbach (1986)

Duhem P (1908) Aim and Structure of Physical Theories, translated by P Wiener, Princeton UP, Princeton, N.J. 1954

Eyler EE, Melikechi N (1993) Near-threshold continuum structure and the dissociation energies of H_2, HD and D_2. Phys Rev A **48**:R18

Feigl H, Maxwell G (eds)(1962) Scientific Explanation, Space and Time. Minnesota Studies in the Phil. of Science. Vol III. U of Minnesota Press, Minneapolis

Feynman R (1967) The Character of Physical Law. MIT Press, Cambridge, Mass.

Franklin A (1987) The Neglect of Experiment. New York

Fraassen BC van (1980) The Scientific Image. Oxford UP, Oxford

Giacomo P (1984) The new definition of the meter. Amer J Phys **52**:607

Grodzicki M, Förster H, Piffer R, Zakharieva O (1988) Profitability of the Combined Application of MO-calculations and Vibrational Analysis on Intrazeolitic Sorption Complexes. Catalysis Today **3**:75

Heitler W, London F (1927) Wechselwirkung neutraler Atome und homöopolare Bindung nach der Quantenmechanik. Z Physik **44**:455
Hempel CG (1965) Aspects of Scientific Explanation. Free Press, New York
Hertz H (1965) The Principles of Mechanics. Dover, New York
Herzberg G (1970) The dissociation energy of the hydrogen molecule. J Molec Spectr **33**:147
Herzberg G, Monfils A (1960) The dissociation energies of the H_2, HD and D_2 molecules. J Molec Spectr **5**:482
Hirschfelder JO, Meath WJ (1967) The Nature of Intermolecular Forces. Adv Chem Phys **12**:3
Holton G (1973) Thematic Origins of Scientific Thought. Harvard UP, Cambridge, Mass.
Holton G, Brush SG (1985) Introduction to Concepts and Theories in Physical Science. Princeton UP, Princeton, N.J.
James HM, Coolidge AS (1933) The ground state of the hydrogen molecule. J Chem Phys **1**:825
James HM, Coolidge AS (1935) A correction and addition to the discussion of the ground state of H_2. J Chem Phys **3**:129
Kolos W, Roothaan, CCJ (1960) Accurate electronic wave functions for the H_2 molecule. Rev Mod Phys **32**:219
Kuhn TS (1961) The Function of Measurement in Modern Physical Science. Isis **52**:161
Kuhn TS (1970) The Structure of Scientific Revolutions, The U of Chicago Press, Chicago, Ill.
Lakatos I (1974) Falsification and the Methodology of Scientific Research Programmes. In: Lakatos I, Musgrave A (1974)
Leinfellner W (1967) Einführung in die Erkenntnis- und Wissenschaftstheorie. BI Mannheim
Ludwig G (1978) Die Grundstrukturen einer physikalischen Theorie. Springer, Heidelberg
Martens E, Schnädelbach H (eds.)(1986) Philosophie - ein Grundkurs. Rowohlt, Hamburg
Maxwell G (1962) The Ontological Status of Theoretical Entities. In: Feigl H, Maxwell G (1962)
Parr RG, Donnelly RA, Levy M, Palke WE (1978) Electronegativity: the density functional viewpoint. J Chem Phys **68**:3801
Peierls RE (1955) The Laws of Nature. Allen and Unwin, London
Planck M (1983) Vorträge und Erinnerungen. Wiss Buchges, Darmstadt
Popper K (1962) The Logic of Scientific Discovery. Hutchinson, London
Przelecki M (1969) The Logic of Empirical Theories. London
Quine WvO (1951) Two Dogmas of Empiricism. Reprinted in: Harding SG (ed) Can Theories be Refuted? D Reidel, Dordrecht, Boston 1976
Rothman MA (1972) Discovering the Natural Laws. Dover, New York
Slanina Z (1981) Chemical isomerism and its contemporary theoretical description. Adv Quant Chem **13**:89

Sneed J (1971) The Logical Structure of Mathematical Physics. D Reidel, Dordrecht
Stegmüller W (1973) Probleme und Resultate der Wissenschaftstheorie und Analytischen Philosophie, vol I–IV. Springer, Heidelberg
Suppe F (ed)(1977) The Structure of Scientific Theories. U of Illinois Press, Urbana, Ill.
Wolniewicz L (1993) Relativistic energies of the ground state of the hydrogen molecule. J Chem Phys **99**:1851
Woolley RG (1978) Must a Molecule Have a Shape? J Amer Chem Soc **100**:1073

Chapter 5
The World View of Physics II: Implications

The relevance of causality, explanations, and the search for truth are discussed against the background that the key objectives of fundamental research in physics are gaining, structuring, and securing knowledge. Necessary conditions for the existence of causal connections are given which prove that causality plays at best a marginal role in physics. Eight types of explanations are defined, and their relevance for physics discussed. A concept of truth is developed that focuses on the capabilities of the physical methods and enables a reasonable definition of the concept of a true theory. Finally, various steps are elaborated which establish physical reality in such a way that this reality is a consistent combination of construction and discovery.

5.1 Aims of Physics

Like most human activities, research in physics is goal-oriented in the sense that thinking and acting are determined by goals and interests. Consequently, the objectives of physical research belong to the factors contributing to the characterization of physical methodology and knowledge. These objectives are largely determined by cognitive interests and application interests. Although these are not strictly separable, it is generally assumed that the aims of fundamental research are primarily, if not exclusively, determined by cognitive interests. These aims are closely correlated with views about what physical knowledge and methods ought to achieve and what they are actually able to attain. These two ideas are rarely separated, and no distinction is usually drawn between wishes and reality. In addition, objectives and cognitive interests are not constant factors in research work, but have undergone transformations during the course of the historical development of physics which cannot be explained solely by advances in the cognitive domain. As a result, there has virtually never been a consensus about how to define the aims of physics. Therefore, a brief presentation of the historical development

M. Grodzicki, *Physical Reality – Construction or Discovery?*,
https://doi.org/10.1007/978-3-030-74579-0_5

of these aims will be instructive, since it offers a survey of possible aims and the ways they have changed over the course of time (Cassirer 1957, Losee 1972).

The scientific objective of ancient Greek natural philosophy consisted in recognizing the deeper nature of things by uncovering a systematic order as the proper reality hidden behind phenomena. The prevailing consensus was that this deeper nature and the principles that would establish this reality cannot be found in the observable phenomena. Knowledge about nature should thus not be constrained to the mere description of patterns of events and the coexistences of these phenomena. The nature of these principles remained controversial, especially the question of what the phenomena should be reduced to. The three conceptual approaches with the greatest influence on physics may be identified as the atomistic, idealistic, and causal views. The atomistic view goes back to Leucippos and Democritos and was elaborated later by Lucretius. This reduces phenomena, properties and changes in the perceivable reality to the motion and interaction of tiny material particles, viz., atoms, that were considered to be indivisible and possess only the properties size, shape, impenetrability, motion, and ability for mutual interaction. The qualitative properties of the perceivable reality were supposed to be explainable in terms of the quantitative changes of this atomic world. According to this view, the world of atoms with their interactions and motions through empty space is qualitatively different from the world of phenomena. The colour of things, for example, does not reduce to the idea that atoms are already coloured. Here for the first time the idea was expressed that there are different levels of organizational complexity in reality and that properties on a level of higher complexity are not explained by those already existing on the lower levels. The conception of emergent properties is thus intrinsically tied to the atomistic world view. Although the close relationship with modern science is obvious, atomism did not exert any essential influence on the physical world view until the nineteenth century, for two main reasons. The first was the associated radical materialism that attempts also to reduce thinking and sensations to the motion of atoms, a notion that is likewise familiar to modern science. The second reason was the missing link with empirical experience. The cognitive ideal of the reduction of observable phenomena to quantitative changes within an atomic world could not be realized, because there was no access to this world, which remained independent of the phenomena to be explained. As a consequence, atomistic explanations were always post-hoc and circular because observations are utilized to postulate properties of the atomic world with the aim of explaining just these observations.

The idealistic approaches represented by the Pythagoreans and Plato viewed the reality hidden behind the phenomena as embodied in mathematical structures such as symmetries. Solely knowledge of these mathematical harmonies would be able to give insight into the fundamental structures of reality and eventually the cosmos. Nature itself was mathematical, so the mathematical representation was not merely a mode of description, but the very explanation of nature. Plato, for example, correlated the five elements fire, water, soil, air, and ether with the five regular "Platonic" solids. When a group of phenomena matched a mathematical relation, this was not by accident, but furnished a proof of the harmony of the cosmos. Criteria for the significance of a mathematical correlation were goodness of fit and simplicity.

The scientific objective of the third approach, represented by Aristotle, was not only of ontological kind, but was also aimed at gaining and structuring knowledge about nature. By searching for the first principles through induction and the subsequent deduction of phenomena from them, Aristotle did not restrict himself to the conceptual design of a world view, but established for the first time the basic features of a scientific method and a cognitive objective with the deductive systematization of scientific knowledge. The two together essentially constituted the transition from natural philosophy to natural science. The ontological component of the Aristotelean world view becomes manifest in the fact that the first principles represent structures of reality which cannot be other than they are. Therefore, they must be true and necessary, whence genuine scientific knowledge as knowledge of these principles is also assigned the status of necessary truths. In contrast to the Pythagorean view that the first principles were mathematical harmonies, Aristotle considered a mathematical description of reality only as a formal representation, but not as reality itself. The aim of science was the search for truth and scientific explanations that implement the transition from factual knowledge to knowledge about the causes of these facts. An explanation must have the properties that the premises are true, unprovable, and better known than the phenomenon to be explained, and they must be its causes. Causation comprised four parts, viz., the formal cause (causa formalis), the material cause (causa materialis), the efficient cause (causa efficiens), and the final cause (causa finalis). For this reason, Aristotle criticized both of the other approaches. He blamed the Pythagoreans for restricting to formal causes, and the atomists for ignoring the final causes. The central cognitive scientific aim of Aristotelian science was the deductive systematization of knowledge with the true first principles as the most general, unprovable postulates at the top, e.g., the first principles of physics are:

- every motion is either natural or enforced,
- every natural motion proceeds towards a natural place,
- forced motion is possible only through the effect of an agent,
- a vacuum is impossible.

Every explanation and every proof must begin with these principles. Albeit these first principles have revealed themselves as an inappropriate basis for physics, this type of deductive systematization has remained one of the dominating cognitive ideals of modern physics.

Irrespective of these dissimilarities in the details, the common objective of ancient natural philosophy was the creation of a unified world view in order to explain the deeper nature and signification of objects and phenomena.

These were to be searched for in an ideal, proper reality hidden behind the phenomena, while the perceivable part of the world was viewed merely as an incomplete mirror image of this ideal reality. The scientific objective focused on nature itself, and was thus of a distinctly ontological kind because the principles of this proper reality were thought to be inherent in nature and necessary truths. With the development of a scientific methodology by Aristotle, these objectives were supplemented by cognitive ones that were directed, not toward nature itself, but to the abilities of the methods to gain knowledge about nature. Therefore, the cognitive ideals and objectives were no longer determined by the human cognitive faculties alone, but also by the specific methods accepted and applied by science. The resulting insight into the limitations of scientific knowledge initiated a debate about the ontological status of the first principles, and this finds its modern continuation in the discussion about the status of theoretical knowledge. In ancient science, this controversy came under the heading of "saving the phenomena", and sprang in principle from Aristoteles' critique of the Pythagorean view of nature. According to this critique, the mere reproduction of the phenomena by mathematical relations had to be distinguished from the explanation in terms of the principles of nature, i.e., one had to differentiate between the deduction of the phenomena from hypothetical-fictitious assumptions and the deduction from the true principles of nature. Only the latter would be accepted as an explanation. The modern continuation of this discussion is the dispute over whether physical knowledge just provides a description of reality or also an explanation.

The subsequent historical development reveals how these three conceptual traditions have influenced the scientific objectives of physics. Copernicus, Kepler, and Galileo, the founders of the new astronomy and physics, put forward a Pythagorean view throughout. Both Copernicus and Kepler searched for mathematical harmonies in the phenomena, because they were convinced that these are inherent parts of nature. With his three laws, Kepler believed he had discovered such harmonies because he felt it could not be an accident that mathematical relations match so well with observations. In the context of his criticism of scholastic natural philosophy in *Il Saggiatore*, Galileo expressed his Pythagorean credo explicitly in a methodological program for the new physics:

> Philosophy is written in this great book – I mean the universe – which ever lies before our eyes, but which we cannot understand if we do not first learn the language and grasp the symbols which it is written in. This book is written in the mathematical language, and the symbols are triangles, circles, and other geometrical figures, without whose help it is impossible to comprehend a single word of it, and without which men stray in vain through a dark labyrinth.

In order to recognize reality, the direct perceptions and experiences must be "decoded" by theoretical analysis in which the geometrical representation plays the central role for Galileo, while nowadays one would speak more generally of mathematical representation. Another conceptual tradition in

Galileo's work is the view that an ideal reality is the proper one since it is only via idealization and abstraction that its mathematical structures can be uncovered. It was due to this attitude that Galileo did not consider a precise quantitative agreement between theory and experiment to be a criterion for the quality of physical knowledge. His ambivalent attitude towards experiments thus becomes comprehensible. On the one hand, he conceded that experiment could have a heuristic function to uncover the mathematical harmonies and to arrive in this way at new knowledge. On the other hand, he did not hesitate to gloss over quantitative discrepancies between theory and experiment because he did not consider them to be problematic or in need of explanation.

The expansion of the aims of physics to those resting on application interests goes back to Bacon, who was most likely the first to recognize the potential for the practical applicability of physical results due to the development of experimental techniques. In contrast, applications did not play any role in ancient natural philosophy, which deliberately distanced itself from any possible practical applications and research. This had its reason in the self-image that understands philosophy as a leisure activity of the freeman, eluding anything that might be earmarked for practical purposes and leaving utilitarian thinking and skilled manual work to the unfree classes. For the same reason, the idea of scientific or technical progress was entirely alien to ancient natural philosophy. This clearly demonstrates how social circumstances influence the cognitive structure of science: the lack of a connection between science and technology and the idea of a "free" science like *l'art pour l'art* can be understood and interpreted only in the context of such a social environment. The same applies to Bacon's formulation of application-oriented goals in his *Novum Organum*, which symbolized humanity's growing self-confidence and a detachment from the dependence on nature, and which established in this respect the ideological basis of the new physics. Most notably, Bacon substantiated the ethical commitment of science, preserved right up to now, to serve the welfare of humankind through the mastery of nature. Since Bacon's central criticism of both Aristotelian and scholastic natural philosophy concerned precisely the lack of this ethical category, the emphasis on applications procured a strong legitimation for the new physics that has prevailed virtually unmodified until today, serving frequently to justify eligibility for research grants.

The emergence of applications was the first sign of a growing trend in physics towards practical experimental activities followed by the development of a codified experimental methodology. This development was of crucial importance for scientific objectives, because quantitative experimental methods largely entail the restriction of physical investigations to the measurable properties that were referred to as primary qualities by Galileo and are here called objectifiable properties. Due to this trend, cognitive interests were no longer focused on nature in its entirety, and the central cognitive interest did not consist in establishing a universal world view, but rather a physical one. This change in the scientific objectives marked the definitive differentiation of physics from philosophy. In addition, Galileo used his separation into primary and secondary qualities to dismantle one of the Aristotelian cognitive ideals, by demonstrating that his explanations and, in particular his first principles, did not explain anything, and thus did not contribute to expanding knowledge. This conclusion found its continuation in a movement where explanations were largely or generally refused, and correspondingly, cognitive interests focused on the description of facts and phenomena, and the central cognitive scientific objective became the construction of empirical classifications. With the increasing weight of experimental experience in gaining physical knowledge about nature, this trend intensified and eventually came to dominate in the natural sciences in the eighteenth century, except for Newtonian mechanics, which, in parallel to this empirical position, established its own cognitive ideals and scientific objectives as the only mathematised science.

These may be extracted from Newton's account of scientific methodology and the explicitly formulated *Regulae philosophandi* that led to the revision and elimination of certain scientific aims of Aristotelian science, although in another direction than Galileo. With the status of his axioms as mathematical principles of natural philosophy, the proof of the necessity of the first principles becomes obsolete. According to Newton, physical methods may enable one to conclude how a group of phenomena is interconnected, but they cannot in principle exclude this connection being of some other kind. While necessary knowledge about nature is unattainable, he still believed true knowledge to be realizable: he understood the explanation of phenomena as the reduction to "true causes". With this position, he only appeared to be continuing the tradition of ancient natural philosophy, because the concept of cause in physics is restricted to causes that should be realized by forces. In contrast, the assumption of final causes has at best a heuristic value, and does not lead to true causes. Overall, the central cognitive ideals of Newtonian mechanics as the paradigm of mathematised knowledge about nature were the ideal of a deductively structured setup of physical knowledge and the explanation of phenomena by forces as their true causes. The structuring and securing of physical knowledge were to be carried out in respect of both ideals. This established a research program whose kernel was the reduction of phenomena to forces acting between tiny particles, where a similar function was ascribed to those particles as to the atoms of ancient natural philosophy. Although with the necessity of theoretical knowledge and the final causes, two essential metaphysical elements of Aristotelian natural philosophy were eliminated, Newton's central scientific objective was still an ontological one, with the explanation of phenomena by true causes: he was convinced that he could explore reality-in-itself, and the search for objective truth remained his dominating aim.

During the eighteenth century, there were thus two views about the objectives of natural science that could be referred to as empirical and theoretical. The theoretical objectives applied to the relatively well structured and mathematised mechanics, while the goals of the non-mathematised natural sciences were adapted to the abilities of empirical methodology. With few exceptions, it seems that, until about the middle of the nineteenth century, there was no controversy between the two positions. One of these exceptions was the dispute about the nature of the gravitational force, especially regarding the ontological significance of actions at a distance. Its introduction by Newton was criticized by many of his contemporaries as slipping back into Aristotelian metaphysics. Due to the impressive successes of mechanics and increasing familiarization with the conceptual system, this criticism fell silent, whence both positions existed in parallel, in a peaceful coexistence and largely isolated from each other.

This state of affairs changed only around the middle of the nineteenth century due to two far-reaching developments in physics that triggered a discussion regarding the fundamental claims, aims, and capabilities of physical research. Until the First World War, this controversy played a central role, known as the "energetics–mechanics dispute" in the German-speaking community, and has not yet come to an end, in principle, constituting the main reason for the lack of consensus about the aims of physics. The first of these developments was the expansion to new research fields like the kinetic gas theory, the wave theory of light, and electromagnetism that are much less accessible to direct perception, which means that the relatively direct contact with the objects of investigation so familiar from mechanics was increasingly lost. Empirical data are obtained virtually exclusively from experiments and require more or less speculative hypotheses to interpret and explain them. The second major change concerned new developments in mathematics, e.g., field theory and statistics, that enabled the stepwise mathematisation of additional areas of physics, viz., electrodynamics, thermodynamics, and statistical mechanics. As a consequence, the cognitive ideals and scientific objectives of Newtonian mechanics were extended to the whole of physics, while the numerous practical successes of this mathematised physics resulted in extremely optimistic assessments of the capabilities of physical methods and knowledge (Helmholtz 1847):

> It thus determines the task of the physical sciences to reduce natural phenomena to inalterable attractive and repulsive forces whose strength depends on the distance. At the same time, the resolvability of this task is the condition for an exhaustive understanding of nature [...] The theoretical natural sciences, if they do not wish to stop half way to a full understanding, have to bring their views in line with the postulate about the nature of the fundamental forces and their implications. This undertaking will be accomplished once phenomena have been fully reduced to the fundamental forces and at the same time a proof can be furnished that the established reduction is the only one permitted by the phenomena. This would then be proven to be the necessary conceptual form of the view about nature, and objective truth must be ascribed to it.

The conviction expressed in this statement is without any doubt the consequence of the enormous advances made due to the mathematisation of all areas of physics that led to the belief that the limitations of knowledge in the past could all be overcome. The actual result, however, was the revival of previous metaphysical positions and the formulation of new ones concerning the objectives of physics. Firstly, by bloating the principle of causality to a universal pattern of explanation, the belief was revived that it would be possible to proceed to the final causes by its complete realization, and thus to furnish a proof of the necessity of physical knowledge. This development culminated in efforts to derive all knowledge about nature from the principle of causality. Secondly, with the mechanical analogy models of Maxwell and Lord Kelvin as the most prominent proponents, the construction of patterns of explanation which do not explain anything was declared an objective. As belatedly constructed analogies, they do not even possess heuristic value, and are thus unnecessary accessories that do not contribute anything to physical knowledge.

Faced with the return of such metaphysical goals, long believed to have been rejected, and which largely ignored the achievements of the seventeenth century criticism of Aristotle, the reaction was not long in coming. Beside a related temporary revival of teleological patterns of explanation which remained without great importance, these causal-mechanistic views of physical knowledge were strongly rejected by the empiricists. The central question of the programme and objective of the physical description of nature had already been formulated by Rankine in 1855 with the subsequent alternative. On the one hand, one may postulate unobservable, and thus hypothetical, entities in order to derive the observable phenomena from them. On the other hand, one may classify perceivable objects and phenomena according to common properties and the relations between them without hypothesizing anything. According to Rankine, the latter method is to be preferred because one remains on the safe ground of empirical experience, and this finds its continuation in Duhem's aim of constructing "natural classifications" (Duhem 1908). Virtually the same view was opined by Kirchhoff who, in his lectures on mathematical physics in 1876, defined the description of nature, as complete as possible, as the central aim of physics, but refused the explanation by causes. The unambiguousness of the description was supposed to guarantee the principle of economy (Occam's razor), according to which the simplest representation must be preferred. The lasting merit of this position consists in the insight that physical laws constitute functional relations rather than causal ones because, due to their formal structure as equivalence relations, both sides of the equation must be viewed as entirely equal, while the designation of one side as cause is not a result of physics, but an externally imposed interpretation that is unrelated to physics. In particular, this argumentation was supported by Maxwell's equations formulated some years earlier. These indicate, for example, that a time-variable electric field induces a variable magnetic field which in turn generates an electric field. Cause and effect are thus arbitrarily interchangeable, so these laws are not causal, but coexistence relations.

The empirical position of Rankine and Kirchhoff that ties in with the tradition of saving the phenomena and aims to eliminate any metaphysics from science and to build up a "hypothesis-free physics", finds its radical continuation in the "phenomenological physics" associated with the names Mach and Ostwald, which was one of the historical roots of logical empiricism. Mach viewed reality as the sum of all sense data, while the objective of physics would then consist in their description. Physical laws were nothing else than a summary of facts and did not provide more information than contained in the data they described. Their value lay solely in their convenience of use, i.e., it was purely economical. Just like laws, theories were not considered to possess any cognitive value, but were the mere reflection or cataloging of factual knowledge. According to Mach, human intellectual ability for abstraction and theoretical thinking was to be downgraded from a strength to a cognitive weakness, because an intelligence with the ability to register and store each single fact separately would not require this detour through abstraction: abstraction and idealization were a necessary evil, rather than methods for obtaining novel knowledge. This is certainly the most extreme form of theory-hostility, identifying the elimination of metaphysics with the total elimination of theory and autonomous theoretical methodology from physics. At least, two positive achievements may be ascribed to this radical variant. Firstly, Mach pointed out the non-empirical character of both the classical concept of substance and Newton's concepts of absolute space and time. Secondly, he proved that the "truths" of physical knowledge were historically grown, and thus unmasked the search for truth as a metaphysical objective. Despite these merits, this attitude with regard to the objectives of physics is obsolete since, due to its fundamental misjudgement of the capabilities of theoretical methods and knowledge, it not only implies an unnecessary reduction of the actually realizable scientific objectives, but also entirely ignores the fact that fundamental physical research is primarily about structures, rather than the collection of pure facts.

In summary, the historical development of science exibits the following basic pattern. Beginning with cognitive interests determined by the ideal of an all-embracing understanding of nature, the scientific objectives of ancient natural philosophy were initially of ontological type, since they focused on nature itself. In the second stage, following the development of methods to realize these aims, cognitive ideals and aims supervened that were considered to account for human cognitive abilities. The different assessments of these abilities resulted in the formation of two fundamentally different positions that may be called empirical and theoretical. The first is based on the tradition of saving the phenomena and aims to build up a hypothesis-free physics which should be realized by the elimination of theories considered as metaphysical speculations. Physics is viewed as the mere description of reality, and knowledge is strictly denied necessity, truth, and explanatory value. The other position does not adhere to this minimalist program, but assiduously attempts to realize the original objectives of ancient natural philosophy by searching for methods to justify the claims of physical knowledge regarding necessity, truth, and explanatory value. The general pattern of the historical development was an oscillation between the two extremes, with a tendency to the reduced empirical position during times of great change. With the growing duration of a phase of normal science, with improved mastery of methods and rising confidence in the abilities of the paradigm, the demands on the aims are increasing again. This very human behaviour must be taken into consideration when assessing different opinions about the aims of physics. Beside the different assessments of the abilities of physical methods to gain knowledge, this is the second reason for the lack of consensus regarding the aims of physics. Thirdly, objectives resulting from cognitive interests and application interests are not usually distinguished. A unilateral orientation toward the latter tends to reduce the drive to explain and predict facts and phenomena, while the structuring and securing of knowledge play a subordinate role or none at all.

From this basic pattern, it follows in the first instance that the definition of realizable objectives of physics requires as a necessary premise a differentiation into metaphysical and cognitive ones. The cognitive objectives focus on the current state of physical methodology and knowledge, so they are closely linked to the scientific development. In contrast, the metaphysical goals, as externally imposed ideals, exhibit a distinctly normative character and are independent of cognitive developments, in the sense that they are not adapted to them. Consequently, between the desire to realize these normative visions and the reality of human cognitive abilities and physical methodology, there is an "epistemological gap" that is not readily bridgeable. The objectives for which such a bridging is basically impossible must be judged as genuinely metaphysical. Therefore, they always remain decoupled from cognitive developments and are out of the question as realizable aims of physics, e.g., the proof of the uniqueness and necessity of knowledge, the recognition of a reality-in-itself, and the search for absolute truth. Some of these metaphysical aims, however, contain parts that can be transformed into cognitive objectives. An example of such a cognitive transformation is the reformulation of the Aristotelian goal of explanation by first causes into the objective of deducing the phenomena from a set of general, theoretical fundamental principles, as already described by Poincarè around 1900 (Cassirer 1957):

> From a physics of images, modern physics has changed to a physics of principles. The development of physics in the nineteenth century is characterized by the retrieval and the increasingly sharpened formulation of these principles. A "principle", however, is not a mere summary of facts, and nor is it only a summary of particular laws. It contains as inherent claim the "always and overall" that can never be justified by experience as such. Instead of extracting it directly from experience, we are using it as a guideline to understand and interpret them. The principles constitute the fixed points we need for orientation in the world of the phenomena. They are less statements about empirical facts than maxims according to which we interpret these facts.

Initially, this aim of reduction to first principles may look like that of the Pythagoreans or Aristotle, but such a cognitive transformation did actually occur. For the status of the principles is no longer ontological, but epistemic, because they are statements *about* structures of reality rather than being a part of it, in contrast to the mathematical harmonies of the Pythagoreans or the first causes of Aristotle. Bohr expressed this essential dissimilarity in the remark:

> The idea is wrong the task of physics would be to elucidate the constitution of nature. Physics is rather concerned with our statements about nature.

The abandonment of ontological claims of validity in combination with a realistic assessment of the capabilities of physical methodology that does not unnecessarily reduce the cognitive abilities is the presupposition for the transformation of the original, metaphysical goals of natural philosophy into cognitive objectives of physics without epistemological gaps between objectives and cognitive abilities. Such a transformation does not at all mean the elimination of an autonomous theoretical methodology and knowledge, and must not therefore be identified with the reduction of the aims of physics to a mere description of reality, e.g., by empirical orders. The construction of explanatory and in particular unifying theories representing autonomous knowledge is not a metaphysical goal, because between empirical and theoretical knowledge there is no unbridgeable epistemological gap, as has been demonstrated by the analysis of the multifaceted relations between theory and experience in Chap. 4. On the other hand, the fact that these connections establish neither the uniqueness nor the truth and necessity of theoretical knowledge merely unmasks the metaphysical character of these aims, without implying that theoretical knowledge in its entirety is metaphysical and must be eliminated.

The elimination of metaphysical goals does not mean that the original cognitive interests from which these goals derive really did vanish from physics. Elements that are not transformable into cognitive objectives appear, at least partially, in the form of tacit assumptions as heuristic principles of physical methodology, as described in Sect. 1.6. However, their status and character must be carefully distinguished from the cognitive objectives in order to avoid misunderstandings. Considering the fact that physical knowledge represents throughout structured and secured knowledge, the three cognitive aims of fundamental research in physics may finally be identified as gaining, securing, and structuring knowledge about nature:

- Gaining means, in the first instance, the extension of factual knowledge both by experiment and theory, but also new insights into interrelations by virtue of abstraction. This aim corresponds to the cognitive ideal of a comprehensive knowledge about nature, and its metaphysical root is the desire for omniscience.
- Securing of knowledge has its metaphysical root in the search for truth and corresponds to the realizable aim of gaining secured, intersubjectively approved knowledge. What is assessed as secured is defined exclusively

by the methods of physics, viz., predominantly the consistent adjustment of theoretical and experimental knowledge, and the application of proof-theoretical procedures. It is a fallacy, however, to believe that one can arrive in this way at objective, true, unequivocal, or necessary knowledge. Consistent adjustment and intersubjective approval do not provide criteria for that. These original metaphysical goals have experienced a partially cognitive and partially sociological transformation, according to a consensus principle which is recognizable, e.g., by the fact that the search for alternative theories virtually ceases as soon as a closed theory exists for a certain universe of discourse.
- Structuring of knowledge corresponds primarily to the construction of unifying closed theories. Its metaphysical root is the desire for a unified, comprehensive world picture enabling the explanation of all phenomena of nature. Such structuring aims toward the ideal of deductive systematization that should be realized by an axiomatization adapted to the specific properties of physical theories. In contrast, the construction of empirical orders as structuring at the empirical level is at best a temporary goal. In addition, structuring must provide the connection between theoretical and empirical knowledge without any epistemological gap. Therefore, it contributes to securing knowledge, but also furnishes the proof, firstly, that the construction of theories is not a metaphysical goal, and secondly, that physics presupposes the existence of a uniform, subject-independent reality. According to the results of Sect. 3.5, the construction of a "theory of everything" resting on a unifying, and thus purely quantum(-field) theoretical conceptual system is unattainable and must be classified in this respect as a metaphysical goal.

All things considered, as the cognitive objectives of fundamental physical research, the gaining, securing, and structuring of knowledge correlate with the construction of a uniform *physical* world view resting on consolidated, factual, and secure intersubjective knowledge. Other views about the objectives of physics go back either to their metaphysical roots or to application interests, as exemplified by:

- the discovery and study of causal relations,
- the explanation of observations and the deduction of predictions,
- the proof of the truth of physical laws and theories,
- the proof of the existence of a reality-in-itself.

Three issues, viz., explanation, causality, and truth will be discussed in more detail below because in a certain respect they correspond to the aims of gaining, structuring, and securing knowledge, respectively. On the one hand, their relevance for physics will be studied. On the other hand, it will be shown which conclusions can be drawn with regard to the corresponding epistemological problems against the background of the results of the preceding chapters. Such a restriction enables a considerably more specific discussion of these issues, but does not of course claim to be comparable with a general philosophical analysis. This should be kept in mind when reading the subsequent sections.

5.2 The Problem of Causality

One of the basic human thought patterns is beyond dispute causal thinking, having psychological, action-theoretical, and linguistic roots. Firstly, overcoming the feeling of being at the mercy of the forces of nature presupposes the belief that natural processes do not occur entirely at random. This belief has a real basis due to the existence of perceivable regularities, but goes back most notably to the need to build structures, because gaining one's bearings in a chaotic world would be impossible. In this respect, the construction of causes and analogies is the favoured auxiliary means, with causal thinking being assigned greater importance due to the seemingly lower probability of error compared with thinking by analogy. In addition, it is considerably more relevant for pragmatic purposes: where causal thinking fails, a feeling of helplessness frequently emerges, because there is then no possibility of exerting any influence. Accordingly, the second root of causal thinking is the direct, pre-scientific experience of influencing something by actions. In this way, it is immediately and vividly experienced how one's own action can become the cause for changes that would not otherwise have occurred. In physics, experiences acquired due to such actions are contained in the tacit assumptions of experimental methodology and form the basis for planning, executing, and interpreting experiments. These repeatedly practiced patterns of action, together with their directly experienced repercussions, subsequently determine thinking to such an extent that perceptions and experiences are ordered and explained according to the cause–effect principle even in the case where they are not generated by one's own actions and the causes are not directly cognizable. The frequent actual or supposed success of such causal explanations of singular events is the third root of the widespread causal thinking that eventually becomes such a self-evident basis of human activity and reasoning that the assumption of causes appears as the reflection of objective-causal structures of reality, and thus as enforced by nature.

Subsequently, the way is paved for the transformation of both individual patterns of action and singular causal explanations into a causal principle of universal validity which considers the cause–effect relation as a general and necessary feature of all events and changes. This generalization to a causal principle imparts an ontological dimension to the problem of causality, while the mere construction of causal explanations referring to single actions and events is without ontological relevance. In this ontological meaning, the causal principle appears in the Aristotelian conception of scientific methodology by ascribing to each event a complex of causes, and the search for the first causes hidden in nature becomes the most important goal of science. For this reason, the problem of causality is a traditional issue of philosophy and epistemology, with a wealth of literature that is far too extensive to discuss in detail here. Therefore, this section can neither aim to provide a survey of the current state of the philosophical discussion, nor address all the issues associated with the problem of causality. Instead, the following discussion will deal exclusively with the question of its significance in physics. The fact that, even with this restriction, there is no univocal answer is demonstrated by the broad spectrum of opinions, ranging from being entirely irrelevant to being very important. On the one hand, the influence of causal thinking and reasoning in gaining physical knowledge is so clear that it might look as if physics aims to search for causes and provides a causal description of natural phenomena and events. In the second half of the nineteenth century, this led to attempts to add the causal principle to the principles of physics and to build the whole of physics upon it. On the other hand, even at that time, a number of prominent physicists advanced the view that the problem of causality is without any significance for gaining, structuring, and securing physical knowledge, in that the search for causal explanations and the construction of cause–effect relations do not belong to the objectives of physics. With respect to the background of this controversy, the question arises whether physics may be able to provide indications for a universal causality of nature, and to what extent physical knowledge and methods may contribute to the clarification of controversial epistemological problems; and this all the more so as the adherents of universal causation frequently attempt to justify their conviction with arguments from physics.

The facets of the causality problem as far as they are relevant for physics concern the question of whether the relation between causality and physics is ontological ("nature is causal"), epistemic ("physics provides a causal description of nature"), or methodological ("causal thinking and acting are necessary preconditions to gain knowledge"). A reasonable discussion of these questions requires two premises. The first is the differentiation of the causality problem into three levels of meaning, represented by the terms "causal principle", "causal law", and "causal explanation" (Stegmüller 1970). The causal principle postulates the causation of *all* occurrences corresponding to the conviction that every event has a cause (*nihil fit sine causa*). A causal law is understood as describing a general cause–effect connection. It is often claimed, for example, that the laws of classical mechanics are causal laws. However, when in the literature one sees reference to *the* causal law, it concerns, in the terminology used here, the causal principle. Finally, a causal explanation is the explanation of an event by specifying its causes. As for the second presupposition, one must determine the properties an event is ascribed to in order to constitute a cause–effect relation. Corresponding to the dynamical correlations, as defined in Sect. 2.6, there are structures of reality that suggest a causal interpretation, since they conform largely to the common-sense notion of causal connections. With reference to these correlations, criteria can be itemized which, although each is insufficient in itself, may serve in combination as the definition of a cause–effect relation.

1. Criterion of time-ordering: the cause must precede the effect in time. This conforms with the pre-scientific experience of the temporal succession of an action and its impact. Conversely, due to this experience, causal connections are also constructed when only a succession is perceptible. Accordingly, the mere temporal succession of two events is taken implicitly as a sufficient condition for the existence of a causal connection. Pursuing this fallacy, the initial state is then taken as the cause of the final state, e.g., the presence as cause for the future. The criticism on this act of equating post hoc with propter hoc has occasionally led to the radical conclusion that any type of causal connection must be considered as pure fiction due to the fact that it is inspired by causal thinking, while what is objectively provable is only the temporal succession.
2. Ontological status: a causal relation concerns natural events, so it is part of the object domain. Cause and effect must have the same ontological status. A relation of the conceptual domain is not a causal relation, but is called a reason–consequence relation. The same applies to relations between the object domain and conceptual level. Constituents of the conceptual domain cannot be causes for events occurring in nature: reality cannot be affected by the conceptual description.
3. Criterion of relevance: the causing influence must effectuate something, i.e., the initial and final state of the system under study must be well defined and dissimilar. A perturbation that does not change the state of the system cannot belong to the possible causes for the incidence of the final state.
4. Criterion of locality: the perturbation must propagate in space and time with finite speed, and the final state must not influence both the initial state and the cause. This restriction to local interactions is crucial for the assessment of the epistemic significance of causality in physics because it definitely excludes actions at a distance as impossible to describe causally. It does not imply, however, that the causal connection is deterministic, because it may be that the initial and final states are defined by statistical distributions while the transition between them may nevertheless be reconstructed causally.

This definition of the causal relation rests on both the properties of dynamical correlations and the results of the special theory of relativity. In this way, a conception of causality is obtained that is meaningful for physics, but severely constrained compared with the meanings in philosophy. This has the following implications:

- Only events lying within the light cone of the causing influence can be causally interpreted: space-like correlated events are never causally interpretable. The same applies to statistical correlations, as defined in Sect. 2.6.

- A sound discussion of the causality problem in physics requires definition of the state and change of state of a system, and these both depend on the underlying theory, specifying, in addition, what is in need of explanation.
- As conceptual descriptions, physical laws and principles can never be the cause for occurrences in nature: nothing happens due to physical laws. The opposite claim belongs to naive realism, which does not distinguish between reality and its conceptual description.
- The empirical proof of a causal relation cannot be furnished by specifying the cause alone: the causal connection ("causal chain") is rather established by a physical interaction, and is thus empirically provable, in principle, because beside the causing influence, the transmitting interaction can also be identified and studied experimentally.
- In combination with the interaction, the causal relation enables a law-like description: the same causes produce the same effects. This corresponds to Hume's regularity hypothesis and shows that the regularity rests on the interaction, but cannot be justified when restricted solely to the cause.

Further consequences also demonstrating the need to define the state and change of state may be exemplified with a class of physical processes whose general structure has also contributed to the view that the study of cause–effect relations ought to be an objective of physics. A system residing in a stable state ("ground state") may be subject to an external perturbation, so that the system goes into an excited state. The experimental detection of this change of state takes place by the measurement of effects. An example may be a metallic wire whose ends are connected to a voltage source. This induces an electric current and this change of state is detected by effects as the deflection of a magnetic needle or the heating of the wire. If the perturbation lasts a finite time, the system will in general return to the ground state. The cause for the first change of state is obviously the external perturbation, and a central part of the interpretation of such experiments is the explanation of what happens to the system due to this perturbation. The reaction of the system to the external perturbation is thus in need of explanation. In contrast, the second change of state, i.e., the transition to the ground state is not explained in physics, or at least not in the same vein. The sole explanation provided by physics is that every closed system being left to itself will eventually return to the state of lowest energy and maximal entropy. This, however, is not a causal explanation, but a rationale, within the conceptual system of physics, by subsumption under a universal principle. The transition of a system into an energetically more favourable or into a more probable state does not require a cause within the framework of physics. More generally, this applies to all processes occurring spontaneously.

The need to define the state and change of state is also demonstrated by Newton's first law, which defines uniform rectilinear motion as the dynamical ground state. Consequently, this state is not in need of explanation, because it is a motion without change of state, in the sense of mechanics. Therefore, the theory does not provide any cause to explain how this state comes about, nor does this belong to its objectives. Only changes of state, i.e., deviations from this ground state, are in need of explanation. Both a theory of motion assuming the state of rest as the ground state and the adherence to a universally valid causal principle must postulate some (not yet known) cause for this type of motion. This shows that a serious discussion of the causality problem in physics necessitates the definition of the ground state and change of state, and this of course depends on the underlying theory.

Additional aspects are illustrated by certain processes that can only be described quantum theoretically, such as radioactive decay. As the name suggests, these are situations where the nuclei in certain substances convert spontaneously, i.e., without external influence, into other nuclei. In physics, this process is characterized by a half-life, specifying the time after which half of a certain amount of the substance has decayed. The half-life is independent of the amount of the initial substance, so in the same time interval, the same fraction of the initial substance decays. This determines the number of decayed nuclei and the average lifetime, but it is impossible to specify how long an individual nucleus will survive until it actually decays. The fact of the decay is explained as above, viz., a radioactive nucleus is in an excited state, so after a finite time it turns into a more stable state on its own, e.g., into the ground state or another, more stable nucleus. In a similar vein the lack of knowledge of the lifetime of a particular nucleus is substantiated by one of the fundamental quantum theoretical principles, viz., the indistinguishability of identical particles. Accordingly, the question of the lifetime of an individual nucleus is meaningless, because two radioactive nuclei of the same substance and in the same state do not possess any physical properties by means of which they could be distinguished. More generally, this indistinguishability implies a number of experimentally detectable properties of an entity of identical quantum systems that can only be interpreted as statistical correlations, whence they are not causally interpretable, as a matter of principle.

Subsequent reasoning about the universal validity of the causal principle depends crucially on how the status of this indistinguishability is being assessed. If one follows the commonly accepted position in physics, that this is an objectifiable property of an ensemble of identical quantum systems, the belief in a universally valid causal principle is disproved. Alternatively, the adherents of this principle could argue that the indistinguishability and the associated statistical correlations are either shortcomings of the quantum theoretical description or only experimentally produced, instead of being an element of a reality-in-itself that is not subject to observation and experimental manipulation. Insofar as empirical verifiability is not postulated as a necessary condition for a causal connection, such an argument cannot be refuted per se, but rests only on the metaphysical belief in this principle, and is thus irrelevant for physics. For the subject matter of physics is reality as an object of perception and cognition, i.e., as recognized by observation and experimental investigation, while reality-in-itself is a physically meaningless thought construct.

In a first summary, one must conclude that the causality problem is irrelevant for physics in its ontological meaning, since physics neither presupposes a causal principle nor furnishes a proof of its existence. In contrast, physics provides distinct indications against a universally valid causal principle due to a number of processes that cannot be assigned a cause and a causal connection within the framework of physics. Nevertheless, its adherents may still argue that either the physical description of reality is incomplete, since the cause for these processes is missing, or that these processes are not elements of a reality-in-itself. Irrespective of one's view, it is in any case not an objective of physics to furnish a proof in favour or against the existence of a universally valid causal principle.

The next issue concerns epistemic relevance, i.e., whether physics provides a causal description of nature irrespective of whether the natural events and phenomena are causally structured throughout or not. Accordingly, the concern is now whether physical processes are described as cause–effect relations, whether the reduction to causes is a part of the formation of physical theories, and whether physical laws are causal laws. Whether this is actually the case may be exemplified by Newtonian mechanics, which was long considered as the scientific specification of the concept of causality and as the empirical consolidation of the causal principle. For in combination with the pre-scientific notion of forces as causes, Newton's programme of attributing all changes to the effect of forces complies exactly with the program of reduction to causes and the belief in objectively existing cause–effect relations. Forces have been viewed as causes of the various types of motion and the laws as causal laws, i.e., as the mathematical representation of cause–effect relations. If it could subsequently be proven that a force exists as cause for every change, the empirical proof would eventually be provided that nothing happens without cause. The causal principle would have received an empirically consolidated basis, and the representation of law-like correlations as cause–effect relations would in turn constitute a part of theoretical methodology.

Newton's first law defines uniform rectilinear motion as the dynamical ground state. Therefore, as elaborated above, the theory does not provide a cause to explain how this state comes about, because it is a motion without change of state. Next, as shown in Sect. 2.6, Newton's second law is inconsistent with a causal interpretation because it violates the criterion (1) of time-ordering. Consequently, the relation between force and acceleration or change of momentum corresponds to a coexistence relation, rather than a causal one. Finally, Newton's third law violates the locality condition (4) since it applies only to actions at a distance, e.g., Newton's gravitational force, but not to local interactions exhibiting retardation. Accordingly, the statement "the planets revolve the Sun because they are deflected from the rectilinear motion by the gravitational force exerted by the Sun" does not describe a causal connection, but merely gives a reason within the conceptual framework of Newton's theory of gravitation. Therefore, as already cited in Sect. 4.2, the gravitational force is not the cause, but the result of the description of central motion by that theory (Hertz 1965). Analogous conclusions hold for those applications of the force concept in mechanics where a dynamical correlation does not exist, i.e., where the relevant forces, like constraining and inertial forces, are mere conceptual constructs, as exemplified by the extension to non-inertial systems. All things considered, the force concept of Newtonian mechanics is substantially more general and not reducible to the description of causal relations. For these reasons, classical mechanics does not describe cause–effect relations and nor does it construct them. This conclusion is also confirmed by the alternative representations of the Lagrangian and Hamiltonian schemes, where the possible types of motion are deduced from extremum principles, suggesting rather an interpretation as finality. The potential question of "true causes" for an object traversing a certain path is physically and epistemically meaningless, since the choice of the mathematical description is a convention determined by the particular problem. Overall, what matters in physics is not the description of causal connections, but the deduction of empirically comparable statements from laws and contingent conditions, both being ruled out as causes according to criteria (1) and (2).

In view of these obvious counter-arguments, the question arises as to why Newtonian mechanics was so long considered as the empirical consolidation of the causal principle. Actually, there are a number of reasons for this erroneous belief:

- The first is the widespread, but nonetheless inadmissible identification of causality with determinism, criticised, e.g., by Born (1957), or occasionally the fact of equating it with lawfulness (Mach 1926). Accordingly, causality is understood in the following way, especially in mechanics. Provided that all initial conditions are known, the future behaviour of the system can be predicted with certainty. In a subsequent step, the initial state is then taken as the cause for the final state. Structurally, equating causality with determinism corresponds once again to identifying reality with its conceptual description, as is typical for naive realism. While causality refers to natural occurrences, determinism is a particular type of conceptual description.
- A second reason is the uncritical transfer of pre-scientific notions of forces as causes for changes of state, i.e., the deviations from the ground state of the uniform rectilinear motion. According to Newton, these changes occur exclusively due to external influences, described symbolically as forces. Actually, the physical meaning of the force concept remains undetermined, since it is also applicable to cases where it cannot be interpreted as cause. The identification of a force with a cause may thus be justified in particular cases, but cannot be maintained in general. Consequently, the identification of the physical force concept with causes disregards both the considerably more general meaning and the formal structure of mechanics.

- Thirdly, it is ignored that the empirical proof of a causal connection is not only provided by specifying the cause, but that the crucial role is played by the transferring interaction, which does not exist for actions at a distance.
- The last reason is the adaptation of the notion of causality to the patterns of explanation and to the conceptual system of Newtonian mechanics, in particular, the ground state and changes of state. As a consequence, the concepts of explanation, cause, causal connection, and causality become theory-dependent, because the theory itself defines what is in need of explanation. Therefore, Newtonian mechanics could be considered as the empirical specification of the concept of causality, since conversely, it is only within its framework that it has been specified how to interpret cause and causality, and this restriction has not been recognized.

In short, Newtonian mechanics is a deterministic theory due to its formal structure, but definitely not a causal one. The latter holds for all theories of physics, e.g., classical thermodynamics or the theories of static fields like electrostatics or magnetostatics that do not contain time as an independent variable, so that any causal interpretation is obsolete. For the same reason, the physical laws are not causal laws, as already demonstrated by the analysis of the concept of law in Chap. 4. The law-like description alone does not provide any proof for the causality of the given regularity, since physical laws as equivalence relations are invariant against additional attributes such as causality or finality. The ascription of one side as cause and the other as effect is rather an externally imposed interpretation which cannot be justified by the formal structure of the laws. As already mentioned, cause and effect are arbitrarily exchangeable in Maxwell's equations, and a causal interpretation of the state equations of thermodynamics is entirely meaningless. The same applies to all laws that are invariant under time reversal, because there is no preferred direction of time. Finally, the corresponding experimentally comparable deductions are entirely independent of the existence of a causal connection. For these reasons, physical theories and laws do not provide a causal description of natural events and phenomena, but rationales within the framework of a specific conceptual system. This is the essential reason for the epistemic irrelevance of the causality problem in physics.

With regard to methodology, the relevance of causality in physics rests on experimental practice. In most experiments, effects are generated by actions and interpreted as resulting from cause–effect relations. According to Sect. 2.2, an effect is defined as a reproducible and thus as a law-like event. A representative example is the observation that, by applying a voltage along a metal wire, a nearby magnetic needle will be deflected. By means of physical knowledge, the creation of this effect may be decomposed into several steps:

- the action of the experimenter creating the voltage,
- the onset of the electric current due to this action,
- the emergence of the magnetic field around the wire,
- the interaction of the magnetic field and needle that leads to the deflection.

As the conceivable causes of this effect, one might consider the experimenter's action, the electric current, the magnetic field, or the magnetic interaction. With respect to a purely empiricist position, aiming at an entirely theory-free, phenomenological description of the experiment, only the action and the observed effect are considered to be real. Accordingly, only the action comes into consideration as the cause, while electric currents and magnetic fields, which are not directly perceivable entities, possess the status of conceptual constructs, and these are not the cause, but only conceptual auxiliary means of reconstructing in thought the connection between action and effect. This conforms with the empiricist position, denying any causal explanation, and restricts itself to the pure description of observations. In contrast, according to criterion (2), postulating the reconstruction entirely in the object domain, any causal interpretation presupposes a realistic ontology that does not deny the existence of currents, magnetic fields, and interactions as elements of a subject-independent reality. Remaining possible causes are then the current and magnetic field, while the interaction is not a cause, but establishes, according to criterion (4), the causal connection between cause (current or magnetic field) and effect (deflection of the magnetic needle). Due to the objectifying mode of representation in physics, which abstracts from the acting subject, the action is needed, but irrelevant, in the sense that it is assumed that the effect would also happen provided that the conditions, as designed by the experimenter, are realized in a natural configuration. This was the decisive reason for assigning effects to the empirical basis. The objectifying description, with the transformation of the action into a cause that is no longer ascribed to an acting subject, but to the subject-independent reality, eventually yields a physically meaningful, causal explanation of the observed effect.

Although the realistic ontology in combination with the objectifying representation of physics seeks to interpret the explanation of effects as causal explanation, it does not furnish any proof that the causal relation is actually a structure of reality. This is shown by the fact that, for many effects, the opposite effect also exists (Schubert 1984). A representative example, mentioned in Sect. 4.2, is the piezoelectric effect: in polar crystals without a center of symmetry, electric fields are generated at surfaces or interfaces by a mechanical deformation, while conversely, by applying an electric field, changes in the geometrical structure can be induced in these crystals that correspond to a mechanical deformation. Cause and effect are thus arbitrarily exchangeable, so the causal structure is imposed only by the decision of the experimenter, but does not correspond to a structure of the subject-independent reality. This structure rather consists of the mutual influence of mechanical deformation and electric field. Consequently, if time ordering, i.e., criterion (1), is postulated as a necessary property of a causal connection, the objectifying representation of physics with its abstraction from the acting subject yields the result that causal explanations of effects do not furnish proof of a causal structure of reality. Additionally, this confirms a result of Chap. 2, viz., that causal explanations of effects are hypotheses at the empirical-conceptual level that do not appear in this form as a part of secured physical knowledge after the completion of theory formation. Finally, numerous physical laws possess an implicit action-theoretical content, expressing the idea that anybody whose course of action complies with the instructions following from the law will obtain the desired result. As an example, the ideal gas law definitely does not describe a causal connection. When the volume of a gas is reduced in an experiment, e.g., by moving a wall of the container, the pressure will increase. Accordingly, the change in the volume appears to cause the rise in pressure. Here, as well, the causal structure is imposed only by the experimenter, but is not part of the law itself, let alone part of the subject-independent reality.

In conclusion, even with regard to methodology, causality in no way possesses the central function frequently ascribed to it. Neither is it an indispensable heuristic for gaining and structuring physical knowledge, nor an a priori condition for gaining knowledge. There may be several reasons for this overestimation. Following empiricist views, reducing theory formation to the level of the construction of empirical orders, physics would indeed provide causal explanations and would search for causes. Another reason is a wording of the causal principle that emphasizes the pragmatic and methodological aspect, according to which every event ought to be explainable or ought to possess an adequate scientific explanation (Stegmüller 1970). Without any doubt, the vision "everything is explainable" essentially codetermines, in the form of a tacit assumption, the general research heuristics and methodology of physics. It is precisely for this reason that physical knowledge does not consist in the mere statement and description of observations and effects, but also in their explanation. Of crucial importance, however, is the sense in which physics explains. This question will be dealt with in detail in the next section. However, it has already been shown that physical theories do not provide causal explanations, but rationales with physical laws or principles as premises that are interpretable neither as causal laws nor as causes. It is only due to the objectifying representation of physics that the law appears as the cause of observed effects, since the acting subject, complying with the law, is replaced by nature, which likewise ought to comply with the law. Such a conclusion is a naive-realistic anthropomorphism: as a constituent of the conceptual level, a law can neither be the cause of a real event nor govern natural processes. Accordingly, the feigned de-anthropomorphization of causality which consists in abstracting from the acting subject and making the transition to a subject-independent lawfulness of nature, reveals itself to be the very opposite, namely an anthropomorphization of nature entirely in the vein of mythical world views.

In summary, it has been shown that only in methodological respects can some importance be assigned to the causality problem in physics, while it is entirely irrelevant from an ontological or epistemic point of view. In its ontological meaning as a causal principle, this is the case because the structure and status of physical knowledge, as well as the physical conception of reality, are unrelated to the validity of a universal causal principle. Consequently, it does not belong among the universal physical principles, in contrast to symmetries and conservation laws. From an epistemic point of view, it is irrelevant because physical knowledge and descriptions are independent of whether the processes under study are causal or not. In particular, physical laws are not causal laws, and none of the physical laws has ever been criticised just because it does not correspond to a causal connection. Actually, the reason for the irrelevance is not that causal events and processes do not exist. It is even possible to specify by means of physical methods whether a causal interpretation is admissible and empirically provable. Therefore, the reduction of the general causality problem to action-theoretical aspects would be too restrictive, because certain causal explanations of effects, and especially the existence of dynamical correlations, provide clear indications of its ontological and epistemic dimension. Regarding the objectives of physics and the structure of physical theories and knowledge, however, this is irrelevant: physics does not provide a causal analysis of nature, and nor do physical laws provide a proof of the causal structure of any physical processes. The formulation of physical laws in the process of theory formation cannot be interpreted as a search for causes or causal laws, and the setting up of causal hypotheses is not a part of theory formation, but belongs to attempts to structure empirical knowledge. In this respect, causal thinking and reasoning serve initially as a heuristic basis for the formulation of particular questions, and for the planning and interpretation of experiments, so it has a methodological function rather than an epistemic one. This is reflected in the tacit assumptions about experimental and theoretical methodology, which contain a number of causal aspects. More generally, although the belief in the existence of causal connections is the root of the methodological norm of physics according to which everything is explainable, the concept of explanation underlying this norm is not identical with causal explanation. For this reason, it cannot be concluded that causality is, at least in methodological respects, an indispensable prerequisite for gaining physical knowledge, or even an a priori and necessary condition for any knowledge: it is not causal, but deductive structures that are the basis of physical reasoning.

5.3 Explanation and Prediction

Opinions about the significance of explanations in physics are just as diverse as those regarding the causality problem. They range from being totally irrelevant to being of central importance. The first position goes back to the identification of explanation with causal explanation, understood as an answer to a "why" question. Subsequently, it is argued that "why" questions are never asked in physics, since they are instead transformed into "how" questions. As an example, Galileo's studies on free fall are frequently cited and it is argued that the prerequisite for decisive advancement of knowledge was that Galileo no longer asked for the (unobservable) causes of motion, but tried to describe the (observable) form. This argument is eventually generalized to the assertion that physics does not provide any explanations at all, but just descriptions of reality. The most notable counter-argument against such a position is that the identification of explanation with causal explanation is too restrictive. As a scientific explanation, one must rather consider deductive arguments. With respect to this premise, explanations are not only a central constituent of physics, but providing them should allegedly be its primary goal: "With the formulation of laws, natural scientists aim primarily to provide a theoretical auxiliary means in order to be able to supply predictions and explanations." (Stegmüller 1973).

The confrontation between these two positions shows, first of all, the need for a precise definition of the concept of explanation. In analytic philosophy of science, this problem of definition is reduced to the question of the formal, logical structure that a statement must fulfill in order to be accepted as an explanation. A commonly applicable definition goes back to Hempel and Oppenheim (1948) who believed they had solved this problem with their deductive-nomological or D-N explanation. According to them, explanations are conceived as deductive arguments in which the conclusion is the explanandum E and the explanans comprises a set of premises consisting of general laws L_i and contingent statements C_i about particular facts (Hempel 1966). Although this definition obviously aims at the special case that has been described in Sect. 4.4 as the deduction of experimentally comparable hypotheses within a scientific application, Hempel and Oppenheim initially claimed their schema to be applicable, in principle, to any type of explanation. The subsequent discussion and critique showed that neither necessary nor sufficient conditions which an explanation must fulfill are defined in this way (Salmon 1989). On the one hand, the D-N explanation is too broad, since it does not exclude pseudo-explanations. On the other hand, types of explanations exist that are not encompassed. Reducing the formal structure of explanations to the form "explanans $\Rightarrow$ explanandum", the problems associated with the conception of explanation may be divided into three classes:

- Requirements on the explanans in order to exclude, in particular, pseudo-explanations.
- Permissibility conditions on the explanandum in order to exclude, in particular, senseless questions.
- Conditions for the type of relation between explanans and explanandum that enable one, in particular, to prove the relevance of the explanans for the explanandum.

The discussion of these challenges in philosophy of science is ongoing. There is no generally accepted theory of explanation, nor a satisfying definition of the concept of explanation. Consequently, it cannot be the purpose of this section to describe the current state of this discussion. Instead, the focus here will be on whether and in what sense physics explains something and to what degree it belongs to the objectives of physics to provide explanations. The first prerequisite, especially in order to avoid naive-realistic fallacies, is a classification of types of explanation according to whether the explanans or explanandum are assigned to the domain of objects (S) or concepts (C). This criterion leads in a first step to four classes of explanations, by analogy with the classification of models in Sect. 2.7. Additionally, the type of relation between explanans and explanandum now plays a central role which may be primarily vertical (v) or horizontal (h). This leads to eight types of explanation that may be represented symbolically by an ordered triple $\langle X, r, Y \rangle$. Accordingly, the D-N explanation is a $\langle C, v, S \rangle$-explanation. When $r = h$, one usually has an explanation of meaning or of understanding. Theories of explanation identifying explanation largely with understanding are preferably oriented toward examples of this type. When $r = v$, one is mainly concerned with an explanation in the sense of a deductive argument. If the explanandum is from S or C, it is a matter of an event explanation or the explanation of a term, law, or theory, respectively. In physics, understanding explanations play a subordinate role, with few exceptions. In discussions regarding the significance of explanations in physics, a vertical relation between explanans and explanandum can virtually always be assumed. For this reason, only the vertical types will be discussed in detail here, while the understanding explanations will be elucidated by some examples.

1. The understanding explanation $\langle S, h, S' \rangle$ of one system or process by another takes place mostly by using $\langle S, S' \rangle$-models. Regarding the significance of this type of explanation, essentially the same applies as has been said about this model type in Sect. 2.7. The analogies between electric and mechanical phenomena, e.g., as commonly used in school, have an exclusively illustrative character without any epistemologically relevant explanatory value, and as a rule, they tend to lead to misconceptions. Therefore, even from a didactic point of view, they should be used with caution.
2. The standard example of a $\langle S, h, C \rangle$-explanation is the ostensive definition: an abstract term is made comprehensible by showing a concrete object. Since according to Sect. 2.4 these definitions are not used in physics, the same is true for this type of explanation. Citing the Solar System as an example of a system of mass points can at best serve as an illustration, but does not really consider the scope of this concept as described in Sect. 2.5. Rather, it gives the incorrect impression that such a system is defined by the condition that the size of the bodies must be small compared with the distances between them.
3. A $\langle C, h, S \rangle$-explanation in ordinary terms is the conceptual description of objects or events that cannot be presupposed as generally known, whence it cannot be considered as a mere definition. In the form of the conceptual description of the function and mode of operation of instruments, this type of explanation is significant in physics just as an explanation of how

to measure a certain physical quantity. Accordingly, it plays some role for experimental practice and is closely related to the notion of operational definitions.

4. A $\langle C, h, C' \rangle$-explanation is, in the first instance, the explanation of the meaning of a term by a predicate: an equilateral triangle is a figure with three equally long sides, and a rigid body is a system whose dynamical state is completely determined by its translational and rotational degrees of freedom. This type of explanation may be referred to as definitional elucidation or explication. Regarding the extension to conceptual systems, the relation between the various realizations of a theoretical model is of this type, where the simplest one constitutes the explanans. As shown in Sect. 2.7, the use of such analogies, although theoretically precise, does not explain anything and is thus without epistemic significance. The identification of certain common properties does not permit conclusions regarding the physical nature of the system as a whole. The existence of a formal analogy between mechanical and light waves may contribute to the understanding of some phenomena of wave optics, but does not indicate that light propagation is a mechanical process.

Common to all h-explanations is the elucidation and explanation of a less well known S or C by relating it to a better known or more familiar one. Accordingly, the most important criterion of acceptance for h-explanations is to guarantee this better understanding. In contrast, the epistemic significance, i.e., whether they provide a contribution to improved knowledge of nature, plays a subordinate role or no role at all. They may then lead to subjectively new insights, but are on the whole epistemically irrelevant and not structurally different from patterns of explanation in religion or myths, whence they must be rated as pseudo-explanations. For this reason, the majority of h-explanations may serve as didactic or methodological auxiliary means, but it should be remembered that it is not the objective of physics to construct explanations of this type.

An example of general type for a $\langle S, v, C \rangle$-explanation is the explanation of thought patterns by patterns of action, described in Sect. 1.5, if actions are assigned to the object domain. Relevant for gaining physical knowledge is the inductive derivation of empirical correlations from experimental data and that of probability distributions from statistical events. For example, the form of an empirical correlation is explained inasmuch as it is justifiable by experimental results. Since one remains in the empirical-conceptual domain, this type of explanation plays only a minor role in physical theory formation. The $\langle S, v, C \rangle$-explanation shares the property with the horizontal explanations that the explanans ought to be better known or explained than the explanandum. In contrast, the opposite is the case for the other three vertical types of explanation. Although they are usually considered to be the genuine scientific explanations, the explanans is rarely better explained or more precisely defined than the explanandum. This means that facts or concepts are "explained" through incompletely or unexplained premises. Therefore, one must question in what respect these three types of explanation explain anything, what are the criteria of acceptance, and what goals are pursued with these explanations by unexplained patterns of explanation.

The prototype of a $\langle S, v, S' \rangle$-explanation is the causal explanation. One attempts to explain events or actions by establishing a connection with other events or actions. According to the criteria given in Sect. 5.2, the plausibility and thus the acceptance of such an explanation is based on two conditions:

- The proof of existence for S: the explanans, i.e., the cause, must be an element of reality, not a conceptual construct.
- The proof of relevance for S: the existence of a connection between S and S' must be provable, i.e., S must be identifiable as a cause for S' and must not be solely an accompaniment of S'.

Examples of non-physical causal explanations are the explanations of individual actions and singular events, because neither can be elements of the empirical basis of a physical theory. The paradigm of a causal explanation in physics is the explanation of physical effects. However, in Sect. 5.2, it was shown that in most cases the causal structure is created only by an acting subject, rather than being a structure of reality. Actually, most explanations in physics are not causal explanations, but of an entirely different type. This is shown by the conceivable answers to the question of why a stone falls when it is dropped:

- because all objects fall down,
- due to the law of gravitation,
- because the Earth attracts the stone,
- due to the gravitational force,
- because all masses attract each other.

The first answer is the explanation by subsumption of a particular case under a fact accepted to be universally valid. In the spirit of a causal explanation, this is a pseudo-explanation since a particular event does not happen just because a certain class of events occurs according to a general pattern. The second answer is not a causal explanation since nothing happens due to conceptual descriptions like physical laws, and the computability of an event does not explain it. The same applies to the last three answers, all of them referring directly or indirectly to mass attraction. They are pseudo-explanations because the phenomenon of mass attraction is not better explained than the fact of the stone falling that one is seeking to explain. Usually, this is not recognized because the principle of mass attraction is generally considered as secured physical knowledge which is no longer in need of explanation.

All in all, the explanation of effects in physics does not exhibit the structure of a causal explanation, but rather that of a deductive argument, and thus has the structure of a D-N explanation, i.e., an explanation of the type $\langle C, v, S \rangle$. In order to be accepted as meaningful, such an explanation does not need to meet either the acceptance criteria for causal explanations or the requirement that the explanans must be better explained than the explanandum. Rather it suffices to subsume it under the patterns of explanation of a physical theory in the form of its laws. This underscores the relativity of such explanations. For anybody who is not convinced of the correctness of the relevant physical laws required for the explanation, such a deductive argument does not represent an acceptable explanation. In addition, it should be mentioned that a $\langle C, v, S\rangle$-explanation is not a strictly deductive argument. Hempel and Oppenheim themselves already pointed out that it is not the effect itself, but a statement *about* this effect that is deduced. Therefore, a $\langle C, v, S\rangle$-explanation contains the implicit assumption that the deduced statement and the observed effect agree with one another. According to the results of Chap. 4, however, the verification of this agreement is not at all self-evident and may lead in the case of discrepancies to modification of the premises. Consequently, unlike a causal explanation, a $\langle C, v, S\rangle$-explanation may very well be of importance for gaining and securing physical knowledge, instead of being a mere application of knowledge, as is the case for a causal explanation or for a strictly deductive argument. All things considered, the problem of acceptance for $\langle C, v, S\rangle$-explanations comprises three aspects: the premises must be accepted, the formal deduction must be logically correct, and the agreement with the observed effect must be well-founded. Acceptance of the premises presupposes that the relevant laws are considered not to be in need of explanation and that the contingent conditions are realized in an appropriate experimental design.

The fact that both the D-N and $\langle C, v, S\rangle$-explanation do not refer initially to the object domain, but to the conceptual one, might suggest interpreting the D-N explanation as a special case of a $\langle C, v, C'\rangle$-explanation. This view appears to be supported by a special class of $\langle C, v, C'\rangle$-explanations, viz., the deduction of empirical correlations from the model assumptions of a theory, like the derivation of Balmer's formula and Moseley's law in Bohr's theory. On the one hand, as a part of the empirical-conceptual domain they are closely correlated with experimental data, while on the other, they are rarely distinguished from physical laws due to this deducibility. However, taking the formal structure of the D-N explanation seriously, assuming the premises to be a combination of general statements, e.g., physical laws, and contingent conditions, the $\langle C, v, C'\rangle$-explanations are structurally dissimilar, because the premises do not contain contingent conditions due to the law-like character of C'. This is already apparent for the empirical correlations and becomes obvious for the deduction of physical laws within a theory since these are entirely independent of any particular empirical circumstances. Representative examples of $\langle C, v, C'\rangle$-explanations are accordingly:

- The realizations of the theory frame by its scientific applications: according to the theory conception developed in Chap. 3, this type of explanation is a central part of theory formation in physics since, without the existence of scientific applications, comparison between theory and experiment would be impossible.

- The relation between the structural principle and its specializations within the theory frame, which comprises both deductive and unifying aspects since it may be interpreted as subsumption of specializations under a unifying theory.
- The relation between explanatory and descriptive theories, although according to Sect. 3.3, such a relation is not strictly deductive, but contains also horizontal structural elements.
- The reduction to universal, supra-theoretical patterns of explanations or principles like conservation laws and the associated symmetries: due to their normative character, these principles are accepted as not being in need of explanation in physics.

These examples show on the one hand that, by means of the $\langle C, v, C' \rangle$-explanations, the abstract statements of the theory frame are explained in the sense that they become more comprehensible through the empirically relevant realizations, and on the other, that the specializations and realizations are explained by subsumption under universal principles considered not to be in need of explanation. Accordingly, the $\langle C, v, C' \rangle$-explanations play the central role in the structuring of physical knowledge, through both deductive systematization and unification. These types of explanations, however, do not explain anything in the sense of either an h-explanation or a causal explanation. The laws and effects of wave optics, for example, do not become more comprehensible due to the possibility of deducing them from Maxwell's equations, and nor do Maxwell's equations provide the "proper causes" of the optical effects.

The broad spectrum of meanings of explanation as demonstrated with these eight types makes any statement about their significance in physics arbitrarily vacuous as long as the concept of explanation is not specified. The same applies to all problems discussed in this context in epistemology and philosophy of science, e.g., the question of acceptance criteria for an appropriate explanation, demarcation criteria between scientific and non-scientific explanations, and those between proper and pseudo-explanations. The possible answers depend first and foremost on the goals that are pursued with an explanation. If it consists in the reduction to something more familiar or comprehensible, as in the case of h-explanations, epistemically irrelevant or even incorrect explanations may be acceptable. In this case, the dominating reason for acceptance is that the explanans is better known than the explanandum, and that the relation between the two should be plausible and illustrative, even when it is a matter of pseudo-explanations in the spirit of v-explanations. In textbooks, one even finds understanding explanations contradicting confirmed physical knowledge, adopting the attitude that a false explanation may be better than none at all.

Exactly the opposite occurs with the v-explanations. If a statement "p explains q" is accepted as an explanation only if p is also explained, most explanations of this type do not explain anything. In particular, this applies to the physically most relevant $\langle C, v, C' \rangle$- and $\langle C, v, S \rangle$-explanations, because the relevant theories, axioms, and principles are not explained, but must be accepted either as not in need of explanation or as not further explainable. Accordingly, the relevant explanatory framework is provided exclusively by the theory that first defines what is not in need of explanation, and then determines the patterns of argumentation serving to explain those phenomena that are in need of explanation. This does not exclude the possibility that these patterns of explanation may occasionally be explained by means of other theories, as described in Sect. 3.3. However, even such $\langle C, v, C' \rangle$-explanations always end with patterns of explanation that are not further explained: no physical theory provides an explanation for its own patterns of explanation.

Analogously, the question of the demarcation between pseudo-explanations and proper explanations can be answered only with respect to a certain type of explanation. In the sense of h-explanations, most of the v-explanations are pseudo-explanations, and vice versa. With respect to causal explanations, $\langle C, v, C' \rangle$- and $\langle C, v, S \rangle$-explanations are pseudo-explanations since theoretical statements cannot be the cause of natural phenomena. Consequently, the acceptance of causal explanations rests only marginally or not at all upon the approval of theoretical knowledge; it rests rather upon the proof that causes and causal connections are parts of reality. In particular, causal explanations are not just a variant of a D-N explanation resting on causal laws. In this context, the introduction of the concept of a "theoretical cause" is absurd, because this implies the identification of causes with reasons, i.e., the identification of reality with its conceptual description once again. Therefore, in the spirit of constructive realism, one must distinguish carefully between $\langle S, v, S' \rangle$- and $\langle C, v, S \rangle$-explanations and between causal explanations und D-N explanations, respectively. Another aspect of the demarcation problem is the distinction between scientific and non-scientific explanations. An explanation, understood usually as an event explanation, is frequently considered to be scientific when the results of a science serve as premises. However, this is mostly a matter of applications of science, so it is more appropriate to rate as scientific only those types of explanation that contribute to the objectives of gaining, structuring, and securing scientific knowledge. In this respect, only the $\langle C, v, C' \rangle$- and the $\langle C, v, S \rangle$-explanations are scientific, insofar as S refers to reproducible events. In contrast, none of the other types of explanation fulfill this condition and they must be classified as non-scientific in this respect.

Finally, the status of predictions in physics can be discussed, especially against the background of the claim that the central purpose of physics is to provide predictions. If one initially understands a prediction to be the prognosis of singular events, it is not a part of physics. Although it is undisputed that singular facts and events can be explained and predicted by means of physical knowledge, they are not a part of the empirical basis, so this type of prediction does not belong to physics as science. More generally, one could define as prediction any theoretically deduced statement, but this does not change its status. Either it is again a matter of mere application of physical knowledge or, as a scientific $\langle C, v, S\rangle$-explanation, it is a hypothesis that is subject to refutation in the comparison between experiment and theory. A prediction, however, is commonly not a hypothesis, but a *confirmed* statement that is deducible within a secure scientific application of a closed theory and does not require any empirical corroboration because it is undisputed. Due to this property, predictions do not even contribute to the acquisition of knowledge, and their production is neither an auxiliary means nor an objective of research in physics. For the same reason, it is easy to refute the idea that explanations and predictions are structurally equivalent in the sense that they differ only because, in the former case, the explanandum is already known, while in the latter case, it is not. Such a claim is meaningful only if explanation is identified with the explanation of events, and it seems to be sensible just because many studies in the philosophy of science are restricted to this type of explanation. But simple examples already prove that predictions are not even structurally equivalent to explanations of singular events, provided that they rest upon physical knowledge. Through skillfully constructed algorithms, equally good or even more precise predictions can be obtained than by physical laws. However, algorithms do not advance any explanatory claim since their construction does not require any understanding of the process under study. This is exemplified by the very precise astronomical calculations of the Babylonians. Consequently, a prediction may be the mere calculation of a datum that does not presuppose any physical understanding and is thus epistemically irrelevant. In contrast, the explanation of an event does not consist in the pure prediction of a fact or datum, but raises at least the claim that the premises contain those causes or reasons that are relevant for the occurrence of the given event or contribute to its understanding. Overall, a structural equivalence can only be claimed if one gives up any relevance condition for the premises. In the end, this does not lead to a reasonable concept of of explanation, since an important demarcation criterion against pseudo-explanations is abandoned.

The relevant aspects of the explanation problem in physics, the significance of the dissimilar types of explanation, and the extent to which they belong to the objectives of physics may be summarized as follows:

1. None of the horizontal types of explanation are a part of physical knowledge, and nor are they an objective in the acquisition of knowledge since they are irrelevant, even misleading, with regard to their epistemic content, as demonstrated by the mechanical explanations of non-mechanical phenomena which have been eliminated from physics precisely for this reason. Subjectively, they may lead to an improved understanding or may be useful for didactic purposes in particular cases, or again they may have some heuristic value, but they do not contribute to gaining, structuring, and securing knowledge, so they cannot be counted as scientific explanations.
2. In contrast, the $\langle C, v, C'\rangle$-explanations are essential constituents of physical knowledge due to their decisive contribution to structuring and occasionally to securing. Moreover, understanding structuring as the de-

ductive systematization and unification of knowledge, the construction of $\langle C, v, C' \rangle$-explanations is a central objective of theory formation in physics.

3. Similarly, the $\langle C, v, S \rangle$- or D-N explanations are scientific explanations insofar as they refer to reproducible events. Firstly, as deductions of statements about observable effects within a scientific application, they establish a necessary prerequisite for the securing of knowledge via the comparison between theory and experience. Secondly, the subsumption of effects under a small number of model assumptions has unifying aspects, and thirdly they can contribute to the extension of knowledge. Due to these features, they are non-eliminable parts of physical knowledge. Although their significance does not rest on the fact that it would be an objective of physics to provide such explanations, they are an indispensable methodological means. In contrast, as explanations of singular events, the $\langle C, v, S \rangle$-explanations are not a part of physics, because their study is not the purpose of physics. Their "scientificity" consists at best in using scientific findings.
4. Causal explanations are not $\langle C, v, S \rangle$-explanations by causal laws and do not belong to the physically relevant explanations, because they are without significance for gaining, structuring, and securing physical knowledge. In constructing causal explanations, the proof of existence of both the cause and the cause–effect connection may be furnished by means of physical knowledge, but this is an application of physics rather than an objective. The aim regarding the explanation of effects is not the detection of causes, but deduction from a theory.

In summary, physics provides explanations in the sense of (2) and (3), and is by no means restricted to mere descriptions. However, physics does not explain reality in a theory-free or ontological sense: neither does it provide causal explanations, and nor does it assign *ontological* significance to its patterns of explanation in the sense that they are part of reality. The physical explanations are epistemically significant, theory-dependent rationales because the patterns of explanation are integrated in intersubjectively accepted theories. Consequently, the challenge of acceptance for explanations in physics focuses on the question of the conditions that the explanation patterns ought to fulfill as premises of the $\langle C, v, X \rangle$-explanations. According to Hempel and Oppenheim, this condition should be their truth. Accordingly, the further objective of physics beyond the construction of $\langle C, v, X \rangle$-explanations ought to be to prove the truth of the explanation patterns and thus to achieve the highest degree of secureness. Whether such a search for truth belongs to the objectives of physics will be discussed next.

5.4 The Problem of Truth

Since securing knowledge is one of the aims of physics, the question arises as to how secure physical knowledge can be. The highest degree of secureness of knowledge is without any doubt its truth. Therefore, securing is frequently identified with establishment of truth, which is one reason for the belief that an objective of physics is the search for truth. That not only physics, but science in general would search for truth is a widespread conviction that is in Germany even legally fixed in the Basic Law according to the commentary of Maunz and Dürig which is generally acknowledged as the prevailing view (Scholz 1977):

> The concept of science in terms of the constitutional law is thus to be summarized as follows: science means the autonomous intellectual process of systematic, methodical, and self-dependent search for findings of factual objective truth, as well as the communicative transmission of such findings.

Actually, securing knowledge has as little to do with establishing truth as secure knowledge has to do with objectively true knowledge. So what led to the opinion that physics might provide findings in the form of "factual objective truths"? One reason is certainly the successful applications of physical knowledge and its predictions, which, although not correct in all details, are sufficiently informative for practical purposes; or in the words of Lenin, practice proves the truth of physics. However, this is not a criterion for truth: one may very well deduce empirically correct results from physical theories that are false according to the current state of knowledge, so successful applications of knowledge alone do not provide a proof for truth. A concept of truth oriented toward practice has its roots rather in the commonplace understanding of truth as the agreement with facts: the statement "there stands a tree" is true if and only if a tree actually stands at the designated site. Therefore, the criterion "agreement with facts" seems to provide both a definition of truth and a procedure for stating it. One checks whether it behaves as claimed. In this context, facts are usually understood as particular objects or singular events, as elements of a fixed, predetermined reality which does not depend on statements about it. Accordingly, one could define: "Truth is the agreement of a statement with an element of reality." The element of reality is the verifying element or "verificator", the claim as a part of the conceptual domain is the carrier of the truth value, and the question of truth concerns only the claim, but not the verificator. Reality is not true, but, according to Aristotle, falsehood and truth are not in the things, but in the thinking. Consequently, the truth problem consists in the proof of the correspondence between a linguistic expression and a part of the nonverbal reality:

$$\text{truth-value carrier} \hookleftarrow \text{correspondence} \hookrightarrow \text{verificator} \qquad (*)$$

A possible definition of truth requires the specification of all three constituents of this relation, thereby providing the framework for sensible questions, such as:

- What are the possible truth-value carriers: single words, statements about facts or events, generalizing statements like physical laws, or entire theories?
- What comes under consideration as the verificator: things or facts in the form of singular events, or also reproducible events like physical effects?
- What does the correspondence mean: is it an identity, a reflection, or a similarity in the sense of a correspondence concerning selected aspects?

Regarding the discussion of the truth problem in physics, these questions also constitute a suitable starting point. Although less comprehensive than the corresponding explorations in philosophy (Ritter et al. 1971ff., Skirbekk 1977, Puntel 1978, Künne 1986), it will be more specific. For instance, due to the fact that physics generates knowledge *about* something, viz., nature or reality, each concept of truth relevant for physics corresponds to a binary relation in accordance with (*). As a result, philosophical truth theories considering the property "true" as a unary predicate (non-relational truth theories, Künne 1986) do not play any role in a truth discussion regarding the objectives of physics. Similarly, the entire speech-analytical dimension of the truth problem is irrelevant for physics. Accordingly, the philosophically influential redundancy theory (Puntel 1978, Künne 1986) going back to Frege and Ramsey can also be discounted.

Another indication about which kinds of truth are relevant in physics comes from the circumstance that physical knowledge about reality is obtained by two mutually independent methodological approaches, viz., by empirical and theoretical methods. The question is whether securing only concerns the relation between empirical and theoretical knowledge, or whether it also concerns the relationship between physical knowledge and the external reality of arbitrary objects and non-reproducible events as a whole. In order to make the subsequent analysis more precise, three meanings of truth should be distinguished (Grodzicki 1986):

- logical mathematical truth as the prerequisite of the (internal) consistency of theories or aggregates of statements within the conceptual domain,
- empirical or epistemic truth as a term describing the consistency between results of the theoretical and empirical domains,
- non-epistemic or ontological truth, with the assumption that the properties and constitution of the external reality as a whole are independent of its conceptual description, so that the type of correspondence is unimportant.

A non-epistemic truth concept underlies both the vernacular meaning of truth and the relation defined by (*). The various forms of the correspondence theory of truth (Puntel 1978, Künne 1986) essentially attempt to make these common-sense notions more precise. They are classified as relational and non-epistemic (Künne 1986) because truth is assumed to be a binary relation and objectively statable: the truth of a statement does not depend on being held for true and nor is truth a matter of arrangements, but is exclusively determined by the subject-independent reality. For this reason, the correspondence theory of truth is closely related to realistic ontologies, and criticism of it largely resembles the objections against naive realism. By analyzing the constituents of the truth relation (*), it can be shown to what degree this critique is justified, and how the notion of truth in physics has changed from a preponderantly non-epistemic attitude to a more epistemic one.

In the first step, a correspondence theory of truth must clarify what may be considered as a verificator, i.e., as the element of reality that may serve as a measure of the truth of statements. As already mentioned, the universe of discourse of physics can only be a reality as constituted by human cognitive faculties, but not a reality-in-itself. In addition, it is usually assumed that this is a directly and predominantly passively perceived reality, comprising sufficiently many attributes that are not influenced or changed and, in particular, that are not generated only through perceptions. Although it is not denied that this reality contains subject-dependent aspects, because every perception is structured and occurs against the background of previously acquired knowledge, the possibility of a sufficiently clear separation between the two parts is presumed. Furthermore, the passively constituted part of reality is considered to be so large that the assumption of a perceptually independent and in this respect subject-independent reality of facts is justified, where facts are considered as particular objects or singular events.

Such a definition of facts is unsuitable for the analysis of the truth problem in physics. Although physical knowledge exhibits many relations with the passively constituted reality, as exemplified by the successful predictions and explanations of singular events or by the design of devices and machines, only elements of the empirical basis can come under consideration. This results in substantial modifications since this basis represents an actively constituted reality of reproducible effects, representative objects, instruments, experiments, and data produced in a goal-oriented way. Although the ontological status is the same according to the tacit assumption (2C), this reality must not be identified with a passively constituted one. Consequently, the question of the truth of claims about singular events that plays the key role in everyday life and forms the basis of many philosophical studies is entirely irrelevant for the truth problem in physics. Most notably, the empirical basis itself already represents physical knowledge that is subject to the question of truth. Moreover, the reproducibility of effects, the representativeness of objects, the display of data sets, and the use of instruments requires linguistic elements originating for a large part from the scientific language. These linguistic elements are indispensable, because on the one hand, they ascertain the communicability of the facts, and on the other, they express verbally properties of the facts that are indispensable for the examination of theoretical statements. Therefore, the previously mentioned problem of the relationship between language and reality arises, i.e., the extent to which verbalization adulterates. In a similar form, the colloquial truth problem is concerned. When speaking about a fact, e.g., there stands a tree, this is usually considered as a mere statement that cannot be questioned. Actually, even the mere statement of a fact is a judgement which implicitly contains the assumption that the fact has been recognized correctly, similarly to an element of the empirical basis. Consequently, two conceptual claims are always compared, while in a strict sense statements and facts as parts of the verbal and nonverbal levels, respectively, are not comparable, so the verificators are not elements of reality. The expression "agreement with facts" then means agreement with evident judgements about correctly recognized facts which are not questioned by anybody. The study of the general truth problem must already assume that the comparison between verificator and truth-value carrier occurs in the conceptual domain.

This fact is made explicit in the empirical basis, in the sense that it has some kind of intermediary function between the two domains. On the one hand, their elements are factual enough to ascribe to them the same ontological status as the external reality, and on the other, it contains a linguistic component that makes its elements communicable and enables a reasonable comparison with theoretical statements. Accordingly, the function of the empirical basis as verificator rests on the assumptions, firstly, that the difference between a passively constituted reality and the empirical basis is judged as unproblematic, and secondly, that the verification of its elements is not the subject matter of the truth problem in physics. Its kernel is rather the issue of whether and in which respect theoretical findings may be designated as true, although the correctness of an element of the empirical basis may be disputed in particular cases. This problem, however, resides on another level, because the elimination of errors in the cognition of facts and the assessment of observations like meter readings or pointer positions is the function of experimental methodology during the transition from the singular reality to the empirical basis, and it is unrelated to the truth problem of theoretical knowledge. For this purpose, it is always assumed that the elements of the empirical basis have been recognized correctly.

In the second step, the potential truth-value carriers must be identified and studied. In physics, these are always verbalized, science-based statements, rather than dogmatic convictions or non-verbalized thoughts. In a correspondence theory of truth, only single statements come under consideration as potential truth-value carriers that are not mere tautologies, usually advance a claim of validity and admit a direct comparison with facts. In physics, this applies to real definitions, empirical hypotheses, and hypotheses of type (H9). The study of these truth-value carriers may use the methods by which their claims of validity are redeemed, i.e., the methods used to secure the knowledge they contain. According to the correspondence theory, this occurs by the proof of agreement with facts, i.e., by empirical comparison. Regarding real definitions, this is directly applicable, because the empirical information contained therein must agree at least enough to enable identification of the relevant object. Similarly, the comparison with experimental data is the best suited method of verification for the deductive hypotheses (H9). In contrast, the claims to validity of empirical hypotheses, e.g., the empirical correlations and generalizations, are redeemed by deduction from a theory, because in particular the epistemic significance cannot be secured through empirical comparison, but only through embedding within a theory. In contrast, according to the correspondence theory, the verification must take place exclusively by comparison with facts. Its shortcomings become even more apparent when the central problem of truth in physics is concerned, viz., the truth of the knowledge contained in the model assumptions, laws, and axioms which are not at all amenable to a direct comparison with facts. In summary, in relation to the physical truth problem, the correspondence theory exhibits the following deficiencies concerning the truth-value carriers:

- As truth-value carriers, only statements enabling direct empirical comparison, i.e., corresponding directly to facts, come under consideration.
- The claims to validity of real definitions and empirically-inductively obtained hypotheses cannot be secured by the proof of agreement with facts. Since they have already been established in accordance with facts, such a methodology leads to a vicious circle.
- Even when a single, empirically comparable statement is verified in the sense of the correspondence theory, this alone is without epistemic significance. A statement becomes relevant as physical knowledge only by being embedded in a scientific application, so that it acquires the status of a deductive hypothesis.
- A fortiori, the truth of statements claiming universality cannot be proven by methods resting on the agreement with (necessarily finite) facts. The truth of model assumptions and physical laws, let alone the truth of theories, as assessed by the question of the truth of the premises of the deductive hypotheses, cannot be answered, and most likely cannot even reasonably be asked, within the framework of a correspondence theory.

The challenges regarding the type of correspondence that is thought to specify agreement with reality are even greater. The correspondence theory tacitly understands a true statement to be something like a merely verbalized reproduction of a fact as an element of reality. This corresponds to the assumption of a simple and unequivocal relation between language and reality. In logical empiricism, this was expressed in the protocol sentences that were thought to be the unproblematic verbalizations of singular facts. However, such a simple structure cannot even exhibit the much more concrete nonverbal relations. For instance, there is an entirely different type of correspondence between a map and a landscape than between a photograph or a painting of the same landscape. A fortiori, this applies to the concepts of physics where, even in the simplest cases, the assumption of an unambiguous relation between concept and object is untenable, and the kind of correspondence may be entirely dissimilar from case to case. Firstly, in the case of real definitions, the identifiability and recognizability of objects or effects is decisive. Secondly, depending on the context, the correspondence between experimental and theoretical data may consist in a qualitative or a precisely defined, quantitative agreement. Thirdly, concerning the physical laws, the referential aspect plays at best a subordinate role, because their claim to validity refers, in the first instance, to non-existent model systems. Similarly, in the case of the model assumptions of a scientific application ScA, the kind of correspondence can be revealed only within the theoretical context and relative to the claim to validity advanced by the ScA.

All in all, the correspondence theory of truth exhibits a number of shortcomings that are generally inherent in it, but become especially apparent when applied to the problem of verifying physical knowledge:

- The verificators are, in a strict sense, not facts, but assessments containing the implicit assumption that the facts have been recognized correctly. A fortiori, this applies to the data of the empirical basis. In an empiricist sense, a qualitative datum may be considered as finally secured, while quantitative data are never ultimate due to the categorical possibility of improving them. Consequently, the standard of comparison as set by the verificators is subject to change and is never absolute, because it depends on the current state of knowledge.
- Even in the non-scientific domain, the verification of statements does not occur exclusively via the mere comparison with facts, but also by examining the consistency with other statements. A fortiori, this applies to those statements in physics that are not amenable to direct empirical comparison. In this case, the agreement with facts is not a criterion that is readily applicable to redeeming their correctness. For these reasons, the securing of physical knowledge takes place by simultaneously embedding in the empirical and theoretical domains.
- What is understood precisely as correspondence or agreement depends on the context. In particular, the options for securing knowledge, as documented by the multifaceted relations between theory and experience in Chap. 4, are not even approximately reflected within the framework of a correspondence theory of truth.

For these reasons, the correspondence theory of truth does not provide a suitable basis for a conception of truth adapted to the structure of physical knowledge, which is usually secured by entirely different methods and criteria than by the mere proof of agreement with facts. In particular, regarding statements not amenable to direct comparison with facts, only the following alternative remains: either the question of truth is not at all meaningful or other truth criteria must be sought. Actually, in Chap. 3, it was described how the extensive vertical and horizontal interconnections within the theoretical domain contribute to securing theoretical knowledge. It is then obvious to exploit these interconnections in combination with the relations between theory and experience as the basis for the explication of a conception of truth adapted to both the structure of knowledge and the potential and abilities of physical methods. This leads to the following criteria:

(TC1) Internal or logical consistency: this is to be understood as a criterion for the internal, relative consistency of a theory. It refers to both the theory frame and the model assumptions and hypotheses in the scientific applications, by restricting the addition of further hypotheses.

(TC2) Compactness: the theory must have a comprehensive scope in relation to the number of its non-provable assumptions.

(TC3) Connectivity: theoretical knowledge does not consist of isolated statements, but constitutes a network established by multiple vertical and horizontal relations.

(TC4) Empirical consistency: within the scope of the theory, there must not be contradictory (reproducible) experiences that are assessed as crucial, and the integration of new empirical experiences is possible within the claimed scope of the theory (empirical openness).

(TC5) Practical relevance: the theory must work when applied to the reality of singular events and non-representative things. Such applications become manifest, e.g., in the explanation and prediction of events of this reality, in the construction of instruments and machines, and most generally in the technical applications of physical theories.

(TC6) Intersubjectivity: the theory must be universally accepted within the relevant scientific community.

These criteria contain aspects of three competing philosophical theories of truth. The criteria (TC1)–(TC3) correlate with the coherence theory (Rescher 1973). According to this view, truth is established and guaranteed by a consistent, comprehensive, and interconnected system of propositions. Coherence is understood only as a property of a propositional system, but does not concern the relation to facts. This connection is established by the criteria (TC4) and (TC5), which in this respect are correlated with the correspondence theory, while (TC6) corresponds to the concept of truth of the consensus theory. The significance of this sociological criterion for the securing and assessment of physical knowledge is often ignored, but must not be disesteemed. Firstly, without such a consensus, it would be difficult to understand that the theory development in physics arrives at a state where alternative theoretical conceptions are no longer pursued, and the discussion about foundational problems of the theory takes place only marginally or not at all. Secondly, some of these criteria rest on assessments about which consensus must be established within the scientific community. Thirdly, in the non-scientific domain, we already reassure ourselves about perceptions or ideas by communicating with other people we consider reliable. In science, the situation is no different, as can attested by the positive aspects of the elimination of individual errors by the scientific community and the progress of knowledge by communication.

On the other hand, negative concomitants may occur, such as the reinforcement of collective errors or the malpractice of authority for the enforcement of convictions. Understanding modern science as an essentially communicative enterprise, the necessity of this criterion cannot be denied.

In accordance with the methods for securing knowledge treated in Sect. 3.2, the criteria (TC1)–(TC3) show that there are theoretical truth criteria beside the comparison with facts. According to (TC2), for example, the theory frame should enable many, also qualitatively dissimilar scientific applications ScA, and an ScA must provide a large number of empirically comparable results relative to the number of its model assumptions. Vertical connectivity in (TC3) means a deductive systematization as far-ranging as possible. Horizontal connectivity refers, firstly, to the different deductive systems of a unifying theory, connected via inter-theoretic relations described in Sect. 3.5 and, secondly, to the model assumptions of an ScA that must be consistent with the model assumptions of other ScA of the same theory frame. Such consistency conditions are important because even the logically correct deduction of a statement, proven to be experimentally consistent, does not constitute a proof of the truth of the premises. Both (TC2) and (TC3) together stand for the unification of knowledge. In Sect. 3.5, it was shown that this is one of the strategies used to secure theoretical knowledge.

The criterion (TC4) must not be identified with the truth criterion of the correspondence theory, postulating agreement with the facts, because it exclusively concerns the consistency between empirical and theoretical knowledge within physics, i.e., the relations between theory and experience, and in particular, the consistency between the two conceptual systems. In contrast, it is obviously not a meaningful criterion if it refers to the consistency with the external reality of singular occurrences, because its elements are unsuitable for the examination of law-like statements. Criterion (TC5) seems to satisfy the truth criterion of the correspondence theory more closely because it refers to this external reality. However, this similarity is at best superficial, since (TC5) does not postulate agreement with this reality in the way that the correspondence theory does, and as (TC4) does regarding the empirical basis. Rather, (TC5) has more the status of an existence criterion, in the sense that some, but certainly not all, singular events ought to be explainable, and even this only in an instrumentalist sense. A theory not related to real life could rarely be accepted as a representation of reality, but would imply instead that there is something wrong with the relationship between the empirical basis of the theory and the external reality. The relationship with practice guarantees the connection with this reality and contributes through the entirety of its practical successes to taking the associated findings for true.

These comments underscore the close relationship between (TC1)–(TC6) and the properties of closed theories (see Sect. 3.4) and the various strategies for securing physical knowledge. (TC1) refers to securing by proof-theoretical methods taken from the underlying mathematical theories, and (TC2) and

(TC3) to the construction of a tight network in the theoretical domain by deductive systematization and unification. (TC4) corresponds to the strategy of reducing the ambiguities between theory and experience by establishing as many relations as possible between theory and experience. (TC5) emphasizes the significance of practical relevance to confirm the findings of physics, and (TC6) accords with the various strategies of intersubjective communication to reassure oneself of the correctness of observations, calculations, or theoretical considerations. Other criteria like evidence or the plausibility of model assumptions, concreteness, elegancy, and especially consistency with the tacit assumptions, i.e., that a theory must fit into the accepted world view, do not establish such strategies for securing knowledge and do not provide any criteria for truth.

Finally, one should ask whether (TC1)–(TC6) taken together define a useful and meaningful concept of truth. In the spirit of a non-epistemic truth concept, this is definitely not the case because:

- internal consistency can at best guarantee the relative consistency of theoretical knowledge,
- the truth of the model assumptions can be guaranteed neither by the successful theoretical deductions of experimentally comparable results nor by the horizontal interconnections and the unification of knowledge,
- a concept of truth resting on (TC4) and (TC5) is largely unrelated to agreement with reality: (TC4) postulates only the consistency between two different types of knowledge, and (TC5) defines at best an instrumentalist concept of truth, which just considers the functioning, but not the truth of the underlying knowledge as a criterion,
- even the best intersubjective communication does not exclude the possibility of collective errors.

Due to these objections, the criteria (TC1)–(TC6) together do not satisfy the demands of a non-epistemic concept of truth and do not furnish a proof for the truth of knowledge. However, a non-epistemic concept of truth is based on at least one of the three following implicit assumptions:

- the existence of absolutely true first or fundamental principles of theoretical knowledge,
- the existence of a reality whose properties and constitution are independent of conceptual descriptions in that it manifests itself in perceptions forming the true and invariable basis of experience,
- the existence of an unambiguous relationship between reality and conceptual description, so that the distinction between them is unnecessary.

The comparison of these assumptions with the schema (*) of the truth relation reveals that a non-epistemic concept of truth sets an absolute measure for the assessment of truth for at least one of the three components. In contrast, the analysis of the structure of physical knowledge has shown that such absolute measures do not exist. Firstly, physical knowledge rests neither on true, theoretical first principles nor on an empirical basis of invariable facts. Secondly, the relation between theory and experience is ambiguous and constrained only by consistency conditions, so that an unambiguous correspondence between reality and conceptual description does not exist: the necessity of selection and abstraction to enable the mathematization of knowledge admits different descriptions of the same part of reality. Thirdly, a concept of truth resting on absolute measures is unsuitable for the assessment of physical knowledge, because only false theories would then result. For instance, a physical theory of space resting on Euclidean geometry is false because the cosmos as a whole is not Euclidean. Similarly, Newtonian mechanics is false, since it yields incorrect results for high velocities and at atomic scales, and by analogous arguments, any physical theory would be false. It would not be meaningful, however, to impose on a science a concept of truth according to which its acquired systems of knowledge are inescapably false. Attempts have occasionally been made to circumvent this obviously nonsensical result by representing the process of theory development as an asymptotic approach to an externally given final truth, and referring to the status of the knowledge on the way to the final truth as "verisimilitude". Since such a notion cannot be expressed in a provable form because this final truth is unknown, this attempt is nothing more than just a modification of the metaphysical goal of the search for absolute truth. Finally, a non-epistemic truth concept becomes entirely obsolete when asking for the truth of statements and conceptual descriptions of parts of reality not directly amenable to human perception, such as electromagnetic, atomic, or molecular phenomena. Access to this part of reality occurs exclusively through experiments, so according to (TC4) the correspondence is constrained to the physics-internal relationship between theory and experiment. Consequently, the properties and constitution of this part of reality are defined in the first place by the physical description. The constitution of atoms, for example, is determined only by the experimental and theoretical methods of physics, without prejudice to the fact that atoms are a part of the external reality. The correspondence with the external reality is established, according to (TC5), only through the practical applications. In contrast, any question about the "true nature" of atoms, i.e., independent of the physical description, is entirely meaningless. In these cases, truth can be defined in a meaningful fashion only through the criteria (TC1)–(TC6). In short, each non-epistemic concept of truth sets a measure for true knowledge that is by no means appropriate to the potential and abilities of physical methods, in particular, and human cognitive faculty, in general. Such a concept may serve at best as a tacit assumption, i.e., as a normative ideal, regulative idea, or heuristic principle, but the search for truth in this sense can never be a cognitive aim of physics.

In contrast, the epistemic concept of truth as defined by (TC1)–(TC6) is both meaningful and appropriate to assess physical knowledge because it is consistently adjusted to the cognitive capabilities of physics, in the sense that the methods of securing knowledge define what is true. Firstly, it accounts for the fundamental limitations of human and physical knowledge. Secondly, the criteria determine for which parts of knowledge the question of truth may be asked, i.e., which are "truth-capable". Thirdly, one can investigate in a sensible fashion how the truth-capable components of physical knowledge may be examined for truth. A necessary condition for the truth-capability of a statement is, in the first instance, that it must at least be arbitrable, in principle, since (TC1)–(TC6) presuppose the truth concept of the two-valued logic. However, this condition is not sufficient, since isolated single statements may be arbitrable, but not truth-capable due to (TC3): the question of their truth is pointless for the simple reason that it is irrelevant for knowledge whether they are true or false. Finally, the theory frame itself is not truth-capable because it can only be assessed by (TC1) and (TC6), while the other criteria are not applicable without inclusion of the scientific applications. Consequently, even with respect to the epistemic concept of truth as defined by (TC1)–(TC6), the theory frame alone can be designated as useful or powerful.

The situation is substantially different regarding the scientific applications containing three types of statements, viz., the model assumptions as the premises, the physical laws as statements with the claim of general validity, and the deduced hypotheses as the experimentally comparable statements. On the basis of a correspondence-theoretical concept of truth in combination with the empiricist concept of experience, only the deduced hypotheses are truth-capable because they alone are experimentally comparable and apparently concerned with single facts. Therefore, they can be checked by empirical testing, in contrast to the not directly testable model assumptions and physical laws that would not be truth-capable statements.

The epistemic concept of truth defined by (TC1)–(TC6) in combination with the physical concept of experience yields entirely different results. Since every experiment represents reproducible and representative experience, it exhibits a law-like character in the sense that it represents, in principle, an infinite number of singular experiences. According to (TC4), physical laws as generally valid statements are then verifiable by appropriately designed experiments. The claim, occasionally encountered in physics, that a physical law has been verified by experiments may thus be incorrect for reasons regarding content, but it does not a priori contravene logical principles, since it is not a singular fact that is faced with a general claim. The crucial condition for the truth of physical laws and deduced hypotheses, however, is the proof of the truth of the model assumptions they are deduced from. Therefore, in the first step, it must be shown that the model assumptions are truth-capable with respect to (TC1)–(TC6). This is indeed the case. (TC1)–(TC3) may be examined and secured by the procedures described in Sect. 3.2, and (TC4) with the aid of the various relations between theory and experience. (TC5) is proven by the existence of representative practical applications and (TC6) by means of standard textbooks and representative references.

The most challenging task in a concrete case is to furnish the proof of the epistemic significance, i.e., that the model assumptions constitute, in the sense of (TC1)–(TC6), epistemic true knowledge about reality, instead of being merely an instrumentalistically understood auxiliary means for the deduction of experimentally comparable data. The strategies providing the proof may in principle be elucidated by the model assumptions of the theories of Bohr and Drude. Bohr's theory of atomic constitution is about charged objects, but cannot be reconstructed in a consistent way as an application of electrodynamics. Accordingly, the model assumptions in the form of Bohr's postulates were largely isolated and appeared as ad-hoc assumptions, acquiring persuasive power, in the first instance, only from their empirical success which did not admit any conclusion about their epistemic significance. Only the comparison with quantum mechanics enables a decision about which of the postulates can be maintained, viz., those that are true with respect to consistency with quantum mechanics, and which must be eliminated as false, such as the assumption that the stability of atoms results from the equilibrium between the attractive Coulomb force and a repulsive centrifugal force. Analogous arguments apply to Drude's theory of electrical and thermal conductivity of metals, which was based on the assumption that the resistivity results from the scattering of the electrons at the ionic cores, assumed to be at rest. The decisive progress in understanding conductivity provided the assumption, consistent with quantum mechanics, that the electrical resistance in metals is only due to deviations from the ideal periodic structure, e.g., lattice vibrations, lattice imperfections, and surfaces. Although the model assumptions of both theories work locally, in the sense that correct data (like the wavelengths of spectral lines), effects (like the Hall effect), and empirical correlations (like Balmer's formula or the Wiedemann–Franz law) are deducible from them, they are false because they advance claims about interactions that are false according to quantum mechanics. Since this assessment is made relative to quantum mechanics, it appears initially as if consistency within the theoretical domain is all that really matters, without connection with reality. Actually, such a connection exists for those model assumptions postulating interactions or mechanisms, i.e., assumptions about the dynamics. Corresponding to the tacit assumption (1B), there exist subject-independent structures of reality that must consequently be identifiable and represented appropriately in the theoretical description. In the framework of physical methodology, these structures are interpreted as results of interactions which are themselves considered a part of reality. For example, a general consensus exists that electromagnetic interactions are a part of the subject-independent reality, rather than mere thought constructs or inventions of physicists. As a consequence, model assumptions of type (H2) concerning the dynamics of physical systems are not arbitrary conventionalist premises whose epistemic status is reduced to working for the deduction of experimentally comparable results. As an autonomous part of physical knowledge about nature, they rather possess independent epistemic significance. Finally, further examples confirm that it is exactly the model assumptions (H2) that are crucial for the true or false understanding of reality. Not only do these model assumptions establish the epistemological kernel of a scientific application, but they are also associated with an ontological claim of validity whose legitimation must be checked in each particular case.

The fundamental importance of these model assumptions for our knowledge of nature has direct consequences for the scientific applications ScA. In Sect. 3.2, an ScA was defined as secure (see Definition 3.11) when it is internal, empirically and theoretically external consistent, i.e., if the criteria (TC1)–(TC4) are met. A secure ScA is then true when in addition the mechanisms or interactions correspond to structures of reality. With it the ontological claim of validity is secured that comes into play through the model assumptions (H2) and the practical relevance (TC5). The vertical and horizontal interconnections are of primary relevance for the assessment of (H2), according to (TC3), and only in the second instance, the empirically comparable deductions postulated in (TC4). This has been elucidated with the locally working model assumptions (H2) of the theories of Bohr and Drude, which are false because they are inconsistent with quantum mechanics, i.e., not theoretically externally consistent. In short, an ScA is to be rated as false when based on false model assumptions. Finally, further indications exist for the epistemic truth and significance of a scientific application ScA:

- An ScA based on false model assumptions can be improved quantitatively only to a very limited extent, if at all, because qualitatively false model assumptions generally lead to incorrect approaches for the dynamics, e.g., for Hamiltonians. Typical of such false theories is a stagnating or even shrinking scope during the course of theory development due to an increasing number of empirical inconsistencies. A false theory will always end up in a "degenerating research program" (Lakatos 1974), and Lakatos himself reconstructed this development for Bohr's theory.
- According to (TC3), a new ScA must be consistent with the existing secure applications of the same theory frame. Inconsistencies in the theoretical domain may be ignored temporarily, but not permanently, on the way to the unification of knowledge. A new ScA with one or more model assumptions that are inconsistent with those of a secure ScA is theoretically externally inconsistent (see Definition 3.10). As a consequence, it remains isolated and will eventually be eliminated from the accepted knowledge. For instance, Bohrs theory, considered originally as an ScA of electrodynamics since it is involved with charged particles, was inconsistent with the ScA describing the behaviour of electrons in cathode rays. Hence, Bohr's postulates remained isolated. In spite of its empirical successes, it was precisely this isolation which prevented the general acceptance of his theory, thereby failing also to meet the criterion (TC6).
- In contrast, inconsistencies with single empirical facts, i.e., anomalies according to Kuhn (Kuhn 1970), rarely represent serious objections to an ScA, because local discrepancies commonly occur, and their consequences

for assessing the theory cannot be estimated a priori. Due to the historical experience that such discrepancies frequently vanish of their own accord as time goes by, they are not usually considered to be problematic, and thus rarely play a pathbreaking role in theory development. Moreover, taking into account the fact that a false ScA may work locally, in the sense that some data, effects, or empirical correlations are predicted correctly, the comparison with experiment or with facts in the sense of the correspondence theory does not provide a sufficient criterion for truth. This demonstrates once again that agreement with experiment is by no means the sole criterion for the assessment of an ScA, and frequently not even the decisive one.
- A fortiori, this applies to the quantitative agreement with experimental data as a criterion for the truth of an ScA. A quantitatively more precise prediction does not necessarily mean a higher truth content. On the one hand, this is illustrated by epistemically insignificant algorithms offering very precise predictions, but no knowledge of reality, while on the other, a theory may rest upon highly simplifying, but essentially correct model assumptions. When it is later quantitatively improved, as in the case of the theories of the ideal and the van-der-Waals gas, the less precise theory is not rated as false.

In the case where the theory frame of an explanatory or unifying theory possesses so many mutually consistent (TC1–TC6)-true ScAs that their number and diversity is rated in accordance with (TC6) as sufficiently large, the concept of truth as defined through (TC1)–(TC6) may be transferred to the entire theory. In this sense, one may finally speak of true grand theories. Accordingly, Newtonian mechanics is a (TC1–TC6)-true theory for the classical, non-relativistic description of systems with few degrees of freedom, and this applies analogously to the other classical grand theories (statistical mechanics, thermodynamics, and classical field theory). Their combination, however, does not constitute a (TC1–TC6)-true description of reality. The reason is not that there are domains, viz., relativistic and quantum theoretical phenomena, to which this description is not applicable, but that their combination, as shown in Sect. 3.5, cannot be made globally consistent. This result, which seems paradoxical in relation to a non-epistemic concept of truth, demonstrates that the combination of two theories may lead to inconsistencies, and is thus not a purely additive or cumulative process, but an autonomous step towards secured and unified knowledge. Nevertheless, the various theories do not become false within their respective scopes.

Altogether, the metaphysical goal of the search for truth is transformed into a meaningful and well-defined cognitive aim of physics if the truth of theoretical knowledge is defined with reference to the methods used to secure this knowledge. The search for truth according to the criteria (TC1)–(TC6) is essentially the securing of knowledge within a certain theoretical framework, and (TC1–TC6)-true knowledge is secure knowledge supplemented with ontological claims to validity, but without claiming that reality is identical with its physical description. First of all, since truth concerns this description, it is context-dependent: whether momentum or a trajectory can be ascribed to a moving particle as a strongly objectifiable property cannot be decided in an absolute sense, i.e., irrespective of the theoretical context. Secondly, even true knowledge contains numerous constructive aspects and rests partially on agreements with other scientists, because any quantitative conceptual description contains non-eliminable conventions where consensus must be established about their arrangement, e.g., concerning the choice of units or reference systems. Accordingly, the demand for invariance of knowledge under conventions must be understood as a normative ideal that is never entirely realizable. In contrast, agreements about the truth of a statement are inadmissible with respect to a non-epistemic concept of truth, because it is determined exclusively by the properties and constitution of reality. Thirdly, this epistemic concept of truth provides a differentiated characterization of the epistemic status of theoretical knowledge. For instance, in spite of their context-dependence the model assumptions regarding the dynamics are by no means epistemically and ontologically irrelevant auxiliary constructions for the deduction of experimentally comparable data. Consequently, a theory may be false even if it works in an instrumentalist sense: truth is just as much determined by the theoretical context as by the comparison with experiment or reality. In contrast, a non-epistemic demand for truth on theoretical knowledge necessarily results in an instrumentalist conception of theory. Since the question of a context-independent truth of theoretical knowledge cannot reasonably be asked, in no way does a theory constitute a truth-capable propositional system. Accordingly, one cannot speak of false or true theories, but only about useful or powerful ones. Other elements of non-epistemic views about truth are echoed as heuristic principles in the tacit assumptions. In this way, they still influence views about true knowledge in physics, especially the notion that the search for truth is the objective of physics. Based on the concept of truth defined by the criteria (TC1)–(TC6), the development of physics is directed toward the construction of a unified, secured, theoretically and empirically consistent world view without associating with it the metaphysical idea of an asymptotic approach towards an a priori fixed, absolute truth.

5.5 Physical Reality

Since the origin of the new physics in the seventeenth century, views about the relation between physical knowledge and reality have changed considerably. Initially, the discussion focused on the reality of an absolute space and the gravitational force, and afterwards on forces in general. Due to the development of statistical mechanics and electrodynamics in the second half of the nineteenth century, the discussion shifted to the existence of atoms and the ontological status of fields, and in particular, the problem of the carrier of electromagnetic waves, which was thought to be a material entity. In the twentieth century, in the context of the theories of relativity, the reality of a matter-free space and the concept of simultaneity became issues, as did the status of the gravitational force once again. Following the interpretation of the photoelectric effect by Einstein, the question of the nature of light and electromagnetic radiation arose in exacerbated form, and ultimately, with the development of modern quantum theories, so did the question of the properties and constitution of matter in general. Fundamental findings like wave–particle duality, the uncertainty relations with the resulting question of the status of dynamical properties, and the indistinguishability of identical particles highlighted the limitations of our ability to understand reality at the (sub)atomic level, which became even more apparent towards the end of the twentieth century with the systematic study of entangled systems.

The root of physical views about reality is the pre-scientific, vernacular meaning of reality that is identified with the entirety of what actually exists in the sense of physical material existence. In this context, it is not important whether natural entities are concerned like trees, stones, or mountains, or human-made objects like buildings, cars, or aircraft. On the one hand, this establishes the demarcation from potential or fictitious items and also from conceptual descriptions where only the question of the relation to reality makes sense. On the other hand, the reference to concreteness emphasizes the difference, for example, with phenomenalism, which considers only the perception as real, but not the perceived. The question of the objective existence of directly perceivable, macroscopic entities like chairs, houses, or trees may be a challenge for some philosophers, but at least for the physical layperson, it is not obvious how to justify the idea that entities like atoms, elementary particles, or electromagnetic fields, which are not directly perceivable, are constituents of a subject-independent reality, i.e., independent of perceptions, consciousness, and cognizance. Doubts regarding their objective existence are found likewise in empiricist philosophies, which consider all assumptions about things that are not directly perceivable as hypothetical and suspect. Similar views were advanced in physics during the nineteenth century regarding the existence of atoms, although there the debate turned primarily, not around the existence of a subject-independent reality per se, but merely around its extent. Scientific realism is that variant of realism that also accepts the existence of objects that are not directly perceivable ("theoretical entities"), irrespective of any theoretical description.

After the challenges of the existence and extent of a subject-independent reality, the next question concerns its properties and constitution. Due to the fact that different descriptions exist for the same object, e.g., for a piece of metal or for light, a number of questions arise (Willascheck 2000):

- If two descriptions of reality deviate widely or entirely from one another, is it then feasible to assume that they are descriptions of the same reality?

- Can one ever get knowledge of the properties and constitution of the differently describable reality itself?
- Since any attempt to describe reality-in-itself results only in another description of reality, is it then reasonable to speak of a reality that is independent of cognizance and description?

On the other hand, the idea of a description presumes, at least implicitly, that something is being described. Furthermore, since incorrect descriptions exist beside correct ones, their correctness seems to depend not only on the chosen representation, but also on what is being described. The basic thesis of philosophical realism based on these considerations thus reads that there exists a subject-independent reality-in-itself in the sense that both its existence and its constitution are independent of perceptions, consciousness, and cognizance. Disregarding the fact that this reality-in-itself cannot be the universe of discourse of empirical science, but only a reality as recognized by the human cognitive faculty, this idea leads to the following problem (Willascheck 2000): the more independent of the human cognitive faculty reality is thought to be, the more difficult it is to explain the possibility of gaining knowledge about that reality. In order that the "real object" should not vanish as an unknowable thing-in-itself behind the "scientific object", the ontological thesis of independence is to be supplemented with an epistemic thesis of accessibility, postulating that the subject-independent reality must be cognizable.

According to these considerations, reality R may be displayed symbolically as a function of a subject-independent reality R_0 and a recognizing subject S (observer) who, within the scope of his cognitive faculties C and embedded in a socio-cultural context, interacts with R_0:

$$R = f\,(R_0,\, C,\, S)$$

This symbolic representation permits a rough classification of the various notions about R. The naive or common-sense realism presumes that perceptions and observations yield a view of R_0 that is unbiased, in principle, by the human cognitive faculty and entirely independent of S, whence R is virtually identical with R_0. In addition, R_0 is usually thought of as a reality-in-itself. Accordingly, physical knowledge itself is assumed to be a part of R_0, or at least its unproblematic and unequivocal description, so that the properties and constitution of R are largely identical with its physical description. Physical knowledge is objective and the physical reality will be discovered by observations and experiments. In the weakened variant corresponding to a critical or scientific realism, R is thought to be some function of R_0 depending on the limitations of the human cognitive faculty, but the properties and constitution of R are still essentially independent of the observer, i.e., $R = f(R_0,C)$. R is viewed somehow as an image, reflection, or reconstruction of a predetermined reality R_0, irrespective of the extent to which R_0 itself is knowable and comprehensible. Physical knowledge establishing R still possesses the status of objective knowledge. According to the realistic assumption of the existence of a subject-independent reality, it is assumed that perceptions have a counterpart in the subject-independent external world, e.g., in the form of a material carrier, although this notion is not unproblematic, as illustrated by optical phenomena like rainbows and mirages. In generalizing such examples, antirealistic positions assume that the influence of the observer S on R must be rated as essentially non-eliminable. In this context, it remains undetermined whether S is conceived as an individual or as a "representative" observer, or again as a scientific community. In the latter case, R is viewed as a reality resting on intersubjective knowledge which is either determined only by the cognitive faculty of S or additionally by the socio-cultural context.

If the subject matter of physics were a fixed, predetermined reality of direct perceptions, arbitrary entities, and phenomena, the questions of the existence, cognizability, and constitution of physical reality R_{phy} would be largely identical with those of the philosophical dispute on realism, and a complex of challenges specific to physical knowledge would cease to exist. Actually, the main focus of the debate on R_{phy} is entirely different. Firstly, as shown in the context of the comparison between experiment and theory, the existence of a subject-independent reality that is independent of experimental and theoretical investigations must be assumed for methodological reasons. This necessary presupposition appears as the tacit assumption (1A), which is not subject to refutation. Accordingly, it does not possess the status of a "useful hypothesis" because it is neither part nor objective of physics to furnish a proof for it. Secondly, the central subject matter of physics is the study of law-like describable structures. Consequently, corresponding to the tacit assumption (1B), the structures, contained in class terms, regularities, reproducible events, and connections, are assigned to the subject-independent reality without prejudice to the fact that they cannot be identified with individual objects. The two assumptions together imply initially a weak realism, in that the extent and constitution of this reality remain unspecified, and its comprehensibility and cognizability are not postulated. However, they transcend a mere entity realism, because structures are assigned to this reality. Thirdly, the subject matter of physics can only be a reality that is, irrespective of the use of instruments, accessible to the human cognitive faculty, i.e., reality as object of perception and cognition, i.e., as scientific object. In order to establish the connection between real object and scientific object in the sense that physics is somehow related to the subject-independent reality, the tacit assumption (1C) postulates, by analogy with the above-mentioned thesis of accessibility, that this reality is by and large correctly cognizable, so that there is an at least partial agreement between epistemic and ontic categories. The constraint "partial" rests on the circumstance that comprehensibility and cognizability initially concern only the directly perceivable part of reality. The proof of the permissibility of this assumption is likewise not the subject of physics, but it may be made plausible by reference to the evolution of the human organs of perception, since without adaptation to the environment, survival would be rather unlikely.

Most clearly, the dissimilarities with the philosophical realism dispute become apparent in the methods which serve to establish the properties and constitution of the physical reality R_{phy}. These dissimilarities result from the circumstance that the mathematical and experimental methods of physics are unsuitable for investigating and describing the directly perceived reality R. This was already recognized by the ancient Greek natural philosophers (Detel 1986). According to them, physics as a mathematized, exact science of nature is impossible because random and in many ways perturbed natural events cannot be described by mathematically precise laws. Similarly, any goal-oriented experimental intervention in nature is pointless because this constitutes only another disturbing factor for the random natural events. Accordingly, human interventions in nature can never be natural, but must be rated as an unnatural finessing of nature. Therefore, an exact science can only deal with an "ideal", in the sense of a law-like describable reality which is hidden behind the directly perceived reality.

With regard to this background, the methodological revolution leading to mathematized natural research by Galileo consisted in a detachment from the directly perceived reality and the construction of such an ideal reality grounded on idealizations and thought experiments. As a consequence, natural research changed to a physics in which thought experience prevails, rather than sense experience, and the central subject matter of fundamental research in physics became structures whose mathematical description rests upon non-existent model systems that are theoretically precisely defined by selected properties, as shown in Sect. 2.5. Since the objects of theoretical physics are now non-existent model systems, it is not obvious, in the first instance, how laws and theories that are exactly valid only for these model systems could be related to reality. The first contribution to establishing this connection is provided by the experiments performed in the laboratory. Firstly, they provide realizations of the non-existent model systems that are as good as possible, and secondly, they serve to demonstrate the general relations as described by the physical laws. This shows once again that mathematical and experimental methods mutually condition each other. In particular, with the subsequent development of a systematic experimental methodology, a reality is constructed by laboratory physics in such a way that it becomes amenable to mathematical methods. Associated with this development, the claim to record reality in its all perceived variety is dropped, but it persists in the reduced form that, in accordance with the tacit assumption (2C), the reality created in the laboratory possesses the same ontological status. This applies in particular to those domains of reality which are not accessible to human sense organs and can be made accessible only by physical methods, so that R becomes virtually identical with R_{phy}.

As a consequence of this methodological revolution, the discussion about the reality of physical knowledge begins in principle with the abstraction that is necessary for the mathematical description, but is accompanied by a loss of the direct relation to reality. In mechanics, this problem had not yet been addressed because, firstly, its objects of investigation were relatively directly amenable to perception, and secondly, no distinction was made between the non-existent model system and real object. The need for an extended view of reality followed only in the second half of the nineteenth century, from the extension of physics to domains like statistical mechanics and electrodynamics that are no longer accessible to direct perception. Due to this development, not only is direct contact with the objects under study largely lost, but there was also an enormous expansion of reality due to the development of new instruments, in combination with new experimental and theoretical methods. The second problem for the debate about physical reality becomes perfectly apparent here, viz., the universe of discourse of physics is not a fixed predetermined reality. Instead, in physics, reality is expanded and created on widely different levels by constructive activities. This becomes even clearer with the developments in modern physics, where the central results of the theories of relativity and quantum theories of matter are fundamentally at variance with views familiar from classical physics and direct experience. As a consequence, views about the structure and properties of atoms, for example, are to a large extent, if not exclusively, theory-determined, notwithstanding the fact that the existence of atoms as a part of the subject-independent reality is not challenged. The question then arises to what extent a physical theory about atoms is related to reality. In short, the idea that physical knowledge is the result of the reconstruction of a fixed predetermined reality is obsolete.

The third challenge follows from the circumstance that, at the outset, the subject matter of physics is a reality R_{rep} that is constrained to reproducible events and representative systems. This constraint is necessary for reasons of consistency, because gaining, structuring, and securing knowledge as the aims of fundamental research in physics refer to R_{rep}. According to the realistic attitude toward physics, in combination with the tacit assumption (2C), R_{rep} is also part of the subject-independent reality R_0 in spite of its predominantly constructive nature, in the sense that the ontological status is the same. In particular, entities like atoms, electromagnetic fields, and elementary particles that are not directly perceivable, but accessible only via experiment and theoretical reasoning, exist as subject-independent objects, and are thus part of R_0, rather than being mere conceptual constructs to describe and interpret experiments. Finally, establishing the empirical basis of a theory consists primarily in determining that part R_{rep} of R_0 that is relevant for the given theory. On a more general level, independently of particular theories, the issue is to determine the extent of the reality. This leads to the question of the criteria enabling one to decide whether a certain object belongs to R_{rep} or R_0.

According to Sect. 2.2, the proof of existence starts from certain observations or effects that are interpreted by defining properties which are ascribed to an object that is postulated as the material carrier of these properties.

However, a property defined in this way is not necessarily associated with a carrier. For instance, one may ask whether the blue colour as a property of a cloudless sky provides a criterion for its existence. Without physical insights, one might perhaps affirm this, but with such insights, one can no longer be sure. Since the blue colour comes about due to the scattering of sunlight on the molecules of the atmosphere, that colour necessitates a source of white, i.e., polychromatic light, as well as particles of an appropriate size. If the molecules of the atmosphere were as large as dust particles, the sky would not look blue, but grey, as it does over major cities or industrial areas. Accordingly, the blue colour is an emergent property of the sky, because it appears as the result of the molecules of the atmosphere being illuminated by a polychromatic light source. Based on this clarification of the situation, the entire meaning of the empirical term "sky" is challenged. Is it to be interpreted as a certain perception, as the Earth's atmosphere, as some imaginary background defined by the stars, as a physical effect, or as a combination of all those? This example illustrates how eventually, when an object is thought of as a carrier, only properties remain, so that the alleged entity is actually created by observed properties, although its empirical counterpart seems to be objectively existent at first sight: that the sky exists is assumed unthinkingly to be as clear as the existence of a tree or a mountain. Analogous considerations apply to optical phenomena like rainbows and Fata Morgana, or to acoustic phenomena like sounds and noise. As a result, a proof of existence resting on the registration of observations is structurally nothing else than the justification that the Moon exists because it is seen in the sky, and does not contribute to clarifying the philosophical ontological problem of existence.

A universally valid conclusion from perceptions and observations on a necessarily existing material carrier is inadmissible because it cannot be excluded that only a complex of mere sense impressions might be involved. If the circumstances were that simple, the more than a hundred year long dispute in physics about the existence of atoms would not have occurred. Actually, the problem of the conclusiveness of the proof of existence of an object due to observed effects does not only come up in the context of the formation of theoretical concepts in physics in the case of entities that are not directly observable. It is just that it is much more perspicuous because the confirmation takes place exclusively in this way, while in the pre-scientific and empirical formation of terms, this problem does not seem to occur. In this case, it focuses mainly on directly perceivable objects whose existence is taken for granted irrespective of the conceptual description, and this definitely without good reason, as exemplified above by the sky. On the other hand, the conclusion about the existence of a carrier based on the chain of construction "effect $\rightarrow$ property $\rightarrow$ carrier" has frequently contributed to the extension of knowledge about reality, e.g., by the interpretation of particle tracks in cloud or spark chambers, in such a way that the given particle has been detected and proof for its existence has been provided. Although the conclusion from observed effects to a really existent material carrier does not establish a universally applicable method, it offers an exceedingly successful heuristic to establish existence hypotheses about elements of R_0. In contrast, the ontological scepticism rejecting such a conclusion about the existence of a carrier as inadmissible metaphysics rests primarily on occasionally occurring fallacies, as in the case of the phlogiston, caloric, or an aether as the material carrier of electromagnetic waves. According to such a sceptical attitude, the carrier is considered solely as a conceptual summary of properties or as a purely conventionalist construct for the conceptual description of observations. On the other hand, precisely the fact that terms like "phlogiston" and "caloric" have disappeared from science provides a substantial argument in favour of a realistic ontology. When an existence hypothesis about a carrier turns out to be erroneous, and is thus falsifiable in principle, the corresponding conceptual description is not a mere convention or an arbitrary representation, because the falsifiability rests on a structure of the subject-independent reality. There is thus no reason to disqualify such a heuristic as metaphysics due to occasional errors, and to consider entities as fictitious just because they are not directly perceivable and knowledge about them can only be obtained via experimental and theoretical methods. Direct perceptibility is not a necessary criterion for elements of R_0.

In an exacerbated form, the existence problem of structures poses a challenge. According to the tacit assumption (1B), they are a part of reality in the same way as objects, but their identification requires constructive operations because they do not correspond to single perceptions. The problems occurring in this context are the same as in the comparison between statistical and dynamical correlations, discussed in Sect. 2.6. Accordingly, a structure that is proven to be the result of a physical interaction, such as the structures of molecules and crystals, may be considered as a part of R_0. These are structures corresponding to certain properties of objects. Considerably more involved is the following challenge: in which sense and according to which criteria should the structures corresponding to the physical laws be considered a part of R_0? By analogy with the chain "effect $\rightarrow$ property $\rightarrow$ carrier", one might conceivably construct a chain "observed regularity $\rightarrow$ law $\rightarrow$ structure of reality", and consider this as a heuristic for the discovery of structures of reality, in the same way as the first chain is a heuristic for the discovery of objects. Disregarding the fact that it is already hard to furnish a proof of existence for an object as the carrier of a theoretically introduced property, one is faced with two additional difficulties regarding the structures of reality as postulated by laws. Firstly, the claim of strict validity refers only to non-existent model systems, while the range of validity for real systems is not fixed a priori since it is system-dependent. Secondly, laws are functional relations between properties which concern the constitution of reality. In a first step, conceivable reality criteria for properties must thus be identified.

The general question of the existence of a subject-independent reality already assigns the key role to the properties which enable one to identify and recognize the elements of this reality. Accordingly, existence hypotheses, especially concerning objects that are not directly perceivable, are always replaced by hypotheses about properties. Therefore, it must be clarified whether and according to which criteria the properties, and thus the constitution of R_0, may be assumed to be subject-independent. In logical empiricism, the reality criterion for a property is that it is directly perceivable and independent of the recognizing subject, i.e., objective in this respect. In addition, such a property rests on the assumption that it can be ascribed to an object as a pre-existing system attribute, entirely independently of any context. Size and shape, for example, are then suitable predicates to describe directly perceivable objects. Apart from the fact that directly perceivable properties are necessarily of a qualitative kind, a number of difficulties arise. Firstly, it has been shown previously that direct perceptions do not provide reliable information about reality, e.g., the directly observed planetary motions have little to do with the actual structure of the Solar System. Secondly, direct perceptions refer only to a very small part of that reality known until now. Thirdly, even the seemingly objective properties are context-dependent, e.g., the length of a bar appears shortened at very high speed, and whether a particle may be ascribed a trajectory depends on the theoretical context. Finally, properties are intrinsically tied to conceptual descriptions that are per se not subject-independent. A particularly simple example of context dependence is provided by aggregate states as qualitative, macroscopic properties of matter. The fact that water may be described by the predicate "liquid" is valid only within a certain range of temperatures. Consequently, the existence of a property is tied to various conditions that are identified, especially in empiricist philosophies, with the ranges of temperature and pressure prevailing on the Earth's surface and forming the frame of direct experience. Accordingly, the resulting understanding of objective properties is not only tied to the existence of these standard conditions, implicitly assumed to be absolute and often to be extrapolated arbitrarily, but also to the assumption that these properties do not change, or only slowly change over time with respect to these standard conditions. In a further step, those properties viewed as objective are ascribed the function of serving as a proof of the existence of both a subject-independent reality and objective knowledge (Schaff 1984). However, since one cannot conclude with certainty to the existence of a material carrier from perceptions and observations establishing the basis for the definition of properties, the conclusion regarding its existence may be justified in particular cases, but properties do not provide a universally valid proof for existence. Hence, the epistemological view taking the question of the existence of objective properties as meaningful, in the sense of entirely context-independent properties, rests on a naive realism identifying empirical experience with the direct experiences made under standard thermodynamic conditions.

These views become entirely obsolete with respect to physical properties. Firstly, the assumption of properties that ought to be independent of any experimental and theoretical investigations is physically meaningless. How a system or a property is conditioned if nothing is observed or measured is obviously empirically undecidable. Secondly, physical quantities correspond to quantitative, measurable properties that are not directly perceivable. The size or shape of a body may be directly perceivable, but the physical quantities "length" or "volume" are not: the statement "this table has a length of 95 cm" or "this body has a volume of 1 m^3" does not correspond to a direct perception. According to empiricist views, only qualitative properties could be considered to be real. Thirdly, measurable properties are based on conventions that cannot be eliminated, and they are also theory-dependent. This is exemplified by the macroscopic, measurable properties of matter, e.g., mass density. The statement "silver has a density of 10.5 $\mathrm{g\,cm^{-3}}$" implicitly assumes a continuum theory of matter, because the property "density" can be defined in a meaningful way only within this conceptual framework: the statement that a silver atom has a density of 10.5 $\mathrm{g\,cm^{-3}}$ is obviously meaningless. As a consequence, a measurable property cannot be considered to be subject-independent in a strict sense. On the other hand, density is a material property enabling the characterization and identification of a substance and must be seen in this respect as subject-independent. The general question of the ontological status of properties was discussed at length in Sect. 2.5 with the result that objective properties in an absolute, i.e., entirely context- and theory-independent sense, do not exist. The properties and constitution of reality can then be understood as subject-independent only in an intersubjective sense and are based on those properties that are strongly objectifiable with respect to a theory T, in the sense that the assumption that they are pre-existing system attributes does not lead to inconsistencies within T. Precisely these properties are the ones that are used either for the representativeness, identification, and cognition of real systems or for the definition of model systems. In this way, they define the constitution of physical reality. Accordingly, the subsequent criteria according to which physical quantities may be treated as "physically real" are of fundamental importance:

- Existence of a material carrier: if a property can be assigned to a material carrier, it is confirmed, at first, that the corresponding effect, leading to the definition of the property, does not present a mere complex of perceptions. The term "material" must be understood somewhat liberally, in that radiation fields or particles with vanishing rest mass are also admitted as material carriers.
- Measurability and operational definition: the measurability of a property like the mass density of a substance or the wavelength of light adds to it a theory-independent meaning, even though in a strict sense this property is only defined with respect to a certain theory. The conventions necessarily contained in operational definitions are considered to be unproblematic with respect to their status as real. In this sense, what is measurable is physically real.
- Strong objectifiability: if a property can be assumed to be a pre-existing system attribute of a material carrier without running into contradictions

within the framework of the theoretical description, it appears as intrinsically tied to the system and may serve for its identification. The inference to consider strongly objectifiable properties as independent of the experimental and theoretical context corresponds to the notion of reality in classical physics, as expressed by Einstein (Einstein et al. 1935):

> If, without in any way disturbing a system, we can predict with certainty (i.e., with probability equal to unity) the value of a physical quantity, then there exists an element of physical reality corresponding to this physical quantity.

The development of quantum mechanics has shown that such a classical view is untenable. According to Sect. 2.5, the status of strong objectifiability of a property is actually theory-dependent.

- Epistemic significance: if a property corresponds to a structure of reality in the sense that it expresses, e.g., an invariance, as in the case of the density, or a reproducible and thus law-like behaviour (dispositional predicates), it acquires an objectifiable aspect of meaning. Similarly, the representativeness of objects, and especially the reproducibility of effects, implicitly contain law-like aspects reflecting structures of reality. Accordingly, abstract terms ("universals"), e.g., class terms, correspond to something real, in that they reflect certain structures of reality.
- Existence of universal constants like the speed of light, the elementary charge, or Planck's constant: crucial in this context is the physical content associated with their existence. The fact that an upper bound exists for the transfer rate of energy and information or a finite lower bound for actions and electric charges corresponds to properties of the subject-independent reality. Although the precise numerical values are of great practical importance, they play a subordinate role in this context, because they depend, for example, on the choice of the unit system. In particular, fundamental importance is assigned to dimensionless ratios of the universal constants like Sommerfeld's fine structure constant or the ratio of the masses of the electron and proton. These become directly manifest in structures of reality.

The reality of physical properties rests on the consistent combination of experimental, empirical, and theoretical aspects. As a consequence, the nature and constitution of physical reality established by these properties is fundamentally different from non-physical notions of reality. Firstly, the status and number of these properties are theory-dependent. This was exemplified by the property "trajectory of a particle" and is transferred to the status of dynamical properties like the position, momentum, or energy of systems. The subsequent inference that the existence of the carriers of these properties ought likewise to be theory-dependent is without empirical justification and has only led to subjectivistic interpretations of quantum mechanics and antirealistic positions that turned out to be untenable long ago. Secondly, the universe of discourse of physics is not a directly perceived reality, but a reality R_{rep} that is constrained to reproducible processes and representative systems. Thirdly, since the conception of experience in physics rests on instrumental experience, the instruments in combination with the experimental, theoretical, and consistency-generating methods M_{phy} contribute crucially to establishing physical reality. Finally, properties and constitution, and thus the view and understanding of R_{rep}, are determined by the secure physical theories. Hence, physical reality is also a function of these theories:

$$R_{phy} = f\,(R_{rep}, M_{phy}, T, S)$$

Here R_{rep} is assumed to be subject-independent, but not R_{phy}. Although physical knowledge is independent of an individual subject and intersubjective to a high degree, it is not objective in the sense of knowledge entirely detached from a scientific community S. This would exclude the possibility of collective errors, for example, that have occurred at times in the historical development of physics. In addition, both physical knowledge and the associated physical reality R_{phy} are the result of a development process of consistent adjustment, or in symbolic form:

$$R^{(n+1)}_{phy} = f\,(R^{(n)}_{phy})$$

Accordingly, physical reality is itself a function of R_{phy}, albeit on another level, corresponding to the state of knowledge of the scientific community $S^{(n)}$, which depends on instrumental and theoretical developments. This accounts for the changes in views about reality since the seventeenth century mentioned above. On the one hand, the conception of reality in modern physics differs substantially from that of classical physics, and on the other, it is built thereon, as may be exemplified by the changing views on the nature of light. It begins with Newton's particle theory followed by the mechanical wave theory and the electromagnetic theory, and ends up in the modern conception where, depending on the experimental conditions, light exhibits properties that correspond to those of either classical waves or classical particles, a dualism that lies beyond human comprehension. Analogous developments have transformed views of matter from the classical concept of substance up to the indistinguishability of identical particles. Therefore, what is regarded as physically real depends on the state of knowledge, and primarily the theoretical state of knowledge, and thus acquires a historical component. Irrespective of the fact that the various theories $T^{(n)}$ are closed and represent secure knowledge within their particular scope, the associated views of physical reality are subject to change and are in this respect always tentative. If, for example, $R^{(0)}_{phy}$ symbolizes the reality of classical mechanics $T^{(0)}$ with its causal deterministic notions, $R^{(0)}_{rep}$ corresponds to that part of the subject-independent reality R_0 that was both known at that time and relevant for this theory. Both $R^{(0)}_{phy}$ and $R^{(0)}_{rep}$ differ significantly from contemporary insights and theoretical notions, while the status of $T^{(0)}$ as definitive is unaffected.

Consequently, concerning the question of physical reality, entirely different issues are relevant, because it is not a question of the existence and cognizability of reality per se, but rather the range, properties and constitution, and comprehensibility of reality as recognized and established by physical methods, as well as the status and significance of the theoretical description and the reasons why theories are not arbitrary constructs, mere conventions, or fictions:

1. Ontological status of R_{rep}: although as the entirety of objects, effects, data, and instruments, R_{rep} may not be subject-independent from an empiricist point of view, the human-made fabrication of representative systems and the reproducibility of effects, even when based on theoretical reasoning, does not change anything regarding its ontological status. This complies with the pre-scientific understanding of reality that considers buildings, cars, and planes to be as real as trees or mountains, in spite of requiring theoretical knowledge. In this sense, the empirical basis as a part of R_{rep} has the same ontological status as the reality R_0, understood as object of perception and cognition that is itself independent of whether it is known or not.
2. Range of R_{rep} and R_0: it is an undisputed fact that, not just physics, but all the natural sciences have contributed through constructive activities to both the extension of our knowledge of reality by new discoveries and to the enlargement of its range. In this context, it is unimportant whether this enlargement occurred by experimental or theoretical methods because frequently objects that were initially considered only as theoretical constructs later turned out to be elements of reality, e.g., electromagnetic fields and atoms. In particular, according to the tacit assumption (1B) that reality is structured, the observable regularities are ascribed to the reality R_0, and so, too, are interactions that are considered to be causes for structures and regularities. Although in physics, new objects and effects are constructed and discovered as elements of R_{rep} and R_0, the further inference that physics can provide a proof of existence of a reality-in-itself is inadmissible. For there is no way of knowing from observable effects alone whether knowledge is actually gained on such a reality or only on the capabilities of the human cognitive faculty in combination with physical methods. A proof of existence in a particular case cannot provide a universal proof for the existence of reality, and for this reason, such a proof does not belong to the objectives of physics.
3. Properties and constitution of R_{rep} and R_0: the nature and constitution of reality is determined by the properties serving to identify and characterize its constituents. While the constitution of R is based largely on qualitative properties, the constitution of R_{phy} rests on physical quantities, i.e., on quantifiable, measurable properties. Since objective properties, in the sense of being entirely context-independent, do not exist, the status of properties as pre-existent system attributes and thus the nature and constitution even of R_0 depend at least on the external conditions, and the constitution of

R_{phy} also on the theoretical context. As a result, statements regarding the status of properties are meaningful only if the relevant conditions of their definition and investigation are specified. Moreover, it is unimportant in physics whether this concerns experimental or theoretical conditions. The two are not usually separable because the measurement of properties rests on theoretical presuppositions. Since properties form the conceptual basis of physical theories, and physical laws are functional relations between properties, the theory dependence of R_{phy} pertains predominantly to the constitution as defined by the properties. This makes the essential difference with R. Consequently, the nature and constitution of both R_{phy} and R can no longer be assumed as strictly subject-independent, but only as intersubjective. While physics can at least contribute to the determination of strongly objectifiable properties, the epistemological view assuming not only the existence of reality as subject-independent, but also its properties and constitution, reveals itself as naive realism.

4. Cognizability and comprehensibility of R_0: the agreement of epistemic and ontic categories according to the tacit assumption (1C) is partial because it refers only to the directly perceivable part of reality. Limitations of cognizability and comprehensibility exist in domains not directly amenable to the human cognitive faculty. This concerns primarily the (sub)atomic domain, which becomes cognizable and comprehensible only via the experimental and theoretical methods of physics. In particular, due to the numerous reducing aspects of physical methods, it is not clear whether knowledge is obtained about the subject-independent reality. Since the mode of physical description determines R_{phy} but not R_0, the question arises of the connection between R_{phy} and the reality R_0, imagined to be unrelated to physical knowledge. Actually, there are a number of good reasons for believing that physical methods yield R_0-relevant outcomes as well:

   - Instruments and the lab-made representatives of theoretical model systems are not only part of R_{rep}, but also part of R_0.
   - Practice of physics: many working items in daily use are based on the application of physical knowledge.
   - The decisive strength of physical methodology in securing physical knowledge rests on the fact that, with the theoretical and experimental methods, two different strategies are used to obtain two descriptions which can be made mutually consistent because they refer to the same reality.

5. Status of laws as the central scientific statements: the subject matter of fundamental research in physics is primarily structures that cannot be discovered by direct observation alone. The need for making a detour via the (theoretical) study of non-existent model systems rests on the fact that such structures can be detected only via appropriate abstraction and idealization, associated with the decomposition of phenomena and effects

into a universal part and system-specific and contingent conditions. Accordingly, the physical concepts occurring in the laws are not necessarily a part of empirical experience. It must then be asked how laws for such model systems are related to reality, what a realistic interpretation of the laws might mean, and how one might guarantee that physical laws are not mere conventions or fictions. Firstly, a realistic interpretation of laws cannot mean that they are a part of nature or reality R_0. As conceptual formulations in the framework of a theory, they are not a part of nature, but statements *about* nature: nature does not "obey" the physical laws and natural processes are not governed by the laws. Consequently, the view of natural laws that govern nature rests on the naive-realistic identification of reality with its conceptual description. The ontological claim to validity related to a realistic interpretation can be justified only by elaborating the connections between laws and the subject-independent reality:

- The properties defining the model systems theoretically can be assigned to real systems as well, and are thus suitable for their characterization and identification.
- Although the laws are not strictly valid for real systems, they describe their behaviour with an accuracy that may be specified quantitatively by the context-dependent conditions.
- The experimental demonstration of the physical laws provides the necessary connection with R_0 by proving that the general statements of theories about non-existent model systems are realizable in reality, provided that the necessary contingent conditions are established. In this sense, the laws represent structures of reality.
- While the qualitative content of physical laws corresponds to a structure of reality, the mathematical representation may be different, as exemplified by the differential, the integral, and the source-like representation of Maxwell's equations. The physical content has not been changed thereby.
- The conservation laws, e.g., for energy, momentum, or electric charge, are considered as the fundamental structures, not just of the theoretical description, but also of reality itself: each conservation law is related to a symmetry (Noether's theorem). In this respect, the conservation laws represent the invariant, subject-independent structures of reality.

6. Status of theoretical knowledge in the form of theories: concerning the question of the relationship between physical theories and reality, the two conceptual levels of the theory frame and of the scientific applications must be distinguished. Both the relations between theories and reality and the arguments in favour of a realistic interpretation rest on their empirical content, which is established by the entirety of the successful scientific applications. These relations become manifest in the following properties and facts:

- The scientific applications contribute to the enlargement of both the reality R_0 itself and the knowledge about it, by discovering or creating objects and effects through theoretical work.
- The fact that effects and data derived from the model assumptions of scientific applications are consistent with the corresponding experimental results establishes a relationship with reality, while at the same time, examples may be quoted in which empirically correct results are deducible from false model assumptions. Conversely, the fact that not everything that has ever been theoretically developed belongs automatically to the accepted and secured part of physical knowledge provides a substantial argument for a realistic interpretation of theories. That theories like the impetus theory, the phlogiston theory, and Bohr's theory are at best merely of historical interest, and are no longer part of accepted knowledge, the reason is to be found in the fact that they did not provide an adequate description of reality.
- Essential for the connection of the scientific applications with reality are the underlying model assumptions, and especially the mechanisms as the realizations of the abstract structural principle. Associated with at least some of them is an ontological claim to validity to represent not arbitrary constructs, but to be a part of reality. However, realizations of the force concept by inertial forces show that the justification of this claim is not given automatically. Initially, it is therefore hypothetical and must be proven in each particular case, so that the question remains for the methods constraining the arbitrariness of theoretical assumptions. Such constraints are set by the tacit assumptions, as well as by internal and external theoretical consistency, the latter being the most important.
- The theory frame determines the view of reality through the physical content of the structural principle and is thus of crucial epistemic significance. However, since an ontological claim to validity is not even met by all realizations of the structural principle, this applies a fortiori to the theory frame which is thus void of ontological claims to validity. Hence, there is only an indirect relationship between reality and the theory frame of a closed theory, established by the number and diversity of the scientific applications, which is a measure of its capability.
- Finally, the enormous number and diversity of practical applications and successes provide a substantial argument for a realistic interpretation of theories. Although these successes, even with respect to the truth concept as defined in Sect. 5.4, do not prove the truth of knowledge, it appears highly unrealistic to assume that these successes should rest on blatantly erroneous insights.

As a consequence, a theory contributes to the enlargement of knowledge in two structurally dissimilar ways. The first is the enlargement of both the subject-independent reality itself and our knowledge of it, by discovering or creating objects and effects through theoretical insights. The second is the creation of descriptions based on theoretical conceptual systems which cannot exist without theory as a basis and establish a constructed reality in the literal sense. Only the two together establish the physical reality. In the first instance, each theory, advancing ontological claims to validity in any form, defines its own reality. This is associated with the fact that the relation between theory, experience, and R_0 is ambiguous, so that for the same part of R_0, different conceptual descriptions may exist, depending on the associated level of concepts or complexity. This is exemplified by the fact that, in accordance with the results of Sect. 3.5, a description of reality by a unified conceptual system is impossible. In contrast, R_0 represents a uniform reality: the aim of consistent adjustment of empirical and theoretical knowledge is reasonable only relative to the background of a uniform conception of reality that is not decomposed into a reality of phenomena and a reality of ideas or principles hidden behind them. The relationship between physics and reality may then be characterized as follows:

- The existence of a subject-independent, structured, and at least partially cognizable reality R_0 is accepted as tacit assumption. Already for methodological reasons, it is not subject to refutation and is thus not a "useful hypothesis". The assumption of partial cognizability, at least, implies that the universe of discourse of physics is not a reality-in-itself, but reality as object of perception and cognition. This is emphasized by the fact that the securing and structuring of physical knowledge as central aims of fundamental research in physics refer to the reality R_{rep} of representative objects and reproducible events. Although establishing R_{rep} necessitates constructive activities, this does not change its status as subject-independent.
- The range of R_0 or that of R_{rep} as the part that is relevant for R_{phy} undergoes processes of change. On the one hand, new objects and effects are discovered and lead to an enlargement of R_0. On the other hand, elements of R_{rep} are created constructively, viz., objects to be studied and instruments as the auxiliary means for investigations, where the two are not strictly separable. For instance, the study of processes with elementary particles usually takes place by means of other elementary particles. The experimental and theoretical knowledge required for all that has no influence on the status of the subject-independence of the range.
- In contrast to the existence and range, the nature and constitution of R_{phy} can no longer be assumed to be subject-independent, but only intersubjective, because it rests crucially on properties whose status and selection depend on the underlying physical theory. The view of reality is determined exclusively by the physical content of the structural principles and the associated theoretical orders, but not by observations, effects, empirical orders, or mathematical representations. Einstein, for example, accepted quantum mechanics as a means for gaining and predicting factual knowledge, but not the view of reality imposed, among other things, by

the uncertainty relations and the indistinguishability of identical particles (Einstein et al. 1935).

In summary, physical reality is established actively by a combination of discoveries, material and conceptual constructions, and the interconnection, as close as possible, between theories, empirical experiences, and R_0 by means of the methods of consistent adjustment. Physical reality as a supra-theoretical concept, usually referred to as the physical world view, is finally established as a conglomerate of entirely dissimilar constituents:

- the tacit assumptions,
- the experimental, empirical, theoretical, and consistency-generating methods,
- the entirety of intersubjectively accepted closed theories, which determine the view of reality via the physical content of the associated structural principles, irrespective of the particular mathematical representation,
- those scientific applications representing realizations of the structural principles with confirmed ontological claims to validity,
- the supra-theoretical principles like conservation laws, and the physical content of the fundamental constants.

The consistency-generating methods in particular lead to a tight interlocking of R_{phy}, on the one hand, and R_{rep} or R_0, on the other. This is absolutely intended, to make the connection of theoretical knowledge with reality as comprehensive as possible, but must not result in the identification of R_{phy} with the subject-independent realities R_{rep} and R_0. Another fallacy to be found, especially in popular scientific treatises, consists in concluding from the changes in the physical world view that physical knowledge is basically hypothetical and tentative. Actually, as argued at length in Sect. 3.4, the closed theories represent ultimate knowledge. Accordingly, they are neither hypothetical nor tentative. It is just the physical world view established by the closed theories that depends on the actual state of research, is subject to changes, and is tentative in this respect. This again demonstrates that the relationship between theory and reality is ambiguous: the view of reality as imparted by the theory may undergo changes without the theory itself being modified, because this view is also influenced by theory-external notions like the tacit assumptions.

References

Born M (1957) Physik und Metaphysik. In: Physik im Wandel meiner Zeit. Vieweg, Braunschweig.
Cassirer E (1957) The Problem of Knowledge. Yale UP, New Haven
Detel W (1986) Wissenschaft. In: Martens, Schnädelbach (1986)
Duhem P (1908) Aim and Structure of Physical Theories, translated by P Wiener,

Princeton UP, Princeton, N.J. 1954

Einstein A, Podolsky B, Rosen N (1935) Can quantum-mechanical description of physical reality be considered complete? Phys Rev **47**:777

Grodzicki M (1986) Das Prinzip Erfahrung – ein Mythos physikalischer Methodologie? In: Bammé A, Berger W, Kotzmann E: Anything Goes – Science Everywhere? Profil, München

Helmholtz H (1847) Über die Erhaltung der Kraft. Reprint 1982, Leipzig

Hempel CG (1966) Philosophy of Natural Science. Prentice Hall, Englewood Cliffs, N.J.

Hempel CG, Oppenheim P (1948) Studies in the Logic of Explanation. Phil. of Science **15**:135

Hertz H (1965) The Principles of Mechanics. Dover, New York

Krüger L (ed) (1970) Erkenntnisprobleme der Naturwissenschaften. Kiepenheuer und Witsch, Köln-Berlin

Kuhn TS (1970) The Structure of Scientific Revolutions, The U of Chicago Press, Chicago, Ill.

Künne W (1986) Wahrheit. In: Martens, Schnädelbach (1986)

Lakatos I (1974) Falsification and the Methodology of Scientific Research Programmes. In: Lakatos I, Musgrave A (1974)

Lakatos I, Musgrave A (eds)(1974) Criticism and the Growth of Knowledge. Cambridge UP, Cambridge, Mass.

Losee J (1972) A Historical Introduction to the Philosophy of Science. Oxford UP, New York

Mach E (1926) Erkenntnis und Irrtum. Reprint 1976. Wiss Buchges, Darmstadt

Martens E, Schnädelbach H (1986) Philosophie – ein Grundkurs. Rowohlt, Hamburg

Puntel LB (1978) Wahrheitstheorien in der neueren Philosophie. Wiss Buchges, Darmstadt

Rescher N (1973) The Coherence Theory of Truth. Clarendon Press, Oxford

Ritter J, Gründer K, Gabriel G (eds)(1971–2007) Historisches Wörterbuch der Philosophie. 13 vols. Schwabe AG, Basel-Stuttgart

Salmon WC (1989) Four Decades in Scientific Explanation. In: Kitcher P, Salmon WC (eds)(1989) Scientific Explanation. Minnesota Studies in the Philosophy of Science, vol XIII. U of Minnesota Press, Minneapolis 1989

Schaff A (1984) Einführung in die Erkenntnistheorie. Europaverlag, Wien, München, Zürich

Scholz R (1977) Annotation of Art. 5, Abs. 3, Rdnr. 100. In: Maunz et al., Grundgesetz Kommentar, 28th delivery, 1990

Schubert J (1984) Physikalische Effekte. Vieweg, Braunschweig

Skirbekk G (1977) Wahrheitstheorien. Suhrkamp, Frankfurt

Stegmüller W (1970) Das Problem der Kausalität. In: Krüger (1970)

Stegmüller W (1973) Probleme und Resultate der Wissenschaftstheorie und Analytischen Philosophie, vols I–IV. Springer, Heidelberg

Willascheck M (ed)(2000) Realismus. F Schöningh, Paderborn

Printed in the United States
by Baker & Taylor Publisher Services